February 1995

Dear John,
    This is really a celebration of the magical Petrochemical Era, of which we both are veterans. With friendship and respect for a really <u>civilized</u> couple.

David Salutan

CREATIVE TROUBLESHOOTING IN THE CHEMICAL PROCESS INDUSTRIES

CREATIVE TROUBLESHOOTING IN THE CHEMICAL PROCESS INDUSTRIES

DAVID SALETAN

CHAPMAN & HALL
New York • London

First published in 1994 by
Chapman & Hall
One Penn Plaza
New York, NY 10119

Published in Great Britain by
Chapman & Hall
2-6 Boundary Row
London SE1 8HN

© 1994 Chapman & Hall, Inc.

Printed in the United States of America

All rights reserved. No part of this book may be reprinted or reproduced or utilized in any form or by any electronic, mechanical or other means, now known or hereafter invented, including photocopying and recording, or by an information storage or retrieval system, without permission in writing from the publishers.

Library of Congress Cataloging in Publication Data

Saletan, David I., 1927–
   Creative troubleshooting in the chemical process industries / by David I. Saletan.
     p. cm.
   Includes bibliographical references and index.
   ISBN 0-412-98441-5 : $54.95
   1. Chemical process control. 2. Chemical plants—Maintenance and repair. I. Title.
TP155.75.S23   1994   94-9082
660'.281—dc20   CIP

British Library Cataloging in Publication Data

Please send your order for this or any other Chapman & Hall, 29 West 35th Street, New York, NY 10001, Attn: Customer Service Department. You may also
call our Order Department at 1-212-244-3336 or fax your purchase order to 1-800-248-4724.

For a complete listing of Chapman & Hall titles, send your request to Chapman & Hall, Dept.
BC, One Penn Plaza, New York, NY 10119.

Contents

Preface vii

Part 1: Problem Definition

1 Troubleshooting as a Skill in Its Own Right 3
2 The Nature of a Mature Operating Process 9
3 How Processes Develop Troubles: Dislocations, Calluses, and Syndromes 23
4 Role, Attitudes, and Language in Team Troubleshooting .. 31
5 Completing Framing-in of the Problem 45

Part 2: Process Modules for Visualizing Problems or Creating Solutions

6 Classic Problem Elements: Fluid Mechanics 65
7 Classic Problem Elements: Distillation and Other Fractionation Systems 95
8 Classic Problem Elements: Control and Process Dynamics 141
9 Classic Problem Elements: Nonlinear Phase Equilibria 178
10 Classic Problem Elements: Homogeneous Reaction Systems 203
11 Classic Problem Elements: Heterogeneous Reaction Systems 240
12 Classic Problem Elements: Trace Chemistry 284

Part 3: Finishing the Job

13 Diagnostic Aids: Seeing the Unseen 303
14 Thinking Metaphorically 310

15 Turning the Tables: Making Phenomenology Work for You 317
16 Safety and Health: A "Must" with a Price 328
17 The Friendly Arts of Persuasion 354
18 Summing Up 357

Preface

Welcome, reader, to this book on process troubleshooting. As a person who generally gives short shrift to book introductions, I commit a strong statement of my message to this preface with fear and trembling, hoping that readers will absorb it first, before getting into the body of the book. The advice and examples presented come from 40 years of professional activity in troubleshooting and process improvement, from my own experience and from that of many able people I have been privileged to work with.

I have spent most of my chemical engineering career in manufacturing support, working variously in plant locations, a central engineering department, and a research and development laboratory. This book grew out of the disparity I found between the drama, subtlety, and urgency of problems in chemical process operation and the inappropriateness of the borrowed tools with which technical professionals were addressing these problems. Increasingly, as my sympathies were enlisted on the side of those on the front line, I veered from a standard process design posture to seek a mode that matched the method more closely to the need.

This is not a text in the usual sense; nor is it a cookbook where you can find a detailed, quick fix for a specific problem. It is more like a foreign travel guide that teaches you how to pace your activities, preserve your health by avoiding iced drinks, and enhance your language skills to make meaningful contact with the local people. It is primarily concerned with reorientation. My primary goal is to instill disciplinary spine and a sense of pride in the "profession without a name"—the technical tending of ongoing chemical process operations.

This plant support role has often been patronizingly regarded as a career by default for those members of the chemical engineering or chemistry professions who did not become teachers, designers, or researchers. Thus, although a large percentage of chemists and engineers spend their careers in plant support work, that work has lacked an appropriate style, élan, and methodology, in part because of attitudes and in part because there seemed to be no alternative "disciplinary home." That discipline/craft is described here and dubbed *process troubleshooting* (PTS for brevity below).

I expect this book to find a place on the desks of chemist and engineer technologists in the chemical process industries who have been out of school long enough to appreciate how different their situation is from an academic or design environment and who sense a need for a new technical underpinning that can give power and coherence to their everyday activities. Many practitioners of PTS become aware that they are practicing a craft that is very unlike the one their schooling trained them for but have difficulty spelling out the distinctive differences. What are these differences from the design and development crafts?

PTS differs radically from design in its time scale, its set of givens, and the way in which technical tools are marshalled to attack a problem. Instead of having as an entry point a neatly stated problem that establishes the mechanics of the situation and sets the stage for quantitative calculation along a well-defined line, the troubleshooter starts with only a malfunctioning system, whose illness must be defined before any other useful activity can be performed. The problems are more multidimensional than a designer's. There may be several right answers, as in the several distinct modes of operation that are appropriate for many processes, as circumstances vary.

The time scale differences from design are also striking. On the one hand, the PTS practitioner seems to have a shorter string. A problem arises and cries out for resolution, with no time to waste, whereas a designer usually has a 12- to 18-month span in which to marshall an orderly attack on a unidimensional problem. And yet, in a sense, the troubleshooter has more time. He has lived with the process over a span of years. He can see that the immediate problem is but a current eruption of a deeply seated process flaw that has been around for some time. Symptoms of that flaw will recur again, particularly if his solution to the current problem is wrong or superficial. He gets a second chance.

The evolving situation of a plant with time provides other differences from design. An examination of several business cycles reveals a process that is at times stretched to maximize production, with an attendant set of process weaknesses. During times of low demand, opportunities will arise to develop operating modes that capture yield or quality gains unattainable at high throughputs.

Surprisingly, the designer's background information on process chemistry and other textural matters may have been weak compared to what is available to PTS over an attentive span of years. The stresses of new requirements and circumstances cause adaptations that fundamentally change the nature of the operating system from the time of startup. While these adaptations may have been successful in achieving their immediate objective, they often leave the process warped and vulnerable to further circumstantial shocks. When a major operating problem does surface, there may be little context from the original design that is pertinent to its solution. Yet in my experience there is rarely a design-type review/update of a process some years after its startup.

There *is* time for learning, usually spurred by a crisis, rather than as an orderly disciplinary exercise. This learning occurs in the problem definition phase of PTS, which is inherently longer, relative to solution, than in design. Thus the stress in this book on problem definition as the *central* activity of PTS.

Even after problem definition is achieved, PTS technologists often find themselves using avenues of solution that differ sharply from those a designer would use. There are several reasons for this: PTS will often bypass best-practice design approaches because of the many implicit and explicit constraints on the technologist's freedom of action in an operating plant. PTS is rich in options for potential solutions to problems, dealing more in operational variants than in heavy-capital inputs. A change in operating mode can be fast and inexpensive to try; a failure can be readily written off and another option tried. This process is less hit-and-miss than it sounds if based on plant tests that flesh out the true nature of the problem, a mode unavailable to the original designer.

Similarly, PTS can try out long-shot but cheap capital options that the original designer would have rejected. The designer takes maximum-probability mainstream design avenues that are familiar and minimizes risk. His professional reputation is on the line, and the plant must start up on schedule. But the troubleshooter can take risks and failures (as long as he is open in communicating these to management), often learning from a buried failure that last bit of information needed for ultimate success.

The troubleshooter is conscious of the multidimensionality of the system she confronts and her need for outside help, particularly in problem diagnosis. But such help comes more readily from research than design because the flexibility and curiosity of the former is more appropriate to the open-ended nature of PTS problems. The differences between design and PTS approaches, and my growing sense of synergy between research and operations, put into place the last element in my visualization of PTS as a worthy craft. I came to sense that research and operations often required an active, *direct* interface with a lead role in remedial work on operating problems.

These thoughts brought me to an image of a *trilateral* network of relationships (see Fig. 4-1) among operations, process engineering, and research. (This displaces a simplistic straight-line concept whereby research and operations are opposed poles, separated from each other by a mediating process engineering function.) The three disciplines are represented as apices of a triangle. The sides then represent bilateral activities: Process development joins research with process engineering; plant design connects process engineering with operations. Design and development have long been recognized as distinct disciplines, with their own starting points, goals, rhythms, and techniques. It came to me that it was time to recognize an activity, here designated PTS, connecting operations and research, as the heretofore unnamed third side of the triangle, whose methodology I try to articulate in this book.

I do not pretend to have been exhaustive or even-handed in the coverage of technical areas in this book; in many areas of reaction engineering, organic chemistry, fractionation calculations—the standard tools are appropriate and it seemed pointless to duplicate what is already well done elsewhere. In areas such as electrochemistry, solids handling, extrusion, where my background is thin, coverage is minimal. In other areas, such as the application of free-radical kinetics, where literature coverage is thin and I have had extensive experience, coverage is more substantial.

The book falls into three parts:

1. The chapters in Part I give insights into the nature of embedded process problems, plus advice on overcoming psychological and other blocks to achieving problem definition, an essential prerequisite to remedial action. These chapters also demonstrate how a technical team can supply balance and relieve the loneliness of solo troubleshooting. Elements of such a team need not all come from within your own company; they can include outside literature and consultants, balanced by internal resources provided by peers.

2. The chapters in Part II provide a survey of how major technical process elements—fluid mechanics, fractionation, phase equilibrium, reactor configuration, chemistry mechanisms—can both cause trouble and offer answers to a solution. This is the part to scan for technical morsels that may be of specific relevance to a current problem.

3. In the chapters in Part III, some ancillary elements essential to rounding out an effective contemporary troubleshooting activity are sketched: diagnostic tools, use of metaphor to view problems in a fresh way, persuasion of management and others, and ad-

herence to societal constraints to protect the health and safety of workers and the general population. I indicate also how we can use the enlarged understanding of process phenomenology from our troubleshooting activities to enlist effects, previously seen only as sources of process aches and pains, to profitable ends.

> The nature of my message has made it necessary to ignore or even tear down certain totems of the engineering profession that block rather than enhance our understanding of mechanisms, such as the equilibrium stage concept and portions of that current bureaucratic totem, the SI system of units. Atmospheres and cycles/second promote our physical feel for problems; pascals and hertz do not. At the same time, I have gone beyond the staid professional proprieties to convey the magic and excitement that I have found in a career of process troubleshooting and to justify the use of intuition, imagery, and nonrational thought as professional tools of some power.

Despite the inversions of design approaches and the championing of unsung technologists against the received wisdom of school and design establishments, I expect this book to find readers among perceptive teachers, designers, and researchers who can find here some eye-opening exposure to the real texture of living chemical processes and useful points of departure for creative research.

My Value System

Sounding again and again throughout this book like a drumbeat are two cardinal principles that drive me. Perhaps they will seem like moral judgments to you; to me they are common sense.

The worst technical sin (worse than not getting a workable answer to the problem we confront) is *working on the wrong problem*. This flows from the "original sin": belief that visible action always takes precedence over reflection. This is embodied in the familiar plant wail: "There's never time to do it right; there's always time to do it over."

The second worst technical sin also flows from that original sin of worshipping action over reflection: it is working on the right problem with totally inappropriate tools, doing *what we know how to do* rather than *what the problem requires*. In extreme cases this becomes "immaculate disengagement," in which bloodless, truncated versions of the real problem are addressed with special tools in monastic seclusion that allows the practitioners implicitly to avoid the complexities of the real world.

CREATIVE TROUBLESHOOTING IN THE CHEMICAL PROCESS INDUSTRIES

I

Problem Definition

1

Troubleshooting as a Skill in Its Own Right

Computers are useless. They can only give you answers.

Picasso

This book is about troubleshooting as a discipline, perhaps even an art form, an exciting and rewarding technical career. The focus of this troubleshooting is an ongoing, established chemical process with goals, needs, rhythms, and quirks that give it character. I want teachers and design professionals to read this book, as well as the plant technologist to whom it speaks. If I have seemed to "put down" academia and design, it is only to raise the technologist to the collegiality he deserves.

Deduction vs. Induction

The central task of the process support technologist is to react to and correct abnormal and unacceptable situations, which will be referred to as *troubleshooting* in this book. Plainly, this central task involves characterizing the abnormality and drawing *inductively* from the situational facts a cause or constellation of causes, which can then be addressed in a corrective or engineering mode. The problem may have forced itself on our attention by gross effects or it may consist of more subtle weaknesses that will be exposed only by curious and persistent investigation.

The design function, by contrast, proceeds *deductively* from a simple set of assumptions and physical principles that lack the "layering" or contextual richness of a real system. It receives corrective feedback only after startup of the process. Of course, an experienced designer calls on prior experience in starting up and troubleshooting previously launched process systems, but this resource comes from precisely the inductive process we are about to explore.

A typical design task might be: Calculate the area requirement and tube layout for a heat exchanger to cool the effluent from a cracking furnace to 650°F prior to joining other reactor effluents as feed to the crude product fractionator. The inverse (troubleshooting) problem might be: The apparent heat transfer coefficient of this exchanger fails to meet requirements after only 10 days on line (vs. an expected cycle of 6 weeks). Furthermore, there is an adverse shift in butadiene and styrene yield during this 10-day period. What equipment or procedural changes can provide more acceptable performance? A least-cost or least-time answer will require multidimensional thinking, including some shrewd conjecturing on the nature of the fouling process.

Many people think of a design person as the first recourse for troubleshooting, but the stances of the two disciplines tend to be quite different, as the table below illustrates. Troubleshooting is, of course, distinct from research as well, but few confuse the two.

	Design Stance	Technological Support
Goals	Conversion	Selectivity
	Assured performance	Marginal capacity
	Capital cost control	Pushing at constraints
Information basis	Process development program or Licensee feedback	Direct observation Literature culling
Validation	Pilot plant	Phenomenological "corners"
Evolution of know-how	Second-plant or via Licensing	Debottlenecking Quality shifts in feed or products
Methodology	Deductive, feedforward Comprehensive computation Minor alternatives excluded	Inductive, feedback Analogy, order-of-magnitude reasoning Lateral thinking

Nonlinear and Feedback Complications

The distinction of troubleshooting from design is not simply a matter of a "linear" script that the designer reads from left to right, whereas the troubleshooter reads it from right to left. There is a circularity or "tail-in-the-mouth" aspect to a mature, integrated process system; the feedbacks and interactions are so numerous that it is often difficult to distinguish effect from cause. The term *root cause*, often heard in these days of quality jargon, presumes linearity; instead, however, we may have a vicious circle in which it is not at all clear how to snip the chain of causality.

A graphic way to appreciate the kind of interaction that may defeat a simpleminded, brute-force intervention to correct a process is to consider the analogy to a sagging circus tent: A pole inserted inside the tent at one point to raise the ceiling results in the tent's sagging in another, but unpredictable, area. Or we may follow that sage of the media, Marshall McLuhan, in his resolution of the familiar chicken-and-egg conundrum of causation. According to McLuhan, a chicken is an egg's idea of how to make more eggs!

Curiosity and the Unseen

The nature of a designer's task and the closely associated time constraints frequently oblige her to put on blinders. It would be too sweeping to say that design engineers lack curiosity, but the nature of the craft is to keep curiosity bounded. Those who can accept this constraint are happiest in the design activity. In the extreme, this stance can be summed up as follows: "if it ain't on the flowsheet, it ain't there."

In my recent experience, which involves working on reliability and quality issues frequently involving trace components in the 1- to 1000-ppm range of concentration, the norm is that the component we care most about will not be found on the designer's flowsheet material balance. This isn't just a matter of overlooked components, but rather one of overlooked fluid mechanics and chemical mechanisms as well. For the process technologist who has 5 or 6 years (rather than the 5 or 6 months a designer has in a typical project) to assimilate the rich texture and layering of an established process, a different toolkit is needed.

Academic Idealizations

For computational convenience, designers use simplifications that get *their* job done but gloss over the complexities of a real operating system. This gloss is characteristic of schooling as well, for understandable reasons.[1] I have had young plant engineers just out of school say to me: "Do you mean to tell me that in a simple vapor/liquid flash I can't count on *one* theoretical stage?" and I tell them, "No, you can't, at least not for minor components that are sparingly present in one phase." In effect, the equilibrium stage is an idealization justified more by the convenience of substituting algebra for calculus than by its correspondence to reality, and frequently it must be overridden.[2] Its use in multicomponent distillation systems has spawned other problems, such as the inability of most computing programs to handle differing tray efficiencies among major and minor components.

Exploration Follows Design

A lunar landing can be considered a triumph of design. The operative physical laws are all well established; the differential equations for the requisite trajectories comprehensively capture the situation. From the moment the astronauts step out onto the lunar surface, however, the discovery phase begins. They confront the lunar landscape, observing, gathering specimens, seeking to assimilate the internal logic of a whole new system. The U.S. Army puts it pithily: "The map is not the terrain."[3] Discovery is what *process troubleshooting* is all about. The term also conveys, I hope, the excitement, the sense of adventure and creativity, that is an addictive joy for those immersed in troubleshooting for any length of time.

All this is not to imply that troubleshooting is the inverse of design in every respect. After the problem has been satisfactorily diagnosed and a procedural or capital equipment change or add-on is

---

[1] Recognition of this simplifying tendency has made *academic* a pejorative adjective. I have great respect for the contributions university people can and do make, but this use of the adjective is often appropriate.

[2] Computer people have recently rediscovered the transfer unit as though it had been newly minted; older hands never stopped using it. The complications attending the substitution of transfer unit calculus for equilibrium stages are readily met by suitable approximations. In troubleshooting precision will rarely be a problem as long as the problem is appropriately framed.

[3] Taken from a televised press briefing during the 1991 Gulf War.

required to effect a solution, design skills must be called upon to give it form. But even at this stage, the questions posed to the designer by experienced troubleshooters may make her feel uncomfortably constrained in her options. For example, she may be asked to bet on a long shot rather than use the usual odds-on approach (plus a safety factor!) of new plant design, simply because the mainstream option is too expensive or its execution would take too long.

Example: Playing a Long Shot First

I recently made just such an unorthodox proposal to the designers about a problem involving downstream corrosion, which had been diagnosed as being caused by sub-micron-sized droplets of a corrosive fluid escaping capture in a gas scrubbing column. (See Chap. 4 for a more complete exposition of this problem's context.)

The failure to capture such fine droplets is due to their following the gas flow lines at every point in the tray action of the scrubber, with no lateral force to cause them to depart from the gas flow at turns during tray contacting and impact into the waiting arms of the scrubbing liquor. Compounding the problem is the fact that, unlike regular gaseous components, there is not even a Fickian diffusional gradient driving the assorted mist droplets toward the surface of the scrubbing liquid. (Brownian motion may approximate this mechanism, but it applies only to colloidal-size drops even smaller than those involved here.)

Seeking to achieve a rapid and cheap solution, I urged on the designer the expedient of adding a moderate quantity of foaming agent to the scrubbing liquid (short of loading up the column), on the chance that it would increase the gas-side contacting efficiency (N_G in the parlance of transfer units) by breaking up the gas into finer bubbles, thereby securing more complete scrubbing. In effect, the designer was being asked to accept an educated guess. The incentive for him was that if the gambit were successful, not only would the problem be inexpensively resolved, but the edge of his art would be advanced for the next occasion. If the gambit failed (in a full-scale plant test), he could always fall back on a traditional capital-intensive solution such as adding a trayed section.

This "perverse" design approach, based upon needful or wishful thinking, works more often than one might think. In effect, we are taking advantage of the fact that the plant already exists, allowing us to answer questions and test guesses that would be totally ir-appropriate if these remained only an issue in the designer's mind.

In the next chapter, we will try to make clearer what is meant by the *layering* and *texture* of an established process, that is, the complications that defeat simplistic thinking in troubleshooting.

2

The Nature of a Mature Operating Process

All nature is but art, unknown to thee.

Alexander Pope

The techniques of troubleshooting derive from a distinctive way of understanding a process. A mature operating process is a complex web of interactions that gives it an organic nature and a texture. A troubleshooter's perception of this affects both her cause/effect inferences and her selection of curative measures. In both the holistic treatment presented here and the introduction of poetic and metaphorical modes of thought discussed in Chap. 14, we begin the "graduate" education of the technologist, beyond the analytical, the prosaic, the reductionist way of looking at problems.

How an Operating Process becomes an Organic System

A mature process cannot, of course, be an organic system in the same sense that the human body is. Nevertheless, there are enough parallels to justify invoking *medical*, rather than engineering or chemical, jargon in talking about some malfunctions that arise. This chapter explains how this remarkable organicity originates.

An older process, at its inception, typically had poor yields on the raw materials fed; had numerous vents, by-product, and effluent waste streams leaving it; and was profligate of energy usage, intermediate storage tanks, and human surveillance. Automatic control was sparse; deviation of the process from expected norms was frequent and unmeasured. But as economic pressures, environmental regulation, and chemical engineering sophistication tightened the process, innumerable interactions that were not present in the primitive origins of the process came into play.

Example: Trouble From Recycle Buildups

In the sulfuric acid process for indirect hydration of olefins to alcohols, for example, the original once-through use of sulfuric acid has long been replaced by reconcentration of spent acid for reuse, with a minimal purge for control of acid quality. This has made the process more prone to upsets from trace surface-active impurities, many originating within the process from side reactions and building in the recycled acid. The buildup can manifest itself in emulsion/foaming "storms" that spread from one sector of the process to another. This loss of reliability from pursuit of economic and environmental gains requires a new layer of understanding and sophistication to overcome it.

Ingrained Process Interactions

We are rarely aware of the extent of the feedback pathways and interlocks built into a mature process. Patient ingenuity, by countless individuals over the years, has rounded off the rough edges, created the recycles, and closed off the vents (whether driven by environmental law or in pursuit of yield improvement). But when this altered structure is disturbed by a new burden or demand, when we "step on the hoe blade and the handle hits our head," we are suddenly aware of the holistic nature of the organism we are confronting. This first became clear to me in the late 1950s, when a number-crunching technique, *multiple statistical regression*, emerged as a statistical fad (fueled by growing computer capabilities) that promised to uncover fundamental process relationships by cross-regressing *normal* variations in the process against one another. But, in practice, I and others found that built-in interactions among process elements and circularity of the cause/effect chain in an integrated process defeated this naive approach.

Subsequently, I was involved in a more sophisticated attempt to characterize process fundamentals online. This method used a mobile data acquisition trailer and deliberate step perturbations of the process to identify process response functions.[4] Even here, it turned out that considerable technical interpretation was needed before the raw responses could be given physical meaning and used extrapolatively to find a better operating mode for the process.

Respect before Manipulation

What this phrase means, in terms of practical problem solving, is that we must grasp the problem fully and understand its many facets before curative action is tried. This may involve plant tests, getting concentration profiles of key components by analytical surveys, achieving a better grasp of fluid mechanics, or laboratory elucidation of reaction mechanisms or kinetics. Ultimately there will still be a curative addition or deletion, but it will likely come with supplementary measures to counter the possibility that the cure will be worse than the disease.

If we are lucky enough to have problem diagnosis come easily, we should switch our attention to enlarging the field of potential solutions. Not infrequently, our backward glances to the problem as we debate which solution to implement will convince us that our original problem diagnosis was superficial or incorrect.

Inferences and Curiosity

The iceberg analogy applies to process troubleshooting: Often only 10% of the problem is even roughly visible at the outset. With what frame of mind and what kind of toolkit do we set about inferring the shape and location of the other 90%? We must draw on experience, current evidence, and imaginative "patterning."

---

[4] This method somewhat resembled *evolutionary operation* (*EO*), an optimization technique in vogue in the 1960s and early 1970s. EO was a gradient search technique using repetitive small-step perturbations over an experimental design grid to move the process to an optimum operating setting, one that, in principle, might drift with time.

Situations are not repeated exactly, any more than history is. But imaginative linkage of one episode to another over time by the curious troubleshooter can lead to construction of a useful map of real phenomena and their interconnections. This is not a task for dogmatic persons who try to force doctrine onto reality. We must first diligently gather the current evidence; only then can we try to link it to the strands and patterns of prior experience.

Imagination-Stretching Exercises: Depolarization

To maintain flexibility and imagination in patterning our "story," we need mental stretching exercises. *Lateral thinking (LT)* and poetic imagery are powerful aids to problem visualization. The rudiments of LT and an example of its use are given here, with further discussion presented in Chap. 5. Poetic imagery, treated more generally in Chap. 14, is invoked in an extended example below.

There are several obstacles to conceptualizing the full nature of a problem. One is the dominance of the problem/solution axis: We focus our attention on a narrow rut that runs between a (premature) statement of the problem and a (premature) form of solution, screening out a host of peripheral information and possibilities that we should be tapping.

Brainstorming can create a variant of this "rut": conceptual blocking. This may surprise those who know it as a tool for generating a wide range of ideas. The difficulty in troubleshooting is that brainstorming can focus attention on the small part of the problem we can see at the outset. We wander from this "pole," to which we are tethered, foraging for potential solutions before we have explored the full dimensions of the problem.

LT is a technique that "depolarizes" premature patterns that may dominate and limit our mental process. The idea is to introduce into the idea stream an attention-claiming "rabbit" at the edge of our field of vision in order to create a deliberate *diffuseness* in our thinking.[5] There are various methods to create this defocusing correlative. One is called reversal.

Use of Reversal

This technique of LT deliberately inverts a value system. For example, in identifying yield losses in a vinyl chloride process, work

---
[5] Some LT terminology and techniques are appended to Chap. 5.

Fig. 2-1   Metaphorical Conversion: A Leak → A Jailbreak

focused on material balances based on ethylene or chlorine, the costly feedstocks. Reversal was achieved by generating a material balance based on *oxygen*, the low-cost feedstock (to the oxychlorination step) that would normally be ignored. In addition to useful findings on trace chemistry/effluent issues, valuable insights were derived regarding the yield on C_2H_4 and Cl_2.

Attention to the periphery of possibilities, which is fostered by the exercise of linking a pair of seeming incompatibles, creates the mental suppleness to begin to put incongruities and unrelated clues together. While LT may also have value later in generating solutions, this peripheral vision is most useful when gathering phenomenological evidence to define the nature of the problem.

Uses of Poetry

Poetic imagery and metaphors can be used to provide new insights into a problem, as shown throughout the extended example below. *Metaphorical conversion*, which consists of paraphrasing the original problem as a metaphor, may be considered an alternative to LT in depolarizing our thought processes. In McKinney's (1987) example, a search for a critical leak is converted to a jail break-out (see Fig. 2-1 and Chap. 14). Linear thinking and brainstorming may be employed on the transformed problem with far less dominance/narrowing than would be true if applied to the original "prose" form of the problem.

Example: Solving a Complex Process Problem with the Help of Poetic Images

The story that follows provides an example of problem definition in which holistic treatment is combined with alert attention

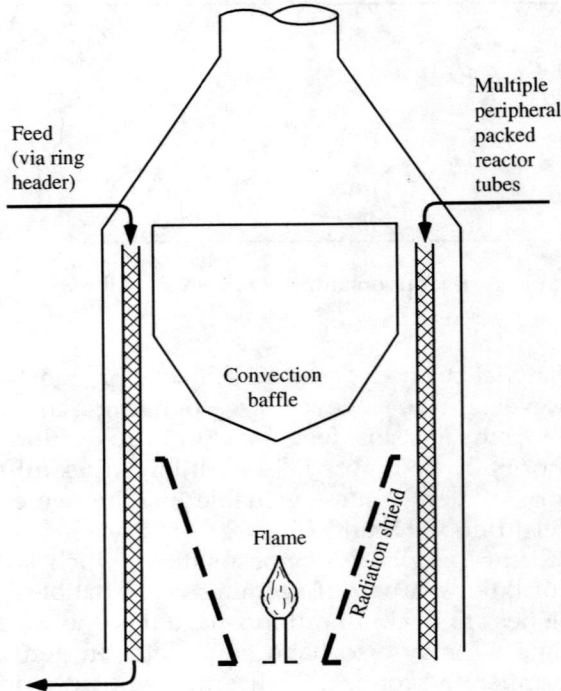

Fig. 2-2 Vertical Layout of a Fired Multitubular Dehydrogenaton Reactor

to all the possibilities in gathering evidence. The ostensible problem was rapid deactivation and plugging of catalyst beds inside externally heated parallel tubes in a dehydrogenation reactor. A cylindrical furnace design had long been used in these reactors, wherein parallel reactor tubes were laid out in vertical downflow runs around the periphery of the furnace, receiving both radiant and convective heat from a central combustion zone (see Fig. 2-2). Elevated temperatures and a consequent rugged catalyst were needed to drive the endothermic reaction toward completion.

Part A: Curing a Down-trend in Catalyst Regenerability

Coke deposits impeded performance both by plugging tubes and by blanketing the catalyst to reduce activity. The standard mode of regeneration was air/steam burnoff. However, experience indicated that catalyst runs between regenerations grew successively shorter.

A holistic approach began when it was recognized that the regeneration procedure somehow contained the seeds of its own undoing. The aptness of this metaphor will be evident shortly.

The Entry Clue A curious plant chemist showed with a magnet that all the coke granules contained an Fe core. This was eventually coupled with laboratory studies indicating that steel catalyzed coke formation in similar gaseous reactant mixtures, but only at particular proportions of water and hydrogen. From this we concluded that the partially reduced form of iron oxide, Fe_3O_4, was the iron species accumulating on reused catalyst and promoting coke formation. With these clues we were ready for imaginative "encirclement" of the real problem by explaining how the Fe "seeds" arose.

How Reactivation Caused Deactivation The bare tube walls, albeit steel, had far too small an area to explain plant coke formation vs. the laboratory results cited above. Furthermore, fine particulate Fe was needed for the seeding effect. After wrestling with these discrepancies for some time, we finally made a connection to the phenomenon whereby rust powder (rouge) is formed by expansion accompanying oxidation of Fe to Fe_2O_3. The air/steam reactivation was oxidizing the inside tube walls—not serious from a corrosion standpoint but creating a rouge of Fe_2O_3, which blew off onto the catalyst particles and was gradually reduced to Fe_3O_4 at the higher partial pressure of hydrogen occurring during the synthesis dehydrogenation reaction.

Each rouge particle, whether adhering to the inside wall of the tube or to the surface of a catalyst pellet, was now seen as a seed for coke encapsulation once reduction to Fe_3O_4 occurred. During each air/steam reactivation, this coke was removed as CO_2 but the original Fe seed remained, to be augmented by new rouge created by the next regeneration cycle. By moving from the image of iron as a "bad seed" to the metaphor "seed of its own undoing," we captured almost perfectly the way the regeneration was undermining itself by stimulating subsequent deactivation.

Now the root cause of poor reactivation had been grasped. Sustained catalyst activity required not just removing coke but also preventing Fe from accumulating. Dropping the catalyst, with water hydrojetting to remove coke plus encapsulated Fe, was initially improvised as an alternative to air/steam regeneration. Later, a chemical procedure to remove Fe selectively from the catalyst surface was devised. A further, more expensive option was developed: to replace the catalyst tubes with a suitable alloy.

This constituted only the chemistry half of the problem. The other half consisted of the physical factors that were diminishing reactor performance and shortening run length. Here alertness to happenstance, even counterintuitive evidence, was the critical ingredient in obtaining a grasp of the problem.

Part B: Failure to Achieve Idealized Flow and Flux Symmetry

The designers who had chosen this cylindrical furnace reactor configuration to provide the needs of the technology were quite pleased with themselves; they had dealt with two obvious stumbling blocks: (1) Radial symmetry and the absence of secondary intervening rows of tubes promised that equal heat flux would reach each tube. (2) The use of sonic orifices to feed vapor to each tube, from a ring header atop the furnace, seemed to promise sustained equal flow to each tube, despite partial plugging that might occur in an individual tube. The combination of the two features seemed to guarantee that the parallel paths would perform identically, or at least that one tube would not foul and plug (the run-terminating event) significantly earlier than its fellows from excessive heat or insufficient internal mass flow.

With the advent of data loggers, observations by curious shift supervisors began to puncture this idyllic picture. Certain tubes were shown to be repeat offenders in terminating runs by plugging. Peripheral exit temperature patterns of the reactor tubes showed that weather events, such as a cold front, caused serious asymmetry of heat input (presumably by wind-induced variations in draft). One day a foreman asked if I could explain why the four worst-coking tubes were the ones closest to the two feed entries into the horizontal ring feed header (see Fig. 2-3).

It seemed that these first takeoff tubes should, if anything, receive more flow than the average tube and be less subject to coking. Rather than reject the observed effect using conventional wisdom, I reviewed the fluid mechanics with experts, including the literature, which showed that it was indeed possible for pressure to rise rather than fall in proceeding along a manifold, depending on the relative diameters of manifold and takeoff lines (Acrivos, et al., 1959). Thus the four tubes in question were indeed likely to be receiving less flow than the others. This finding led to a biasing pattern for the inlet sonic orifice diameters to counter the effect. However, a further and profound cause of plugging, uneven heat input, remained to be uncovered.

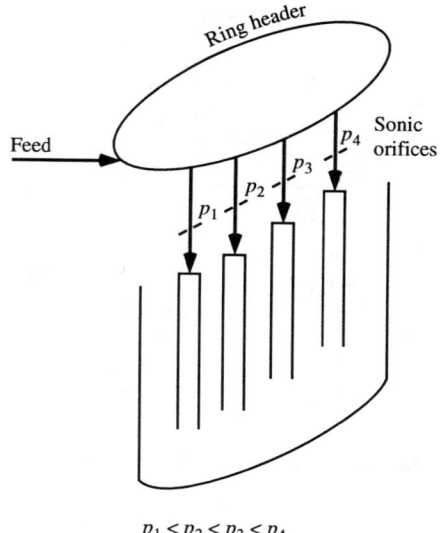

Fig. 2-3 Flow Distribution to a Fired Multitubular Reactor

$p_1 < p_2 < p_3 < p_4$

Part C: Learning from Failure Specimens

Curiosity includes getting as close to a critical phenomenon as possible. Sometimes it takes strong determination to overcome superficial hindrances between you and the undenatured clear-cut evidence needed. The final installment of the story shows how a key failure specimen further altered the kind of imagery we used in troubleshooting tube plugging in the reactor system.

A plugged tube *failure specimen* opened a final avenue to understanding of the total problem. Metallurgists have always used such specimens; I recommend them highly to technologists as well for grasping phenomenology. It is worth fighting shutdown/decontamination procedures to get an "undenatured" failure specimen.

Occasionally a tube was so coked that it could neither pass enough gas for steam/air decoking nor drop its catalyst for external cleanup and replacement; it would be cut out. Up to that point (for lack of looking), technologists had assumed that coking was a deposition on the catalyst surface, as it is in unfired reactors. On this basis, it was difficult to rationalize the axially localized blockage that was the normal pattern of converter tube plugging. However, the critical level of curiosity needed to close the gap between observation and theory

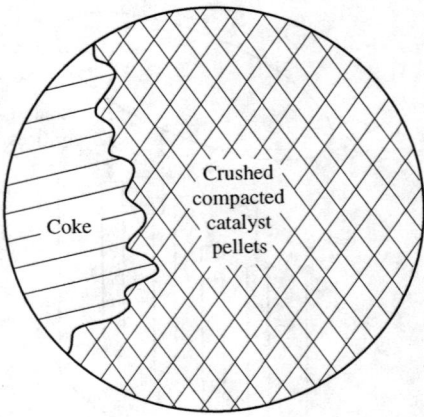

Fig. 2-4 Glacierlike Cross Section of a Coked Reactor Tube

had been lacking. I had one such failed tube sectioned to expose the length and cross-sectional character of the plugging pattern.

Visual inspection of this tube, which was cut across at the point of maximum plugging, revealed a coking pattern very different from the one previously surmised (see Fig. 2-4). The coke was not a snowdrift but rather a solid glacierlike mass invading from the tube wall. Catalyst pellets had been thrust ahead by this coke and ultimately deformed to eliminate voids even in the catalyst-filled space beyond the coke "glacier." Building on this observation, we recognized that the coke had originated at the wall of the tube facing the flame.

The designers who had used a single circle of tubes for uniform heat input had neglected the radiant component, which put more flux into the side of the tube facing inward toward the flame/flue gases. The tube walls were too thin for thermal conduction to even out this unequal peripheral heat flux pattern. Also, the large pellet size of the catalyst relative to the tube diameter guaranteed a voids gap next to the inside wall, which constituted the major resistance to radial heat transfer into the tube. This ensured a major temperature differential between the hot inside tube wall, where coking began, and the working catalyst pellets cooled by reaction endothermicity. Aggravating this situation was a low design mass flow of process gas per tube (to cut reactor height) that further impeded radial heat transport.

Thus, entirely aside from the original insight about the specific chemical effect of wall-originated Fe in catalyzing coke formation, it was essential to understand, using the glacier image, that the thermal effects driving the coking began at the hot inside tube wall facing the combustion zone (presumably from cracking/cyclization side reactions of the feed and/or the reaction product). Once this was realized, it was easy to conceptualize that the initial coke deposit, once begun, constituted a thermal blanket that further increased the temperature of the tube wall where the coke originated. This barrier was, however, porous enough that process components that were coke precursors could still diffuse through to form new coke at a self-accelerating rate at the hot inside wall. This established coke origination, against intuition, as an "inside-out" process, rather than the snowfall-type deposition that was our original mental image.

Summary of the Solutions to the Problem

Note how, once we corrected (via the failure specimen) our original fuzzy conception of the origin of the coke, our background experience or simple analogizing rapidly enriched and enlarged the breakthrough. This cascading effect is a typical climax of the discovery phase of problem definition.

Just as there were many nested deviations and complications causing the reactor's performance problems, the solution was a bundle of complementary actions. We have already indicated elements that can be used to combat buildup of coke-inducing Fe in the catalyst and to even out initial flow inequalities among tubes. To combat heat flux unevenness, two approaches were taken: a cheap expedient and a more elaborate capital-intensive-solution. The first was to remove a screen between the lower (exit) half of the reactor tubes and the combustion flame at the center of the furnace. This design feature seemingly was consistent with needs: preventing excessive heat flux into the tubes from radiation to the side facing the flame. However, what it accomplished was to shift to the entrance portion of the tubes the full brunt of convective plus radiant heat flux. This section, where endothermic reaction had scarcely begun, was far less able to absorb this flux than the section farther down where the reaction was fully underway. The design error of using a screen was in treating the unit as a *heater* instead of as a reactor.

Where the flux peaked above the top of the screen, localized coking had regularly produced plugging. With the screen gone, this plugging was dissipated; both conversion and furnace heat utilization improved. However the short, vertical, natural-draft layout still left two major deficiencies: asymmetrical flux among tubes from wind effects and poor internal heat transfer from low velocities.

Therefore, on a new unit, a horizontal multiburner furnace was built, achieving more uniform heat input and better internal heat transfer, despite using larger-diameter, thicker-walled tubes.

This extended example used poetic imagery and metaphors (glacier, seed of its own undoing) at critical junctures to reach the satisfying problem definition upon which we could build sound solutions. Metaphor is the stronger vehicle; the very inexactness of fit of a metaphor (or analogy) actually increases its power for visualization and for stretching our mind to encompass new possibilities. Over and above specific metaphors we may coin, there is a useful definition of *metaphor*: making the strange familiar and making the familiar strange (MSF + MFS). The above example has illustrated MSF; Chap. 14 presents an example of MFS.

From Diagnosis to Solution

The major portion of this chapter has, quite properly, dwelled on how to secure full problem definition, the beginning and usually neglected phase of troubleshooting. More material on this topic will be provided in Chap. 5. Let us now briefly consider a few dos and don'ts of solution implementation.

> In a holistic approach, perhaps the hardest attitude to acquire is a fundamental respect for the process. One need not be a Mother Earth worshipper to achieve this. What is implied is respect for a subtle and worthy adversary rather than seeing the process as a lump of clay waiting to be tweaked and manipulated.

Example of Counterproductive Tweaking

Acid, generated in situ in a process and wandering destructively throughout the system, is a situation tempting us to do some brute-force tweaking. The problem might be corrosion from HCl, generated in distillation by catalyst residues, reacting with moisture. Here a tweaking approach might add a heavy amine to neutralize the acidity. In several cases I was involved in, however, not only did the amine hydrochlorides decompose in reboilers, to defeat the original objective by regenerating the HCl as a vapor, but the additive fouled the reboilers as well.

An example closer to home is the notion that you should take sodium bicarbonate routinely to counter "heartburn" after meals. The body generates HCl for an essential digestive function; it simply fights back against $NaHCO_3$ by generating more acid.

Less is More

A cure that subtracts process elements, if it can be devised, is usually preferable to one that adds elements. While "less is more" is a principle of art, the preference here is based on the observation that add-ons are more likely to bring unexpected process complications. Their frequent failure to provide a robust solution has earned them the nickname "band-aids." However, some band-aids are more robust than others; one can settle for a robust band-aid if excising the root cause of the problem is too expensive in terms of people, time, or money. The essential thing is to retain awareness of getting by with a patchwork improvisation. "Less is more" should never be an investigative principle; this would narrow unwisely the technical context of a problem.

Example: Complications of a Band-aid Solution

A half-wise technologist avoids dosing NaOH into the overhead accumulator of a light ends recovery distillation in an organic chloride process to mitigate the effect of trace HCl coming up the column. He recognizes the complications that may ensue. Since the problem with HCl lies only in its corrosive effects (both water and HCl are enriched during rectification), the alternative of removing water but not HCl may provide a "less is more" solution. Suppose that our half-wise technologist decides to use an adsorption bed of alumina to dry the overhead condensate.

Avoiding the caustic dosing is certainly sound: Salt deposition and the generation of chemical water in the NaOH/HCl neutralization are only two possible forms of adverse fallout from such a band-aid. However, the environmental costs of spent bed disposal in this case required reuse of the alumina bed. Hot nitrogen stripping, a regeneration procedure improvised after the fact, removed water but not the HCl, which turned out to be appreciably picked up and united chemically with the alumina. Furthermore, elevating the temperature for bed regeneration caused dehydrochlorination of some heavier organic chloride components that had adsorbed on the alumina; the HCl released combined with the alumina to catalyze further tarring and degradation of the new olefinic species.

In such a situation, we must think through the regeneration aspect as soon as fixed-bed sorption is put forth as a solution. For example, by using for drying the hydrogen form of cation-exchange resin (its sorption isotherm for water is only slightly inferior to that of alumina), we can avoid having HCl combine with the sorbent. Sorbed organic chlorides might be less prone to dehydrochlorinate during thermal regeneration, and a workable scheme for in-process drying might be reachable. Starting at the back end (regeneration, replacement, servicing) is a useful screening device before adding a process feature. If adsorptive removal has a very favorable equilibrium, the bed will be difficult to regenerate!

Having considered problem definition at length and solution framing briefly, we will see how process trouble spots arise in the next chapter. They are by no means all designers' errors.

References

Acrivos, A. et al. *Chem. Eng. Sci.* 10 112–24 (1959).
Mckinney, P.T. *Chem. Eng.* 94 (21) 131–3 (1987).

3

How Processes Develop Troubles: Dislocations, Calluses, and Syndromes

A conclusion is the place where you got tired of thinking.
Arthur Block

Troubles in operating processes do not arise solely from accidents or original sins on the part of their designers. We do not live in a static world; changes in the context in which a process operates can render previous functioning no longer acceptable. In discussing the onset of process aches, I have invoked medical jargon to bring out the nature of embedded process problems more clearly than can be done with chemical technological terminology.

Accumulation of Process Defects

It has been said that most doctors today treat *diseases* but that the good ones still treat *patients*. This distinction helps explain why management has difficulty understanding that a 30- or 40-year-old process still has major unresolved deficiencies. A naïve expectation exists that a technical problem can be isolated, defined, and cured by diligent attention once it has ben categorized as a known "disease." But an ongoing organism (and a mature process is an organic entity) can mask symptoms, thwart one-dimensional curative thrusts, and muddle along with the help of band-aids (see Chap. 2) without receiving appropriate help from a skilled specialist. Gans et al. (1991) have captured some of the flavor of a living organism, but the flaws they recount tend to be simple one-dimensional mistakes—the kinds found in the first stages of starting up a new unit.

Because designers miss in certain areas and because poststartup shifts in operating pattern are rarely covered as thoroughly as design and startup by skilled technical people, most processes have or accumulate several *embedded flaws*. If these were catastrophic enough to bring production to a halt, they would have been diagnosed and cured at their first manifestation. But, like minor fissures in the earth, they are bridged over by stopgap measures and productive life goes on. After a few years, operating people become used to expedients like having to dump and manually decoke catalyst after each run. The *callus* phase—an insensitivity to the process "wound" lying buried under the band-aid by both operating and technical support people—has begun.

Example: "Douching" to Cope with Icing from Column Entrapment

A simple callus is the use of alcohol douches to cope with recurrent plugging of an ethylene/ethane splitter column by ill-defined "hydrate" icing. The alcohol causes downstream complications, but the process stumbles along. The designer had put a drier on column feed and thought he had finished his task. The *embedded flaw* is entrapment of trace water between C_2H_4 and C_2H_6 in the column (caused by its nonideal behavior in a hydrocarbon environment) accumulating to create a separate (ice) phase.

Example: A Callus in Living with Reactor Thermal Runaways

An exothermic hydrogenation reactor cokes a few of its parallel jacketed catalyst tubes on every run, to the extent that dumping, screening, and replacement of catalyst are routine after each run, before in-place carbon burnoff is feasible. What causes the coking, and why only some of the tubes plug (with bypassing through the others reducing both conversion and selectivity), are questions no longer asked. Insight into thermal runaway, the *embedded flaw* operative here, and its remedies (see Chap. 11) is absent. Operations plods on with a basically unacceptable situation.

Dislocations Expose Defects

As long as only traditional demands are put on the system, the band-aids hold. But we live amid rapid changes: New stresses are imposed by environmental regulations, feedstock shifts, and quality demands, even while there is pressure to reduce costs and switch smoothly among two or three distinct operating modes. I term such shock events *dislocations*, a term admittedly drawn more from geology than from medicine. A new dislocation exposes the fragile nature of the improvised band-aid, so that the callus no longer anesthetizes personnel to the process's shortcoming.

Example: Impurity Elimination by Massive Purges

The practice of diverting purge streams from a product finishing train to lower-value end uses in order to limit the odor components that would otherwise exit in the prime product, is a callus. In one such case, a serious effort was devoted to identifying the chemistry of the offensive components and devising a clean interception module only after a dislocation resulted in loss of the secondary end use.

Remissions

If this alone were the problem, we could count on the unfolding of crisis events to invoke expert help that gradually excised the original embedded process flaws. Two other considerations, however, interfere with this curative evolution. The first I have termed *remission*, a medical term denoting temporary disappearance of symptoms. This stems from the fact that it may take a combination of stresses to create a dislocation severe enough to expose the inadequacy of the traditional band-aids. For example, a requirement for extra production may be coupled with the need to accept temporarily an inferior feedstock. When one of these stresses disappears, the plant once again muddles along and the pressure to find a solution vanishes.

Since the action of these stresses upon the process is poorly understood at this point, it appears to operations that the problem has gone away. This can frustrate and disperse a newly created curative technical team, particularly since live symptoms, not just anecdotal evidence, are the food such a team lives on. Only after the flaw has surfaced a number of times in the course of the process lifespan will sensitivity to the *pattern* of the deficiency be combined with the dedication to find a better way.

Genesis of New Flaw

The second factor interfering with the curative process is that process evolution embeds new flaws or potential flaws in the process.

Example: Embedding a New Flaw

A homogeneous catalyst system is run at progressively higher temperatures over the course of years, gradually up-rating nameplate capacity, but at the expense of more rapid catalyst decay and increasing vulnerability to feedstock disruptions; sensitivity is gradually lost to these trade-offs. A callus has been created. It takes a new dislocation to expose the wound under the callus: a business recession in which the plant is run below capacity for several years. The throughput advantage of the high-temperature reaction mode has been lost and the debits of catalyst usage and poor operability stand forth as wounds, particularly in the cost-cutting environment that exists at such times.

Problem diagnosis under these circumstances requires continuing contact with the process throughout a number of dislocations over a period of years. Even when the process defect has been grasped roughly, it usually takes a renewed crisis phase before the combination of need, resources, and availability of live evidence from the plant itself makes possible a decisive attack on the problem.

Interaction of Flaws

One additional complication that may also interfere with resolution of a process defect deserves mention. This is a *syndrome*, a constellation of symptoms that interact to aggravate each other. This interactive causality and apparent multiplicity of troubles impede the search for a root cause. An example is corrosion experienced when distilling organic chlorides in steel equipment.

A SYNDROME:
LEAKS AND FOULING IN DISTILLATION OF ORGANIC CHLORIDES

1. Steel catalyzes dehydrochlorination reactions.
2. Trace water nucleates an aqueous HCl phase.
3. Corrosion provides more Fe.
4. Fe generates further HCl and polymerization of olefins.
5. Polymerization/fouling raises skin temperature in the reboiler, increasing dehydrochlorination.
6. Micro-cracks permit water entry even against an adverse pressure gradient.

Fig. 3-1. Corrosion Syndrome in Distillation of Organic Chlorides

Example: A Vicious Circle of Causation in Corrosion

The use of steel has been premised by the designer on the absence of water. HCl is not even shown on the flowsheet material balance. Yet, given a trace of water and some HCl from reboiler dehydrochlorination side reactions (catalyzed by dissolved Fe!), a minor phase of aqueous HCl will form (see Chap. 9). This, in turn, leads to serious corrosion, which puts more Fe in play to catalyze dehydrochlorination (and fouling/polymerization) reactions. The original water presence might have been an accidental startup transient, but the chain of events described may create crevice corrosion, which makes water a regular visitor and sustains a hydra-headed syndrome. The full syndrome is shown in Fig. 3-1.

A fundamental but expensive root-cause solution to this problem might be retrofitting with Ni in the key heat exchangers and column positions. The potential weaknesses of less expensive expedients for this problem were reviewed in Chap. 2.

Example: One Misstep Resonates a Syndrome of Process Flaws

The raw spots in a process can grow to interact with each other over the years, creating an overall syndrome of process wounds. As one example, in the startup of a newly expanded primary synthesis step, a silicone antifoam agent was injected to counter an unanticipated foam-over in a degasser vessel. A devastating deactivation resulted almost immediately in the downstream final catalytic conversion step. In the frantic investigation that ensued, it was found that the silicone, by an unexpected depolymerization, was following the crude primary product through distillation and interacting in several harmful ways with the downstream catalyst. It not only glazed over the catalyst to block its primary function, it also combined chemically with other catalyst ingredients to aggravate several harmful side-reactions.

Unexcised process weaknesses in three areas—managing the native surfactant chemistry in the primary synthesis step, scrubbing and distilling the primary product, and managing the deactivation/regeneration of the heterogeneous catalyst in the final conversion step—were linking together to create a crisis of the first order. To resolve this crisis, it was necessary not only to discontinue the triggering addition of silicone but also to devise new and unique ways of regenerating the conversion catalyst.

This linking of process weaknesses in their subterranean life beneath the calluses is so common in mature technologies that I have coined *Saletan's theorem*:

> Any shock impacting a mature process will cause all the "skeletons" in the process closet to resonate violently.

This skeletal reverberation may sound fearsome and indeed it can be so, to the point of making first-time onlookers wonder how the process ever managed to function at all. However, the positive aspect is that there will be an abundance of symptoms, not normally visible, for identification and even quantification of the phenomena involved. It is a matter of keeping cool, collecting the abundant data, and starting the order-of-magnitude framing-in of the problem, to be described in Chap. 5.

Process Improvements as Dislocations

It is difficult for a member of the technical community dedicated to devising process improvements to admit it, but process improvement itself can be one of the *prime* dislocations imposed on a process. A new and better catalyst, an energy integration scheme, or closing off an undesirable effluent can precipitate a process crisis as the unanticipated repercussions of the change roll.

Example: An Improved Catalyst Makes Things Worse

A proposed catalyst retrofit illustrates some of these complicating side effects. A supported Ni hydrogenation catalyst was giving poor productivity and poor selectivity. An R&D chemist assigned to the problem started with a bias toward contemporaneity. The current catalyst was of ancient vintage; hence a newer catalyst was the answer. Catalyst vendors were all too ready to second this approach. They had catalysts that, while costlier on a per pound basis, gave much more Ni per dollar than the existing catalyst, and spread out on a higher surface area to boot.

The new catalyst was installed in the plant after rudimentary screening in an isothermal laboratory reactor that showed better conversion than the old catalyst at the required space velocity. A laboratory life test was not performed. The flaw in the laboratory research supporting the use of the new catalyst was less in its brevity than in the failure to master the context of the process into which the new catalyst was to be inserted.

A generic "hot spot" in the reactor dominated coking and selectivity problems (see earlier example on living with thermal runaways). Also, the older catalyst had achieved a symbiotic fit with the process weaknesses. Its coarse pores and alkaline modification of the alumina support were virtues in the context of the process. The better dispersion of Ni in the new catalyst was obtained by a higher bare support area, with enough acidic sites to stimulate heavy-end-forming side reactions of the reactor feed.

Not only did the plant revert to the use of the old catalyst, the reputational damage that R&D sustained from this episode fed resistance to a later proposed trial of a graded-activity catalyst bed to address the fundamental hot-spot problem (see Chap. 11).

As one who has been both a victim and a perpetrator of such "improvements," I can say that the moral is *not* to avoid improving existing processes. In fact, Chap. 15 will show some of the many opportunities for grafting new high-tech components onto even ancient processes to restore their competitiveness. What is urged is realistic preparation, canvassing in advance the process context to anticipate possible complications, and startup support personnel who can think on their feet as the situation unrolls, but without pretending or aspiring to perfect foresight.

Retrofits

In an era of pinched returns on new capital investment, there has been much interest in *retrofits*, which retread an older process for new quantitative or qualitative capabilities at small capital cost. Enthusiastic principals may tend to give too little advance consideration to complications from a retrofit improvement. The resulting backlash can lead plant people to quietly "bury" a plant change that could have been salvaged with a modest remedial effort. A converse sin on the R&D side is to worry a new development to death with costly piloting, in the mistaken belief that quality in R&D means never making a mistake, when it might be more economic to fine-tune the development in the full-scale plant.[6]

In this and the previous chapter, we have illustrated the topology of process trouble spots. In the next two chapters, we return to the subject of structuring a proper troubleshooting team activity.

Reference

Gans, M., et al. *Chem. Eng. Prog. 87*(4) 25–9 (1991).

---

[6] Comprehensive and realistic pretesting of a new catalyst in the laboratory may often be more costly than a plant test when we are dealing with an existing plant reactor.

4

Roles, Attitudes, and Language in Team Troubleshooting

Idealism increases in direct proportion to one's distance from the problem.
John Galsworthy

The Plant Technical Support Function

This book speaks to both plant technical personnel and R&D people who may be called in[7] to help them on an intermittent or regular basis. A serious plant problem will need a team approach. This chapter is concerned with the complementary *roles* of typical participants in a balanced team. It is also important to sketch out here *attitude* and *language* barriers that may detract from troubleshooting accomplishment, regardless of individual talents, if they are not bridged. The characterization here of attitudes of the various technical disciplines and of management may verge on caricature in the interest of making the faults memorable.

The designation given to the plant technical function at Shell, where I worked for many years, is *technical support to operations (TSO)*. While the name leaves something to be desired, it recognizes that a plant

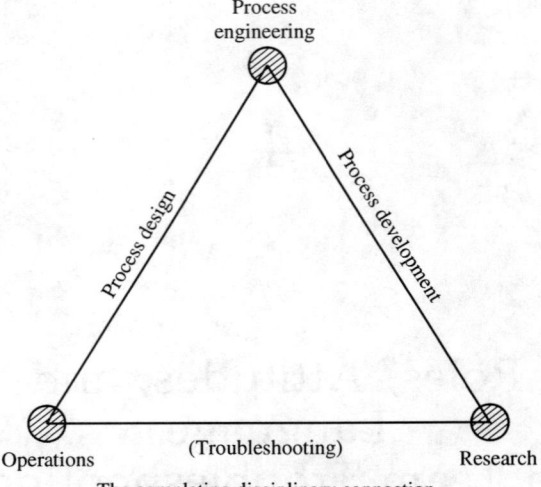

Fig. 4-1 Trilateral Relationships Among Research, Operations, and Process Engineering

engineer does little design and does process engineering only as part of a formalized capital project, a role secondary to the troubleshooting captured by TSO. One problem with the designation troubleshooting is that it tends to connote a crisis-driven agenda. In fact there *is* a craft of preventive troubleshooting (see Chap. 5) with a longer-range pace. The TSO person will usually be a chemical engineer but may also be a chemist. In this book I use *process technologist* as a generic term to distinguish TSOs from their design and R&D counterparts.[7]

Initiating a PTS Team

In what follows we will omit a process engineer/design person as a potential third member of the triad (see Fig. 4-1) making up a troubleshooting team to reinforce what the local technologist can do

---

[7] The implication, here and elsewhere, of a large corporation with separate centralized platoons of research specialists and process design engineers by no means renders the discipline taught here inapplicable to technical support people in isolated plants of smaller ventures. One can always hire a visiting consultant or a bank of outside experts for serious plant problems.

in a full-fledged plant crisis. This truncation reflects a strong belief, introduced in the Preface, that the research/operations interface is the critical one in the problem definition phase[8] of PTS because of the phenomenological curiosity of researchers. Design is not very useful in this phase and cannot enter simply as a buffer or coupler between the operations and research cultures. We are therefore working the bottom edge of the triangle in Fig. 4-2, not excluding the utility of process engineering/design participation at a later implementation phase.

Enter R&D Help

The R&D person, summoned for a one-shot contribution in what will likely be a fully ripened crisis, may look at this TSO work as a nuisance, an unfortunate distraction from the main mission at the home lab. Instead, if the occasion is seen as an opportunity to learn and not simply as an opportunity to put forth a set of prefabricated "solutions," the R&D person is positioned to be both effective and creative. In fact, the knowledge gained in such an effort can serve as a beachhead into a potential second phase at the plant whereby development-type skills are profitably deployed in an innovative/expansive mode.

What are some of the R&D inputs that those creating a problem-solving team have a right to expect? They include the following:

1. Curiosity about phenomena and facile generation of hypothetical scenarios for the problem at hand.
2. Chemical analyses and diagnostic tools of a nonroutine sort.
3. A lateral transfer of know-how derived from work on other processes and unit operations to the issues posed by the process under investigation.
4. The ability to identify and tap other special skills needed and to mediate between the temporary assistance of off-site specialists and plant technologists.

---

[8] Many managements, in a carryover from the early days of the industry (when incremental plants were built so often that support people were expected to learn only enough about the existing plant to clone it) still use the capital expenditure plan and the design function as the "leading edge" of technical surveillance.

Special Strengths of Plant People

What should we expect of plant people as equal PTS team members?

1. Familiarity with process hardware and a realistic sense of what can and cannot be done in a plant setting.
2. Interfacing with operations supervisors and shift workers both for information transfer (in both directions) and to provide an understanding of the capabilities and personalities of particular individuals.
3. A perspective on process history, either through personal experience or by access to relevant documents.
4. Circumstantial information on the timing of problem manifestations and related events that may help generate a causation hypothesis.

The Audit Scenario

A certain subset of management is fond of using R&D as an audit function to define the shortcomings of regular plant operations. This approach creates a bad atmosphere. Plant people feel (rightly) that they will surely be made to look incompetent. The aura of an audit should be avoided in a PTS team. If R&D members can also avoid force-fitting prefabricated solutions to the problem, mutual trust and cooperation will be enhanced.

A Negative Example: A Design-Oriented Product Quality Improvement Team

Production of a major commodity chemical was suffering from continuing odor problems attributed (reasonably) to sulfur contaminants in the raw material feed to synthesis. A team from the central process engineering department was sent out to find a solution. The team was not without skills, but it lacked a member with insight into chemistry. Their main approach was to continuize the NaOH treating system used to remove sulfur compounds from the incoming feedstock and increase the use of NaOH. Previously, caustic had been moved batchwise from the back to the front of a three-vessel treating sequence. The "experts" were unaware that unsteady-state semibatch movement between staged contactors is actually more efficient than continuized flow. They were also unaware of the peculiar chemical kinetics of aqueous NaOH interaction with COS, in which excessive basicity actually impedes the rate-limiting step in absorbing and reacting this particular sulfur impurity.

The team also lacked a perspective on which of the three types of sulfur compounds potentially present was most likely to lead to the formation of the pertinent malodorous impurity in the finished product, and they did not seek a chemical causation thread stretching from feedstock impurities to this compound. Finally, they forced on the plant a costly feedstock analyzer that did not work, both because of unresolved reliability problems as a prototype instrument and because of a poorly thought-out sampling system. In short, this externally imposed team was doing what it knew how to do rather than what the situation required.

Attitudes That Block True Collaboration

Negative and counterproductive attitudes are often present among the members of a troubleshooting team stemming from the bureaucratic, individualistic, and educational attitudes that we pick up almost as a matter of course in contemporary American culture.[9] The reader will recognize familiar elements in the following descriptions.

"Them and Us" (Fig. 4-2)

Organizational "clubbiness" is frequently used as a shoddy substitute for the morale that comes from a true sense of team excellence. We may see this in a smug corporate attitude that "of course, we at Shell do it better than Exxon." Between R&D and an operating plant, this attitude takes the form of a contempt or distrust of the culture of the other organization and failure to hear valid contributions offered by one of its members because of a "screen of disbelief" inserted between us and them. Readers will have heard of this

---

[9] The most damaging of these attitudes is the mushrooming tendency on all sides in American life toward "immaculate disinvolvement," a latter-day monasticism that avoids the complexities and impurities of organic life by working on problems that are esoteric and sanitized. Much computer simulation work is the modern equivalent of counting angels on the point of a needle. This trend is only partly driven by a skewed rewards system; see the later discussion on how we use jargon to wall ourselves off from our peers.

Fig. 4-2 "Them" and "Us"

as NIH—not invented here. Identifying with the problem rather than with the home-base organization will help to shed this attitudinal hangup. It is not professional, but rather most unprofessional, to show this parochialism in a team effort.

"Show Me Your Problem"

This is an R&D mind-set. The underlying presumption is that the problem is grubby, due to the fundamental incompetence of plant people, and only awaiting the arrival of Sir Galahad, the R&D knight bearing prepackaged sets of universally applicable solutions. The difficulty here is not just that *them and us* is being invoked. The solution bearer is actually stunting himself by a false innocence that avoids coming to grips with the real world. He attempts to dedimensionalize and truncate the actual richly textured problem to make it fit his standard solution (as in the Procrustean bed of Greek mythology) instead of adapting to the real situation. Aside from the personal aspect, such an attitude may mean that a solution that might make strong contributions sits on the shelf for lack of that extra bit of adaptation or ingenuity to give it relevant form.

Example: "Customer Service"

This purist attitude is illustrated by the R&D chemist who was assigned, some years ago, to diagnose what elusive characteristic made one of Shell's chemical intermediate products inferior to a competitor's in a particular merchant market end use. In effect the task was to translate a performance property, observable only at the customer's end, into a specification or analytical property that could thereby be controlled and improved at the manufacturing end.

Since the chemist's overriding reaction was repugnance to the assignment, he gave it his least creative response. He scanned the known impurities in our product and invented a threadbare hypothesis that might connect one of the larger impurities to the user's symptoms. He then launched a complex sampling effort at the customer's plant, keyed only to his hypothesis, that went nowhere and used up the customer's limited goodwill for us.

A later effort treated the problem with due respect and was much more successful, although it took some time to achieve a remedial payoff. The operative phenomenon had several facets, which needed to be investigated individually and put together like a jigsaw puzzle to create a rough picture of what to look for in Shell's manufacturing process. An analytical definition of the problem in terms of a newly identified 50- to 500-ppm by-product impurity was achieved, but the chemical mechanism(s) involved in its formation required a literature search, as well as bench-scale and plant experiments, before a multicomponent correction to the manufacturing process could be implemented. We never achieved more than a guesswork conclusion on how this impurity had upset the customer's process.

"Show Us Your Solution"

This exclusionary stance is the counterpart in plant people of the nonproductive R&D attitude discussed above. It conveys cynical doubt that R&D has anything to offer, and a bear-baiting attempt to get the "professors" to put their solution on the table before the full texture of the problem has been exposed, so that it can be shown irrelevant by presenting previously uncited information.

"They" Did It to Us

Another common counterproductive plant attitude is an attempt to blame a source upstream of the process for the problem. This is not to say that there is no sloppy feedstock preparation by suppliers, inside or outside the company. But trying immediately to blame others short-circuits needed initial introspection on the problem, which should focus on the question "What can I be doing wrong in running my (our) process?"

A variant of this ploy is to blame the crew that last serviced the distillation tower for leaving tools in a downcomer to explain away what may be a generic (e.g., foaming) problem involving mechanisms potentially amenable to control by the process operators. There have been many such dodges in my experience, which put off real thinking until a turnaround inspection of the column some time later found nothing physically amiss with the internals. The similarity to the mental laziness of the researcher dealing with the product quality complaint cited above should be noted.

An adult conception of PTS responsibilities will include a realistic appreciation that the person upstream can't/won't care about your troubles as much as you do. Even though an occasional ploy may be available, to enlist the upstream person's concern and even though a desirable root-cause solution to the problem may point to the need for corrective action upstream, it is healthy to start with an attitude of meeting and resolving the problem on home turf. In a philosophical/control theory vein, this is the feedback corrective of a decentralized society instead of the feedforward dictate of a centralized, authoritarian state. The best way not to fall into sterile role-typing and premature blaming is for the problem-solving participants to concentrate jointly at the outset on defining the full texture of the problem.

Classroom Carryovers

Our educational system prepares us poorly for troubleshooting, particularly the problem-defining phase. It is too linear and too given to force-feeding the student. The linear thinking tends to make us start erroneously on a real-life problem by creating a straight-line "project" connection from an incompletely defined problem to an out-of-context, off-the-shelf solution, as offered by the researcher in the section "Show Me Your Problem." The force-feeding tradition makes us try to reduce real-life situations to classroom exercises: a "pat" problem stated by the teacher and a preordained answer, the one path to which we have to reason our way. In real problem solving, a substantial fraction of the total effort involves arriving at a decent verbal definition of the problem. Undue haste to turn the problem into a project that will allow us to race toward completion (all too prevalent in any case) is aggravated by these attitudes carried over from classroom test taking. The following example illustrates the process of enlarging the problem context without invoking sterile poses or a premature implementation phase.

Example: Problem Definition—Reagent Intake Quality Control

Caustic soda used as an in-process treating agent frequently causes unwelcome oxidation side reactions, in addition to its desired acid neutralization or catalytic role. A "classroom" approach might be to assume oxidation meant oxygen content (from dissolved air) and to take a supposedly straightforward engineering approach of stripping the caustic with nitrogen just before use. A PTS approach would consider other chemical sources of oxidation, such as NaOCl or $NaClO_3$ impurities in the NaOH, and would also recognize the technical difficulty of efficiently stripping out oxygen from the caustic. Depending on the results of initial analytical screening to define better the oxidation source, options like substituting more expensive rayon-grade caustic (if usage is small), blanketing storage tanks with nitrogen to prevent air intrusion, or using condensate to dilute incoming 40%w to 10%w NaOH for local use would be considered, in tandem with stripping out oxygen, as potential solutions.

On-the-job experience solving real problems in a constructive team setting is the best corrective to shortcomings in our education. Colleges cannot realistically be expected to provide the texture of realism (even if the instructors had the background) in the class hours available in a 4- or even a 6-year curriculum.

Impedance Matching of Plant People with Outside Experts

Effective team problem solving does not involve a one-way display of expertise. It requires persons with a variety of viewpoints and meaningful communication among them. This is easier said than done. The Tower of Babel is not just an ancient myth; it applies here and now to the use of jargon and the other ways we wall off our craft/specialty from needed collaborators. We have dealt with various aspects of attitudes and roles which close us off from others. Now let us address the language problems—differences in training and the way in which phenomena are conceptualized—between researchers and Plant operations support personnel.

Plant people see and assimilate information on a wealth of secondary phenomena that are screened out of the glossy abstractions of the academic mind. They know that pipes frequently corrode from the outside in, that check valves fail to prevent reverse flow, and that deionized water is corrosive. The more academic minds in research know that diffusion interacts with reaction kinetics in nonlinear ways and that total sulfur poisoning of a metal catalyst can comprise a depth of only one atomic layer.

It is not simply a matter of the number of years spent in graduate school; the difference rests in an accumulation of "trade jargon" and symbolism we use in storing observations in our minds. Communication from different mental starting points and modes of thought has a dim chance of success. What to do?

A Shared Starting Point

First and most important is to establish a common starting point. This is best done by concentrating on jointly reconstructing from available observations the phenomena central to the problem at hand. Only after this has been done should the parties direct their gaze to their respective "home bases." For the plant people, this second phase may involve trying to relate this constellation of misbehavior to physical events in their experience with other units that to some extent may parallel the present case; perhaps to devise brief plant tests that can strengthen the physical picture (particularly fluid mechanics) of what is occurring. For the research members, it may (finally) allow the framing of observations in some theoretical context, with opportunities for extrapolation and interpolation.

The initial collaborative mode desired is for all parties to develop a "working volume" or "grazing patch" where they can exercise their special skills while being jointly anchored to a common starting point: an explicit conception or an image of the problem. It goes without saying that these initial explorations may to some extent redefine the starting point itself in terms of the phenomena that are truly operative.

Building a Shared Inferential Data Base

In terms of the multitubular reactor deactivation/coking problem presented in Chap. 2, we might be trying to assess the relative importance of heat flux and flow inequalities among the parallel reactor tubes. In my experience, R&D participants can almost always find theoretical rationalizations for well-characterized phenomena.

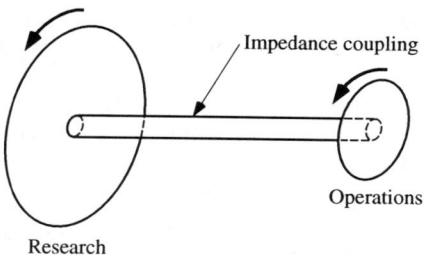

Fig. 4-3 Impedance Coupling of Research and Operations

But the common starting point for the two disciplinary groups will likely rest on accumulation of circumstantial evidence, that is, an inductive data base. Here this means gathering data that establish a pattern of nonrandom occurrence of tube pluggings and rupture around the perimeter of the furnace, against such variables as wind pattern, relative position in the feed distribution manifold, and so on.

The Impedance-Matching Coupling (Fig. 4-3)

To say that operations/research is the proper team for the problem definition phase is not to say that there is no culture clash or language barrier. Even for the bilateral communication of the bottom edge of the triangle in Fig. 4-1, we need a "translation" device. Impedance matching in electrical systems connotes the exchange of power between two dissimilar devices via a coupling device that makes the transfer efficient. As applied to the gap between plant and research personnel (in both language and conceptualization), a mechanical analogy may be more helpful. Think of a transmission that couples a low-torque, high-speed element (operations) to a high-torque, low-speed element (research). This match-up will be greatly facilitated if the plant has a long-resident senior technologist with sufficient sophistication to act as a broker by providing some overlap between the two cultures.

> This desirable overlap conflicts with a common management belief that a sharp separation of capabilities and functions between research and plant avoids "duplication" and cross-contamination of two very different cultures. The two-culture situation can work without a broker only if there have been continuing trust-

ing contacts between the same individuals in R&D and the plant over a number of years. The normal turnover of personnel makes this unlikely.

The premises upon which managements foreclose the long-term presence in the plant organization of brokers endowed with curiosity and better-than-average technical skills seem to be:

1. Promotional needs of rapid growth (no longer the case).
2. The stereotype of an able person as wishing to move on and up as soon as he/she has barely mastered a job.
3. Disbelief in how much there still is to learn about a mature process in order to place it under control in a quality sense.

Example: Team Communication in the Problem—Defining Stage of PTS

A recent PTS collaboration provided positive and negative examples of impedance matching. Continuing long-term difficulty had been experienced in a sulfuric acid–catalyzed process with downstream corrosion from acidic material leaving with the crude product as it was stripped from aqueous H_2SO_4. Despite scrubbing via multistage contacting with NaOH, both gaseous SO_2 and mist droplets of aqueous H_2SO_4 were getting through to an extent incompatible with plant reliability. The technical issues to be addressed and resolved for complete problem definition were thus:

1. How did the SO_2 originate?
2. What limited its capture in the scrubber?
3. How was mist entrainment of acid generated?
4. How did such droplets elude capture in the scrubber?

R&D was able to answer issue 1 with bench-scale demonstrations of the reductive action of organics on H_2SO_4 and its strong temperature dependence. However, neither R&D nor process design personnel consulted gave the needed explanatory insight into issue 2—how much SO_2 removal the plant had a right to expect from seven valve trays of caustic contact. (The limiting factor was in fact gas-side resistance to mass transfer and interfacial resistance to mass transfer by quasisolid surface-active impurities.)

> This hiatus on issue 2 reflected gaps at the sophisticated end of the technical spectrum, in fractionation/treating/fluid mechanics, from a sparse technical "matrix" organization[10] where ex-

---

[10] Specialists in a central engineering function are lumped here with R&D specialists for purposes of characterizing "experts".

perts resided in cubbyholes and few generalists bridged the gaps between disciplines. The treating specialists no longer thought in terms of transfer units (as more fundamental than tray efficiency; see Chap. 1); the fractionation people were weak on the phenomenological texture of contacting; and the fluid mechanics experts dwelled in a classic cubbyhole, needing context translation before they could be relevant.

On issue 4, plant people were actually more expert than the R&D personnel simply because they had seen the problem before, in HCl mists passing through more than one scrubber in series. On issue 3, the support groups offered some rationalizations (as distinct from explanation) of how micron-size entrainment might occur in the acid stripper. Because this did not represent a fundamental grasp of the problem at its source, the team was forced toward downstream expedients as potential solutions. When process constraints blocked three standard hardware solutions,[11] the need for true definition of the fluid mechanics of question 3 was reasserted. The process support generalists in R&D were able to place the acid carryover in a phenomenological context by recognizing the strong Marangoni (surface tension gradient) effect[12] inherently present on the upper trays of the stripper removing product from aqueous H_2SO_4.

This led to further discussion between the plant people and research specialists on how the long-chain alcohols routinely used worked as antifoams, in relation to the Ross defoaming theory,[12] which requires the appearance of a minor second liquid phase that tends to spread at the vapor/liquid interface. In the continuing communication, it became apparent that the long-chain alcohols routinely added as antifoam agents (which manifestly did some calming of the strippers and permitted higher throughputs) were too soluble to qualify as a Ross defoamer in the organic/acid mixture on the top tray, the critical locus for acid carryover. At this point, some older operators in the plant recalled the critical importance, in their experience, of introducing heavy ends from a column elsewhere in the system into the stripper for disposal. These proved to be oligomers of the process raw material, with a molecular weight just high enough to confer insolubility (without becoming a solid) on the organic/acid mixture on the critical top tray.

[11] A demister matrix or venturi scrubber were excluded by unacceptable pressure drops. A 360° full-cross-section spray of aq NaOH that might have given positive interception of mist with only external energy dissipation was not feasible in the large vessel diameter involved.

[12] The material on Marangoni and Ross foaming phenomenology here anticipates a fuller treatment in Chapter 7.

This example illustrates how a proper dialog among the different levels of perception begins to uncover the essential phenomenology of a problem and, from this, some possible solutions, even when a number of standard equipment solutions are not practical. Even more to the point, it shows the value of a "put-it-together" level of sophistication to assist communication between specialists and plant people. Participants in PTS teams, powerless to change management stereotypes on audits and staffing, can nevertheless make an organizational contribution through their personal development in learning to "lean toward the interface."

For a plant technologist, this means not settling for imprecise categories like "polymer" or "fouling" but consciously reaching deeper into the physical texture of a problem. For the R&D specialist, it means approaching a problem from facets other than the one in which she has expertise, consciously increasing its texture and dimensionality. Technical jargon should be avoided, not just because it will intimidate plant people, but also because it has a way of narrowing the focus even for the specialist and introducing unstated limiting assumptions.

In the next chapter, we return to the technical content issues in the late stages of problem definition as the PTS team defers implementation until the groundwork is fully laid.

5

Completing Framing-in of the Problem

Some people never learn anything because they understand everything too soon.
Alexander Pope

We resume here from the point where we left problem definition in Chap. 2. There we dealt with gathering the raw material for this phase. Now, having considered some of the organizational aspects of team building, we address the completion of the rough framing-in of problem definition, both in its own right and as a way of forestalling premature entry into the project phase of solution implementation.

Pressures for Reaching Conclusions and Acting

Premature belief and disbelief must both be held in abeyance by an act of will. Let us recognize that there are external and internalized pressures to truncate this phase of problem solving. There is explicit time pressure from management and the sense of being on the "firing line." But we also have a human need to find pattern in our experience. Wallowing in uncertainty and ambiguities creates uneasiness. Ancient cultures invoked a rain god to explain a crop-destroying drought; we moderns take vitamin C for colds. Lastly, there is in the United States a particular historical emphasis on action (with a minimum of prior thought), of "getting on with it." Suspending judgment and holding on to ambiguities and conflicting threads of evidence while all possibilities are sifted is the difficult posture I strongly recommend.

We have all seen how creating new buildings is more congenial activity for school boards and legislative committees than an in-depth study of the process of learning or the rehabilitation of criminals. The tension-releasing leap from process to project is something we must learn to resist. It is a truism that more hours must go into understanding the problem than into devising the solution. A number of examples follow that show various guises in which the rush to the project phase can occur.

Example: Changing Column Internals to Solve Distillation Problems

When valve trays were a new option on a distillation scene that had been dominated by bubble-cap trays, it became almost a fad, at one plant I am familiar with, to respond to fractionation shortfalls by retrofitting the new valve trays to replace bubble caps, hoping vaguely for efficiency and/or capacity improvements. Many malperformance problems, where the undiagnosed real problem centered in faulty control, a trapped component, foaming, or other fundamental process issues, were "solved" by the mechanical "fad" of valve tray substitution. The hardware cost and downtime of one of these failed projects were only a part of the technical/economic cost: In the wake of the loss of credibility to the plant's technical function from such gimmicky approachs, even a subsequent perceptive diagnosis of the true problem might be denied funds for implementation.

"Horse riding" is a common posture in organizations, despite official cultivation of teamwork; individual players wrap a particular form of presumed solution around themselves as a personal flag or badge and ride "their" horse off into the sunset in search of glory. This posture reflects systems where individuals perceive their advancement as resting on separating themselves from the group. But it also embodies the impatience most of us feel at the slow, inductive piecing together of a story, particularly when the pace is perceived as further slowed by a collaborative effort.

Example of Prematurely Acting Upon a Glib Assumption

This is one I can tell on myself! Cyclopentadiene (CPD) was a key troublesome impurity in isoprene produced by a dehydrogenation process whose development and early operation I participated in. Assuming that the cyclopentene (CP) found in the feed to the dehydrogenation reaction was the precursor of CPD and intoxicated with my discovery that CP's behavior in extractive distillation was unique among the C_5 olefins, I convinced the design team to completely reconfigure the projected extractive distillation unit upstream of the reactor. Only the intervention of a skeptical plant middle manager, who chose to restart an experimental reactor and prove that CPD did *not* derive primarily from CP in our process, stopped my wild horse ride before much damage had been done.

A further obstacle to a properly deliberate defining phase is the tendency to accept the rendition of the problem initially given to the team by the client. Although the latter's seeking aid is testimony to troubles beyond his ability to resolve them, a natural fear of appearing incompetent in his own bailiwick will often lead the client to offer a narrow hypothesis, suppressing details that do not fit the pattern he is forcing onto the problem.

Example: A False Scapegoating of a New Catalyst

A sudden upsurge in corrosion was observed in the fractionation train downstream of a two-stage fixed-bed catalyst system for a liquid synthesis reaction. The plant was almost literally dissolving before our eyes. When the plant people were interviewed by a troubleshooting team, some attributed the corrosion to the new catalyst introduced about a year before. The relationship was circumstantial; the catalyst change was the only thing that had changed in the process in years. There was also a lingering discomfort with the new catalyst system because of the frequent regenerations required, even though synthesis productivity had been enhanced. This made the new catalyst system a suitable scapegoat, in the absence of any obvious villain.

After about 4 months of team study, it became clear that much of the corrosion was related to increased liquid carryover in the second column downstream from the synthesis reactor. The process debottlenecking that was an incidental effect of the catalyst system change had been utilized by operations to boost throughput about 15% above the historic maximum in response to a marketing opportunity. In a roundabout way, the new catalyst *was* responsible for the corrosion problem. It proved possible, however, to unload the entraining column by reboiler and reflux changes without rolling back throughputs or undoing the catalyst change. (See Chaps. 6 and 7 for other problems due to entrainment; it is not unusual for low-level entrainment to cause quality or reliability problems before full column "loading" symptoms are noted.)

Others' Systematized Techniques of Problem Solving

It is worth pausing to distinguish the process of defining the problem presented here from various systems for problem solving presented in recent chemical engineering literature. The term in a college teaching context (Woods 1981), for example, enlarges the problem definition phase but concentrates on psychological and logical reinforcement. H.M. Hill (1989), who comes from an industrial quality control background, tries to reduce a problem to one of three classes—design, control, and research—each with a distinct strategy for solution. Only in the third of these does Hill include a science/curiosity component, needed because of fundamental gaps in our understanding of a process and the causality chains that produce certain effects. In my experience, this last factor is at least a partial presence in all problems; a retrofit with a new process ingredient, which Hill would designate a design problem, often fails from secondary complications unforeseen due to our inadequate grasp of the process into which it is inserted.

Each of these writers has made worthwhile contributions; where they are perhaps most lacking is in describing the motivations and traps that truncate the problem-defining stage in an industrial setting.

Example: Sprucing Up A L/L Extractor (See Fig. 5-1)

Another example of the perils in turning a plant problem into a narrowly framed design problem (paralleling that of the valve-tray substitutions described above) is a recent crisis that claimed my attention: A hydrocarbon-like by-product stream (HC) was facing new and much more stringent quality specifications to permit it to go into the simplest disposal route. The key new requirement was a much more complete extraction (into a countercurrent flow of water) of the process's product, an alcohol (ROH), from the HC. The al-

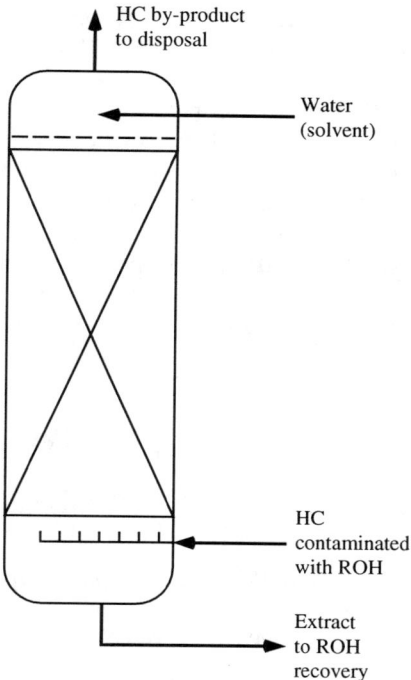

Fig. 5-1 Removal of Trace Alcohol in a Extraction Column

ready existing (L/L) extraction tower contained an old ceramic ring packing, with bottom feed nozzles to disperse the HC raffinate phase in a continuous phase of water extractant.

A process engineer in a central engineering function assigned to the project elected to substitute a superior new saddle-shaped packing, add intermediate distributor plates, and increase the water extractant flow rate. When the revamped column was started up, ROH removal was, if anything, worse than before the modification. A stock answer had been fastened to a process problem with insufficient sifting of context. The excuse for the design engineer's modus operandi was the time urgency after previous delays in confronting the quality issue.[13]

What context enrichment should have been done in a deliberately extended problem definition phase before rushing to implementa-

---

[13] A frequent lament of plant technical people is: "There's never time to do it right; there's always time to do it over."

tion? One important issue was whether it was feasible to invert the contacting, dispersing water in a continuous HC phase, even though the water flow was volumetrically greater. Another was whether the packing was to be wetted by the dispersed phase and in which phase the limiting resistance to ROH extraction occurred before and after the modifications. Flooding issues aside, the increased extraction water flow had the debit of decreasing the number of transfer units in the existing extractor, which was not addressed in any quantitative sense. Finally, the conflict between normal design dicta of dispersing the more viscous phase (in this case, water) and dispersing the volumetrically smaller phase (here, the HC phase) had not been addressed.

Foreseeing Trouble: Asking Good Questions

A first-class Shell chemical engineer, Bob Vincent, has provided a useful foresight device, which I have termed *Vincent's theorem*:

> In trying to guess in advance the kinds of operating problems that may turn up at startup, imagine the presence of one more phase than the process flowsheet would lead you to expect.

If you think that a simple liquid is flowing, imagine the presence of a gas or solid phase as well. If a vapor and a liquid phase occur, imagine a second liquid phase. This is a fine device for enlarging the context.

Example: Hydroclone Array as Process Retrofit

To illustrate the applicability of Vincent's theorem, consider a recent development in which we were planning to inject a small flow of a heavy liquid phase to extract inline certain impurities from a large flow of a lighter liquid phase. All of our attention was focused on the conservative design of hydroclones to disengage the small heavy extract phase from the raffinate. The key problem was seen as avoiding entrainment of fine droplets of heavy phase into the raffinate overflowing the cyclone, which would detract significantly from overall extraction.

Someone using Vincent's theorem said: "What about a gas phase?" Fortunately, this was a retrofit of an existing system, so the question could be answered—in the affirmative. Further investigation showed that bubbles of the minor gas phase were preferentially wetted by the minor heavy liquid phase, effectively reducing the density differential for phase separation of some drops of heavy phase. Addressing the problem at the design stage was much less troublesome than encountering it at startup, particularly since later diagnosis would take place in a time-constrained environment and without the observation tools available with our early insight, in a laboratory setup.

Even when the extra phase turns out not to be present; the exercise has utility as a lateral thinking device (see Chap. 2) to stimulate imaginative framing of a problem.

Example: Near Appearance of a Second Liquid Phase in Distillation

The upper "knockback" section of an extractive distillation (ED) column uses solutes reflux to reduce the amount of (less volatile) extractive solvent that must be processed in the overhead raffinate stream. We can come close to springing a second liquid phase because of the limited solubility of the overhead (solute) component in the ED solvent. (There is a rapid transition from solvent to solute over just a few trays in such knockback sections.)

There was a fear that such a phaseout could be causing the loading observed in one such system. But when the Ross criterion of foaming/defoaming in the vicinity of the two-liquid-phase boundary (see Chap. 4 and the fuller treatment in Chap. 7) was studied in depth, it became apparent that it was the near approach to phaseout, not its actual occurrence, that was causing disruptive foaming. An antifoam was employed as an interim expedient. Ultimately, our ability to pinpoint the few trays at risk permitted a more satisfactory solution by redesign of just those trays.

In view of the pressures most of us feel to move too quickly to the solution-generating phase of troubleshooting, I offer the following exercises to help keep you on the exploratory trail for the needed duration:

Other Procedures for Context Enlargement

1. Move your imagination down through the phenomenological "layers" to achieve a physical feeling for the process. This is more structured, yet less tethered, than brainstorming. The Vincent theorem is one such tool, but it tends to emphasize fluid mechanics. Examples of enriching the reaction context might be questions like "What do I really mean by coking?," "Where does

it originate?," "What are its chemical precursors?" or "What is really going on as the two reactants meet?" and "How is the initial reaction product transformed into the product seen in the reactor effluent?"

2. Avoid sieving each tentative framing of the problem by asking, "What would this imply by way of solution? The use of LT[14] and poetic metaphor to enrich context and insight and to avoid the polarizing effect of a dominating problem/solution axis have already been introduced in Chap. 2.

3. Bring an assortment of skills, not just chemical analysis and process design engineering, to bear on problem definition. Patrol the interfaces between disciplines. Even if none of the team members has expertise in the relevant areas, it may be preferable to wade into them (using reading and expert backup as needed) than to try to force the problem into your own area of expertise.

The virtues of poetry in providing imagery that enriches the context of problem visualization are addressed further in Chap. 14. It may seem farfetched to have metaphor urged upon engineers as a normal tool, but it is a virtual necessity in problem solving. The imperfect nature of the analogy embodied in a metaphor is a strength, not a weakness. Use of metaphor as a discipline for learning and problem solving has been termed *synectics* (Gordon, 1961).

The following example illustrates image building in which one feels almost physically identified with the system under study.

Example: Context Enrichment by Self-identification with the System

In a problem concerning near-loading in a distillation column utilizing dual-flow trays, the context is enriched by avoiding formula thinking, in which "loading" is an equation or a curve on a graph. Instead, identify with the fundamental conflict and jostling in the countercurrent flow of vapor and liquid, where both share the same path. Think in terms of automobile traffic gridlock or use whatever imagery works best for you.

The emphasis is on the fundamental instability of the physical situation, the liquid "mounding" on some portions of the tray, with a counterpart thinning of the liquid in other portions where the vapor is "blasting through," taking the sieve holes to itself, instead of sharing them on an oscillatory, time-shared basis, as in normal oc-

---

[14] A summary of the elements of LT is given at the end of this chapter.

cupancy of a sieve dual-flow tray. By the same token, holes are being used exclusively for downflow in the regions of mounding. One result of this different way of looking at the problem is to visualize entrainment as a localized rather than a uniform phenomenon across the tray. A second finding is to conceive of distributed downcomers and other design variants, which can "act out the logic" of the spontaneously occurring flow segregation.

Even at the point where we are almost ready to move from problem definition to solution framing, it should not be thought that we throw away the extra context that we have generated and go back to the kind of median design exercise presented in our earlier examples of failed project-type approaches. Only at this point does it become good PTS practice to challenge the competing versions of problem definition with the question: If this were truly the nature of the problem, is the solution it suggests capable of being implemented with the resources we can muster? This kind of means test, if not done too early, may lead us to choose a less probable scenario, for which we *can* improvise a cure, as our first try, rather than dwell on a more probable scenario that is nearly impossible to remedy with the resources and mandate at hand. A good design person will be able to leave the norms behind and run with such a strategy, if we have laid a basis in the problem-definition phase.

Assuming that we sustained a curious, flexible mindset in the early phase of problem definition, let us address the systematic lining up of evidence, theories, and information resources.

Marshalling the Facts at the Team's Inception

Overexposure to pat problems in the classroom makes many troubleshooting teams (especially team leaders) prone to begin by lining up the facts and letting them "speak for themselves." Facts do not speak, and the collection of facts needs to be framed around one or more very tentative initial hypotheses. I have seen teams inundate the local, contract, or R&D support laboratory with truckloads of samples as their initial act. Not only is this unfair to the laboratory people, it is self-deceptive, giving the impression of activity when the thinking gears have in fact not become engaged. Rare is the problem severe enough to invoke a corrective action team where there isn't a mass of evidence and clues sufficient to start the framing of hypotheses. What should be avoided in initial hypothesizing is the tendency to merge this problem into the previous problem you solved. Open yourself up to the texture of *this* problem; save the analogizing for a later phase.

Coping with a Disappearance of Symptoms

Not infrequently, the mere appearance of a team on site seems to make the problem go away. I have termed this a *remission*, in line with the medical jargon introduced in Chap. 3. Fortunately, embedded problems can be counted on to recur. A brief hiatus can be profitably used to steep the team in process history and the immediate context associated with the most recent manifestation of the problem. If the remission continues, it is better to fold the tents and return to the respective home bases than to do what a number of teams in recent memory have done: that is, put on their process designer hats and do for the process what it "needs" in a textbook sense. The activities of the product odor team described in the previous chapter exemplify this change of venue when simply going home would have been in order.

Process Emphasis in Building Hypotheses

If the problem continues unabated and hypothesizing is appropriate, a bias toward process explanations over crude mechanical hypotheses (e.g., "they installed the baffle wrong at feed entry to the column") is recommended. It's not that mechanical mistakes never happen. Rather, it is the case that most organizations are well endowed with people given to this kind of physical visualization, and it can usually be assumed that shutdown and physical inspection (plus, today, gamma-ray or neutron back-scatter scanning; see Chap. 13) will have been done well before the persisting crisis caused a team to be assembled.

This bias toward process introspection is also intended to counter the unconscious strategies used to avoid hard thinking cited earlier.

Information in the Literature

It is unreasonable to expect complete guidance via either in-house information or standard textbooks. I have found the open literature, frequently foreign-language (Russian, German, French, Japanese) or obscure publications to be a gold mine of supplemental data. Of course, one cannot do literature searches in the agonies of a plant crisis. Supportive articles from the literature pertinent to an operating process should be accumulated over the years, taking advantage of the extended time scale of a process support effort. Think in terms of dossiers rather than reports. Many of you shrink from this effort because of hesitancy about language or chemistry skills. You will find that there is usually someone with language facility to help you get what you need from an article. (It may involve interpreting just one figure or a table.)

The Russian literature is particularly underutilized. It is true that the quality is quite variable. However, Mendeleev, Semenov, Levich, Kapitsa, and others were world-class. Not only are the Russians gifted in chemistry, but where else can you find their disconnection between basic process information and the strictures of corporate secrecy on commercial processes (as in the capitalistic West) because of the former absence of a market economy?

Computer Models

Another don't is the invoking of computer models, a particular danger today, when a computer gloss is so often used in lieu of rather than as an extension of thinking. An example of the counterproductive use of a computer is a reactor model that misleads because it omits the very level of phenomenology we are trying to find in our team study. But some of you will say that a fractionation model is surely an early ally in a distillation problem. Why be reluctant to let it "do our sniffing"?

I certainly support using a fractionation model to rough out tray loadings at various parts of the columns and the separations of major components we can expect, based on reasonable assumptions about tray efficiency. We use it the same way as back-of-the-envelope, order-of-magnitude calculations: to frame in the dimensions of the problem before we hunt for the primal fault. But distillations that are upset by faulty control, "bulges" of trace components, foaming, anomalous tray efficiencies of minor components, "plunging" of second liquid phases, or fouling side reactions (which are typical of real distillation operating problems) will not be dealt with by the workhorse models of the designers. To linger with them is to fail to get on with the business at hand.

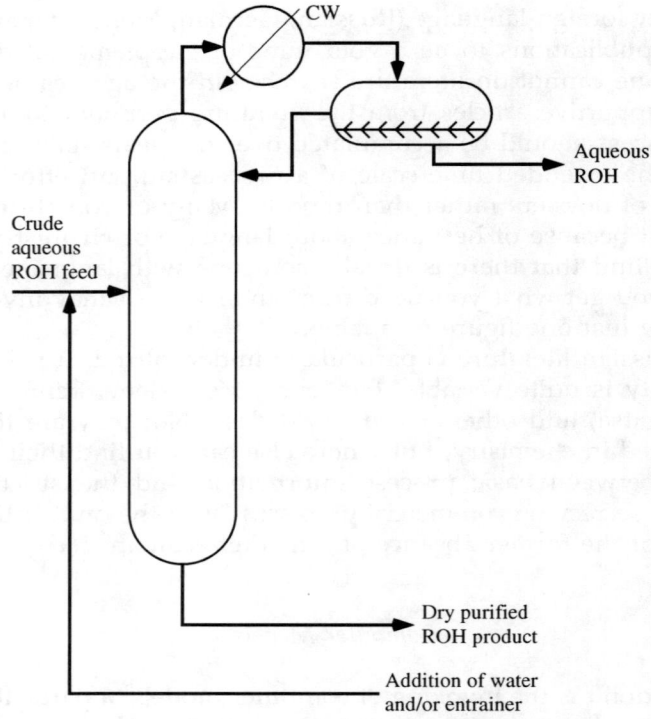

Fig. 5-2 Azeotropic Drying/Purification of Crude Alcohol

Example: Azeotropic Distillation: Commonsense Essentials (see Fig. 5-2)

A brief example of troubleshooting an azeotropic distillation ternary system will illustrate a situation where anomalous tray efficiency of a component (plus other modeling difficulties) makes sophisticated fractionation calculations an unwise entry to the problem. The system, a fairly general one, comprises an alcohol (A), water (W), and a third component (Z) of a nonpolar nature, either present naturally or added as an azeotropic agent to facilitate removal of the W overhead, while the bulk of the A leaves the bottom of the distillation column as a pure, dry product. The interesting interaction in this system, which calls forth the enthusiasm for modeling, is that a certain inventory of Z on the middle trays helps the rejection of W overhead, while a minimum level of W just below is essential to prevent some Z from coming out at the bottom as an impurity in A.

The extreme nonlinearity can indeed encourage modelers to attack this numerically very sensitive calculation. However, when the dust has settled, we will still be left with the simple question of how many stripping trays are required to dry A below the tray where a critical concentration of W has been maintained to ensure that Z goes overhead. This concentration of W can readily be established by estimation of activity coefficients of Z on a pseudo-binary "field" of A-W (see Chap. 9).

It turns out that the critical framework for a solution comprises the control issue of maintaining that W concentration together with the anomalously low tray efficiency of stripping out W from A below the tray where that W level is held. In other words, will that critical control tray be 8, 12, or 24 trays up from the bottom? As long as we have a real column to infer from (in other words, this is not an *ab initio* design problem), it becomes clear that sampling of stripping trays to infer tray efficiency in W/A separation is the proper entry to problem definition.

The Power of Single Digit Estimation

It is remarkable how much can be done with one-significant-figure reasoning to exclude possibilities and narrow the field of search. Brought up as we are to expect at least three-figure accuracy in the result of our calculations, we do not realize how this background restricts our readiness to estimate our way through "mushy" elements of the scenario. If I used to say "Throw away your slide rule" (which gave at best three significant figures), how much more often must I say today "Throw away your hand-held calculator" (with the false assurance of a 12-digit display)? We are literally talking in many instances of power-of-10 operations in your head.

Example: Order-of-Magnitude Framing-in—Catalyst Fouling

This example illustrates the framing-in of a problem of catalyst deactivation potentially caused by feed fouling. Caustic soda is being added to a crude alcohol product, in a dose of about 20 ppmw, to combat trace acidity generation in the subsequent distillation, which is the only overheading of the alcohol before entering a catalytic reactor for conversion to the final product. About 20%w of liquid heavy ends are not overheaded (see Fig. 5-3).

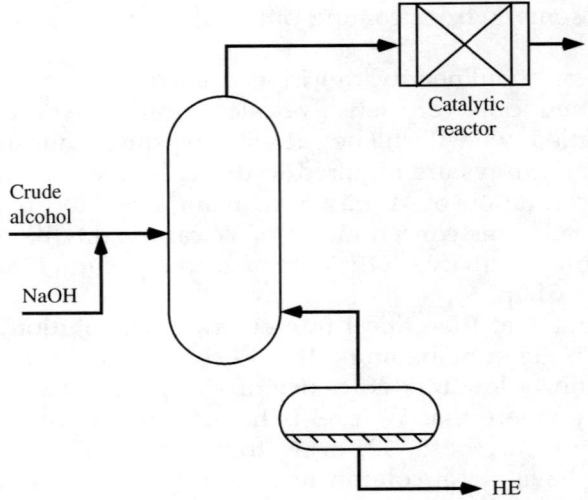

Fig. 5-3 Catalyst Fouling Problem from Caustic Entrainment

The issue is whether deactivation and shortened catalyst runs can be attributed to entrainment of an aliquot of the NaOH into the reactor or whether essentially all the NaOH leaves with the liquid heavy ends. Two questions must be answered before the case can be considered loosely made:

1. What mass flow of entrained NaOH in the reactor feed can account for the observed rate of catalyst deactivation?
2. Is it reasonable that this proportion of the NaOH flow entering the single vaporization will entrain (volatilization of NaOH is out of the question)?

The inexperienced process support technologists will flinch at addressing either of these questions because as design questions they are clearly nonstarters. They have not yet breathed in the heady air of power-of-10 reasoning. But fly with me, you Star Wars pilots, as the Power (of 10) infuses and guides you!

We start by addressing the first issue, because it is slightly more definable, so as to leave maximum maneuvering room on the second. The catalyst is a fairly rudimentary one, whose specific surface is initially unknown. From its chemistry, it is reasonable to presume that deactivation is by simple surface coverage (by NaOH or the compounds to which it will convert), but the key entry question here is: Are we talking about plating out and blocking the simple external surface of the pellets or are we considering physical blockage of the extended active area?

The neophyte will be able to make a reasonable shot at the first issue, the superficial surface of the pellets, by applying a roughness multiplier to the nominal superficial surface/volume, essentially $6(1 - \epsilon)/d$, where d is the pellet diameter (as a cube) and ϵ is the bed's "external" void volume fraction. However, the essential judgment is whether the entrained liquid, concentrated aqueous NaOH[15], will enter the rather rudimentary pore structure of the catalyst. Let us pause to consider this issue, showing how an elementary picture of catalyst topology (available, with some effort, to the plant support person) can be used to complete the alternative causation hypotheses for the problem.

Droplets entrained in normal distillation can be expected to be coarse, at least 100 μm in diameter. For most catalysts, with pore mouths of 5 to 200 nm (0.005 to 0.2 μm), this would essentially rule out extensive coverage of the active pore surface, so that we can assume with some confidence that we are dealing with a glaze that creates pore mouth blockage. Assuming a roughness factor of 10, a bed voids of 40%, and an average pellet diameter of 1.3 cm, the external catalyst surface per bed volume will be 28/cm. (By contrast, the internal surface of this rather rudimentary catalyst, even at a minimal 0.1 m^2/g, will be almost 3000/cm.) Now the gas hourly space velocity (vol. gas/vol. bed hr) (GHSV) for this system is 5 and the density of the feed vapor at the conditions for which GHSV is calculated will be approximately 4.0 kg/m^3.

We assume that five monolayers of NaOH on the external surface is sufficient to create pore mouth blockage, and that this condition is to be reached in 15 days of operation, to explain the rate of deactivation experienced. Our workhorse estimator here is the Avogadro number (6×10^{23} molecules/g-mole) combined with the assumption that a deposited molecule of NaOH occupies 0.2 nm^2 ($2 \cdot 10^{-19}$ m^2 of surface).

On this basis, the mass of NaOH deposited per bed volume after 15 days of operation must be 4.6 g/m^3. The entrainment level E (in

---

[15] The solubility of NaOH is essentially nil in the slightly wet alcohol. (A more sophisticated treatment might allow for the equilibrium of formation of a trace amount of the alcoholate, NaOR, by ROH + NaOH → HOH + NaOR.) Either phase distribution data or estimation of the activity of the water component in the alcohol and caustic phases will indicate that the latter will be concentrated from the initial 10%w to about 40%w by water transfer.

kg/kg vapor) will provide 7200 E kg of deposition per m^3 of bed in 360 hours (15 days) of continuous operation. Solving for E, we find that only 0.6 ppmw of NaOH entrained in the vapor suffices for pore blockage. (If we had used the entire interior surface, but only one monolayer deep, to deactivate all active catalyst sites, this figure would increase to about 12 ppm.)

However, we now must grapple with the second necessary condition listed above: This seemingly small concentration must be justified as a reasonable fractional entrainment of NaOH from the residual liquid in the vaporizer, allowing for the two phases present. Assume that the mass of total entrained liquid relative to vapor ranges typically from 10^{-3} to 10^{-1} kg/kg. [This assumes a load factor $U_v \sqrt{\rho_v/(\rho_l - \rho_v)}$ in a range where droplet settling against vapor flow is minimal.] Taking a geometric mean entrainment of 0.01 kg/kg, we find that 25 ppmw NaOH content in the liquid feed to the reactor (vs. 20 actual) would give the 60 ppmw required in the liquid heel[16] from which entrainment occurs, in order for the vapor to carry about 0.6 ppmw, as estimated above by pro rata entrainment relative to the major liquid components.

Thus, for closure of our scenario about simple fouling of catalyst by caustic added to the crude alcohol, we need not invoke preferential entrainment of the aqueous NaOH second liquid phase.[17] To the extent that this second phase fails to spread at the V/L interface but rather settles out, away from the main V/L "action" (see the discussion of plunging of minor second liquid phases in distillation in Chap. 7), it may be more difficult to make the case for the required entrainment of the caustic phase.[17]

Summation

Obviously, the above example has piled one order-of-magnitude surmise upon another, and we would be wise to use analytical confirmation of the deposits in the vaporizer and on withdrawn deac-

---

[16] A designer's approach would use a one-stage flash equilibration to define the liquid phase from which NaOH is entraining. In fact, vaporization is an integral process, so we assume that the liquid pool averages 40% w of the feed.

[17] For reasons not entirely clear to me (perhaps formation of trace soaps), liquid systems containing free NaOH frequently show disproportionately high NaOH entrainment. In line with this observation, we might profit from a few laboratory benchtop visualization experiments to establish whether NaOH in the system at hand will be under- or overrepresented in the entrainate relative to its bulk concentration in the liquid in the vaporizer.

tivated catalyst to firm up the story. However, I hope it has served to show how order-of-magnitude testing of hypotheses can bring some order out of a rather fuzzy situation. Furthermore, to depend only on samples and analytical evidence therefrom would be a mistake, particularly in an era where occupational/environmental regulations may dictate the kind of purge washout of equipment in shutting down that destroys the critical evidence before one can sample it.

If a first scenario as described above fails to provide even approximate closure, an alternate one should be tried before investing substantially in analytical or other laboratory efforts.

This first part of the book has dealt with troubleshooting as a generic discipline and with framing-in the problem's definition before proceeding to solutions. In Part II, we will provide a smörgasbord of advice and building-block elements for visualizing problems and creating solutions in six major technical areas: fluid mechanics, fractionation, phase equilibrium, control, reactor engineering, and trace chemistry.

Elements of LT

The following material is taken from Edward de Bono, *Lateral Thinking: Creativity Step by Step* (Harper and Row, New York, 1970) and "The Virtues of Zig-zag Thinking," *CHEMTECH*, 1(1), 10–16 (1971).

> ESCAPE from dominating concepts and "tethering" factors
>   by enlarging context
>   by carrying INCOMPATIBLE alternatives
>   by generating imperfect ANALOGIES
> PROVOCATION—movement for the sake of movement
>   Discontinuous JUMPS in thought process
>   Deliberate REVERSAL and exaggeration
>   Use of an INTERMEDIATE IMPOSSIBLE concept
>   Forced bridging to a randomly chosen word

Examples

Confining ourselves to an 8-hour shift in scheduling assignments for an unusual project is a simple example of a blocking or dominating concept at the entry point to a problem that must be sidestepped before creative ideas can be generated.

Giving automobiles square wheels is an example of reversal or an intermediate impossible in provoking original ideas for alleviating traffic congestion.

References

Gordon, W.J.J. *Synectics* MacMillan (1961).

Hill, H.M. CHEMTECH *85*(3) 148–51 (1989); *85*(5) 289–94 (1989); *85*(7) 403–10 (1989); *85*(12) 754–7 (1989).

Woods, D.R. Paper 88c AIChE meeting, New Orleans Meeting (Nov. 1981).

II

Process Modules for Visualizing Problems or Creating Solutions

6

Classic Problem Elements: Fluid Mechanics

How did the great rivers and seas gain dominion over the hundred lesser streams? . . . By being lower than they.

Lao-tse

This chapter provides insights into a few sectors of a very broad field: the intelligent dispensing of energy expenditure in enhancing contact between phases; the management of coalescence of a dispersed fluid phase; flow instabilities and dynamic effects; the effect of phase changes accompanying liquid evaporation; phenomena attending nucleation of microphases; and a general mapping of qualitatively distinct flow regimes in gas/liquid flow in pipes.

Fluid flow is central to chemical engineering. Even if we are not dealing with a continuous flow process, there will be mixing, filtration, phase separation, or other operations involving the physics of fluids. We distinguish two categories of fluid mechanics problems: purely *physical* misbehavior and problems where *chemical gradients* cause or aggravate a fluid mechanics issue. The intent in this and the succeeding chapters is not only to highlight some of the classic phenomena causing operating problems but also to discuss some of the standard ways of dealing with such effects.

In the first category are:

Manifold distribution problems
Sonic orifices for equalized flow distribution
Channeling in flow through fixed beds and open vessels
Unstable boiling
"Seaspray" entrainment; mist-annular flow
Jet stirring of tanks
Bypassing in stirred tanks
Coalescence by droplet/droplet impact
Coalescence by wall impact
Stokes' law and its complications
Anomalous viscosity effects

In the second category are:

Fog formation in condensing
Nuclei "showers" in crystallization
Reactive nucleation of a very fine second phase
Hindered coalescence
Spontaneous emulsion formation in liquid-liquid extraction
"Rag" interphase emulsions
Electrical effects in suspensions
Static electricity

Achieving Plug Flow in Open Pipe

Most of the above problems obviously involve multiphase systems, but let us start with the distortions possible in a one-phase system. Most of our interest in minimizing variation in residence time history concerns flow reactors where shortfalls in conversion and/or selectivity can result from deviations from plug flow. In dealing with scale-up from a small-diameter pilot flow reactor to a larger-diameter commercial reactor, we are usually aware that the change may move us from laminar flow to turbulent flow (since Reynolds number $Re = 4W/\pi D\mu$ will increase as the square root of flow W if constant velocity is maintained). This means forecasting for scale-up is conservative. The error is likely to come in the reverse situation, when we scale down for laboratory or pilot plant tests. We are less concerned here with minor conversion-only shortfalls, which are mostly a matter of the designer's pride and can be covered with volumetric overdesign, than with selectivity surprises.

However, the classic error of designing inadvertently into the laminar flow regime for a pipe reactor is not the only mistake one can make in scale-up from laboratory batch data. Even well-developed turbulent flow may have a long-residence tail, from the laminar sublayer along the wall, which is kinetically significant.[18] Flow pulsing or the use of additives or a nonwetted (e.g., Teflon) surface can sometimes be used to advantage to enforce slip at the wall.

Two extremes in realization of a plug-flow reactor (PFR) deserve discussion. The first is for reactions where even small deviations from plug flow entail serious shortfalls (see the discussion in Chap. 10). The second is pathological deviations from plug flow of mobile fluids when extreme bypassing and flow short-circuiting through vessels may occur from density or viscosity differences between reactants and products, encouraging layering or plunging, natural convection effects, or other serious disruption of uniform flow.

Hooper (1990) has reviewed a number of pitfalls in realizing PFRs or continuous stirred-tank reactors (CSTRs) from such situations. An early method for guaranteeing a plug-flow approximation for mobile fluids in tubular reactors was to force the flow through a series of baffles. Thus, experienced practitioners commonly provide intermediate means of radial remixing of the flow elements that have been experiencing disparate residence time histories. This may take the form of a sequence of large-hole orifice plates or the use of a relatively large length/diameter ratio (L/D) for the pipe reactor to enable radial diffusion to even out the histories. Most of these methods cost pressure drop. In fact, much of the judgment in fluid mechanics involves the intelligent dispensing of pressure-drop energy.

Internals for Remixing

The use of special packing or static mixer elements to remix radially can cause additional complications. In terms of pressure drop expended, it is hard to beat the orifice plate insert. In mobile fluids, static mixers are no cureall and may be justified for mixing only

---
[18] Coiled pipe provides an additional wrinkle for deferring the onset of turbulence and changing boundary layer flow (Schmidt 1967).

when axial length is at a premium for unrelated reasons (see the example of pH control in Chap. 8).

Fingering Instabilities

When a less viscous fluid displaces a more viscous one in an open or packed pipe, plunging or fingering of the former through the latter may occur. This can be a problem in cyclic sorption separations, in pipeline transfers, and in secondary recovery of petroleum from formations. Cooney (1974) shows its interference with separation efficiency in ion exclusion and demonstrates how staying below a critical superficial velocity can mitigate this effect.

Viscous Flow Reactors

In axial-flow tubular polymerization reactors, the enhanced viscosity developing near the wall from extended residence time can create velocity profiles much sharper than a laminar parabolic one, with centerline plunging effects similar to the fingering mentioned above. Heating the wall in what is essentially an adiabatic reactor is one strategy to counter such tendencies. Using a series of CSTRs to approximate a PFR is another expedient; a more imaginative solution is emulsion polymerization systems that suspend the polymer in water or another mobile fluid. In some polymerization reactions, a batch reaction system may have to be retained because there is no adequate PFR realization.

Two-phase Effects

Strong radial variations in the axial velocity profile may occur in nonviscous systems like the solids/gas dispersion of a cat cracker riser reactor, where the gas-phase velocity profile is considerably sharpened from segregation effects and drag of the heavier phase. In an expanded-bed fluidized reactor, the fluid attains a surprisingly good approximation to plug flow, while the solids circulation is close to that of a CSTR. This anomaly causes complications only if there is significant mass or heat partition from the fluid to the solid phase. Another variant, the "downer" reactor, in which both phases flow concurrently downward, is an underutilized expedient in gas-solid systems.

Particularly in countercurrent flow, we should take maldistribution as the expected *norm*, making a special effort to avoid it. This is not to say that process effects cannot aggravate such tendencies, as in an HCl absorber described by Kister (1991), where the nonlinearity of phase equilibrium out of aqueous HCl was unforgiving of an initial vapor maldistribution.

Flow Division

Subdivision of a gas or liquid flow into equal packets is often a necessary prerequisite to a subsequent heat transfer, distillation, or reaction step. It should be viewed as an inherent task of the chemical engineer in countering the debits of scale-up. However, there are potential pitfalls in subdivision:

1. As noted above, pressure drop is usually the price we pay to secure equal division, one that is a significant fraction of the total system pressure drop if it is to be an effective regulator. In Chap. 2 an example was given in which even massive expenditure of pressure drop via sonic orifices feeding a multitubular catalytic reactor failed to equalize flows because of the quirks of pressure loss/recovery along a manifold involving successive flow aliquots drawn off perpendicular to the manifold.[19]
2. The other issue (in a preventive sense) that the technologist should be considering is whether flow maldistributions, once they occur, will be self-correcting or self-sharpening in the system in question. If the latter, a higher standard of initial flow equalization must be set or feedback corrections arranged.

Sonic Orifices

These may be used for flow division, but their basic purpose is to *insulate* the supply pressure from downstream developments, which cannot propagate upstream against a sonic velocity. For example, a reactor's tendency toward oscillatory behavior may be vastly greater

---
[19] The nonobvious relationship between pressure loss/gain and the relative diameters of manifold and takeoff lines should be a standard reference in flow subdivision work (Acrivos et al. 1959).

if changes in temperature or molar conversion can feed back to affect the feed rate. The upstream/downstream pressure ratio will be ≈2 for a sonic orifice, but varying with the properties of the fluid in question. This technique is still very useful for laboratory-scale reactors but can no longer be considered first-class technology in commercial units because of its profligate use of energy.

Channeling

Even viscous pipeline flow can be considered channeling. Normally, it is undesirable for the components of flow to have differing residence time histories (there is usually an adverse shift in reaction selectivity from that experienced in batch development work), but there are cases where this may not be an issue.

Channeling in Packed Beds

One-phase flow through packed beds is the next most complex form of channeling. The presence of the packing does not ensure plug flow; it provides new opportunities for channeling and a short-residence aliquot of flow. Even if the pellets are completely uniform in size, there may be added void space near the wall (or around a central thermocouple sheath) where the fluid will have higher than average velocity. In a deep bed (high L/D) this may be mitigated by radial mixing, as the flow constantly redivides in passing around individual pellets.

The usual recourse in laboratory reactors or commercial multi-tubular exothermic reactions, where tube diameter D is constrained to ≤5 cm, is to keep pellet diameter d small enough (e.g., 0.3 to 0.6 cm) to provide a D/d of at least 8. When this criterion is barely met, the gap at the wall may dominate radial removal of heat from the reaction to the coolant beyond the wall.[20] If the particle size is further reduced, the wall gap no longer dominates, but the overall effective radial heat transfer coefficient may suffer from the pellet size reduction.

In large-diameter reactors, dividing a single bed into a series of smaller beds, with intervening redistributor plates, is the first line of defense for approximating plug flow. The initial distribution should

---

[20] Gunn et al. (1987) have reviewed the effects of both gas velocity and particle size on the thermal resistance of this wall gap.

be regarded similarly to the subsequent redistributions via plates or baffles; it does not guarantee uniform flow very far into the reactor. While the packing can serve to spread out flow from a number of "point" sources provided by the initial distributor, such spreading is not necessarily isotropic in that flow toward the wall of the pipe may be more likely than flow back inward once plunging at the wall (because of its extra voids) has begun. A sparger or inlet distributor (as distinguished from a jet entry) is simply an attempt to prevent the motive power of the feed from stimulating recirculation eddies.

The above situation assumes uniform particles. A dispersity of particle sizes may give greater bypassing, but only if the bed is improperly loaded. Gas/liquid (especially countercurrent) flow can cause severe bypassing and will be discussed further below.

Stirred Tanks

Before passing to multiphase fluid mechanics issues, let us consider the nature of shortfalls in achieving a truly well-mixed, homogeneous fluid in a continuously fed vessel, the classic CSTR. Cost and practicality considerations tend to make power input in large commercial tanks rise less than the 0.7 or 0.8 power of volume required to maintain mixing quality. Our purpose here is not to discuss reactor issues but to depict the dynamics of blending in the real-life tank, anticipating control examples in Chap. 8.

An early investigation with step concentration inputs (Cholette and Cloutier 1959) indicated deviations comprising both a plug-flow shunt of a small fraction of flow straight across the tank and semistagnant zones that participate little in mixing (see Fig. 6-1). However, this shunt flow must have a finite transit time θ (visualizable as the time for flow to pass from inlet to exit at finite recirculation). Even a delay θ that is small compared to the overall residence time may be of considerable dynamic significance.

Jet-Stirred Tanks

Open cylindrical vessels fed axially with an entry jet can sometimes approximate CSTR behavior without other agitation. The effect of L/D and other parameters on deviations from CSTR behavior, in terms of short-circuiting and a long-residence-time tail, have been investigated by Veeraraghavan and Silveston (1971). In their study,

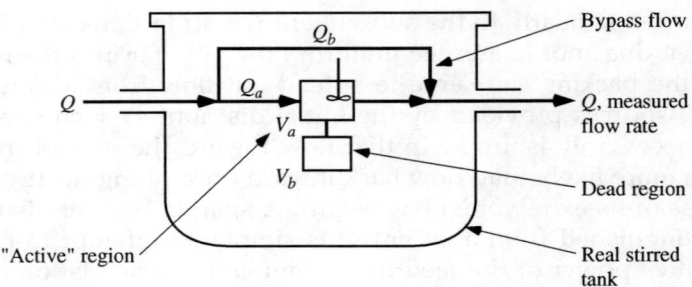

Fig. 6-1 Model of Cholette and Cloutier for a Real Stirred Tank (From Himmelblau 1968)

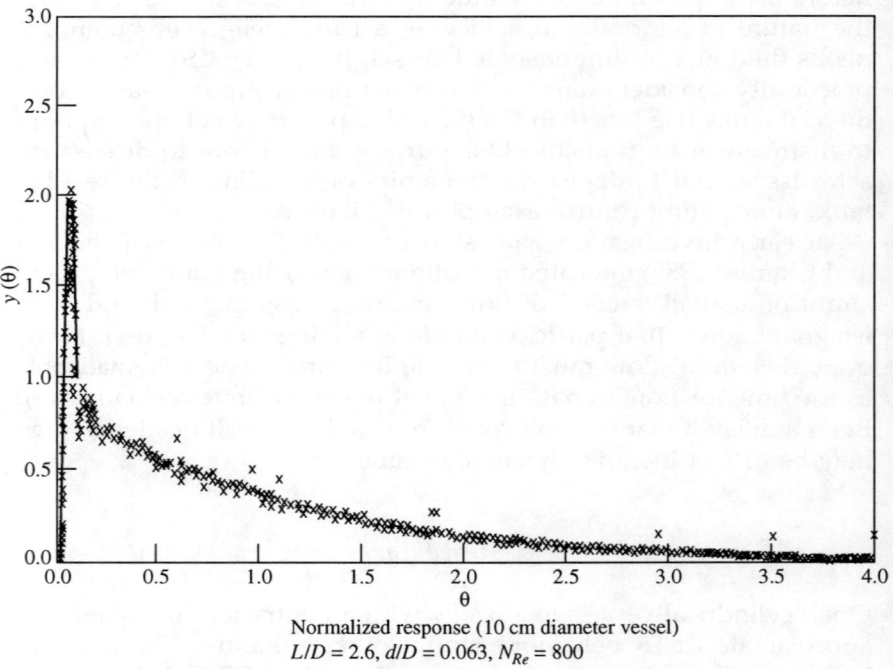

Normalized response (10 cm diameter vessel)
$L/D = 2.6$, $d/D = 0.063$, $N_{Re} = 800$

Fig. 6-2 Pulse Response of Imperfect CSTR with Plug Flow Bypass (From Veeraraghavan 1971)

a superior impulse response technique (see Fig. 6-2) exhibited the early "spike" corresponding to the short-circuit with residence time θ corresponding to about 10% of nominal residence time in the vessel; while the remaining 90% acts as a valid CSTR.

In another context, side-mounted jet recirculation is frequently used for mixing storage tanks, where, for example, the gland of a mounted mixer or some other process risk precludes the use of an internal agitator. The key features of this mode of agitation (e.g., angling the jet entry so as not to break the free surface of the liquid) are now well established (Fox and Gex 1956). Mixing in large storage tanks is usually done batchwise, with the somewhat conflicting goals of minimum mixing time and safety (e.g., relative to static electricity generation). The key parameter, of course, is the effective pumping capacity of the circulating device. The performance of impeller stirrers, mixing jets, and eductors appears to be about the same, relative to energy input, and peripheral issues will settle the choice. An eductor-type mixer seems to have the edge with respect to safety issues.

Agitation for Two-phase Suspensions

Sometimes we are faced with the modest task of preventing dropout of a minor solid phase in a tank. Experimental work has correlated this minimal degree of agitation (Zwietering 1958). For a fairly uniform suspension, more power will be needed. Adding a draft tube to a turbine impeller is usually helpful. Homogenization with wall baffles added has also been correlated (Weisman and Efferding 1960). If gas is being bubbled through a liquid, the task of suspending solids requires faster stirring than in the nonaerated case. Slurry reactors are discussed in Chap. 11.

When mass transfer to or from a solid suspension is involved, extra power will have only a mild effect (1/6 power dependence) on the mass transfer after the initial suspension off the floor of the vessel is achieved. A reasonable correlation of power needs in both the initial suspension and homogenization modes is available (Levins and Glastonbury 1972), using the settling velocity of the particles as a fundamental parameter in predicting particle Nusselt (Sherwood) numbers for mass transfer.

A recent thoughtful review (Villermeaux 1988) compares the extent to which various devices use energy effectively in obtaining mixing or mass transfer contact of a fluid with a solid surface.

Closer Control of the Quality of L/L Dispersion

The objective in passing through a contactor is usually a rapid mass transfer equilibration, maximizing the interfacial area term a by decreasing droplet diameter d. Calderbank (1958, 1959, 1961) was the first to characterize separately area and k_L trends in mechanically agitated G/L and L/L contacting. Our usual goal is to produce a uniform population of droplets just fine enough to achieve the desired mass transfer in a specified time interval, recognizing that both coalescence and surface aging will rapidly diminish mass transfer as soon as we exit the device where energy expenditure is promoting dispersion. An excessively fine dispersion will only complicate the subsequent coalescence needed for decanting the minor liquid phase.

Effect of the Energy Input Field on L/L Dispersion

Correlations for drop size distribution in L/L dispersion systems tend to be based on the theoretical concept of a maximum stable drop size (MSD)[21] with a logarithmic function of diameter that is linear vs. a cumulative probability distribution of volume extending down from the MSD (Mugele and Evans 1951). Such a three-parameter description (e.g., MSD, d_{95}, and d_{50}) holds well between the 10th and 95th percentiles[22] of the cumulative volume distribution of drops (see Fig. 6-3).

Only in layouts where turbulence is generated uniformly throughout the volume involved does the theory apply with any rigor. For many dispersion devices, there are high-shear regions near the impeller or blades where drop breakup predominates and so-called circulation zones where dynamic coalescence takes place. In such devices, a uniform turbulence field is only roughly realized. Similarly, in pipeline flow, realization is best in the turbulent core of the pipe, while other phenomena are operative in the laminar sublayer near the wall. The superiority of a pipeline loop reactor over a stirred tank for maintaining a uniform L/L dispersion (even though both use impeller agitation) has recently been asserted. Such a loop

---

[21] MSD varies roughly inversely with pipeline velocity. Conflicting data give a stronger inverse dependence.

[22] As in most such distributions, the tails outside the 5th to 95th percentile range may not follow the trend line. So-called extreme-value statistics would be invoked to deal with these tails (Gumbel 1958). Of course, by its nature, the MSD relationship can tell us little about the large drops in the 0 to 10th percentile range.

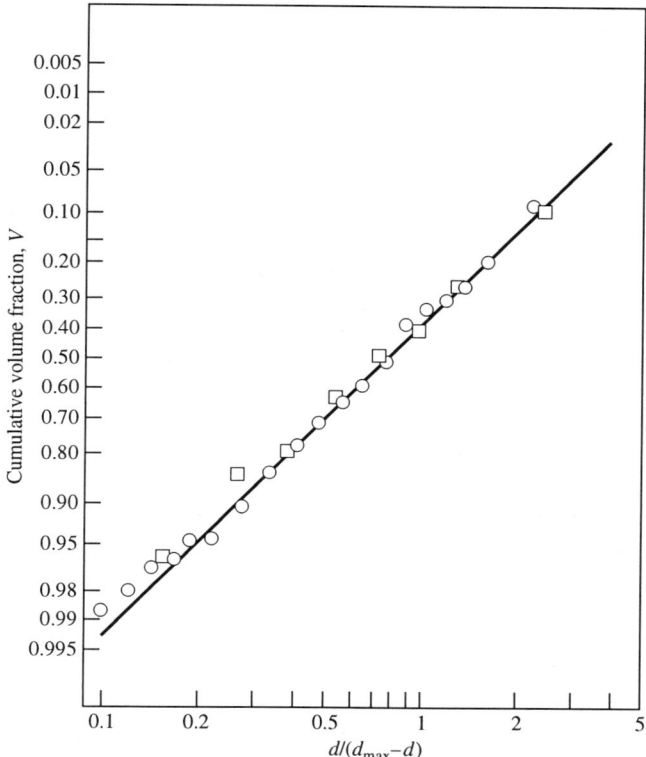

Fig. 6-3 Upper-limit Log/Probability Drop Size Distribution

therefore can give us the residence time characteristics of a CSTR, with better control of the fine structure of dispersion.

There is more theoretical discussion in the literature than good experimental data on L/L dispersion in pipeline flow and considerable divergence (Collins and Knudsen 1970, Karabelas 1978) among existing data. Some of the spread reflects the competing effects of surface tension and viscous forces in breakup and cleanliness, as well as geometry differences affecting coalescence. Fortunately, the offsets among available literature correlations matter less in a troubleshooting mode (e.g., where we are seeking a differential change on an existing pipeline mixer to obtain three instead of two effective transfer units) than in a design mode, attempting to work from first principles. Where the droplet population can be observed directly or by inference, troubleshooting can proceed in an empirical way.

Laboratory characterization of L/L dispersion is most conveniently done in stirred vessels. Translation of such experiments into a field pipeline on the basis of equivalent energy/volume is usually an adequate basis as long as the scale of equipment does not differ widely. However, scale-up (or scale-down) in multiphase fluid mechanics, even absent reaction, can maintain only partial similarity. Zlokarnik (1987) gives a thoughtful review of which parameters to focus on under such partial similarity conditions.

As an example of a situation amenable to calculational approaches, consider a multitray, pulsed extractor column. Here the correlational approach works, largely because most of the interphase transport is associated with the immediate shear field wherein new droplets are created. The rise and coalescence of the droplets between trays have less effect on overall mass transfer.

The newer static mixer elements have no particular advantage over old-fashioned perforated plates or even globe valves as dispersion devices when mobile fluids are involved. The key element is either the velocity achieved in the plate perforation or other dispersion channel or the (closely linked) pressure drop.

Physical Property Effects in Dispersion

The influence of physical properties—density difference between phases, viscosity of the dispersed phase, interfacial tension—has been advanced theoretically but not supported broadly by good data. Lowering interfacial tension via an added surfactant can, in my experience, be useful in achieving smaller drops, provided that it makes sense otherwise in terms of cost, dynamics of action, and process compatibility.

The question often arises: What determines which of two liquid phases will tend to be dispersed when stirred together? There is a wide range of "ambivalence" that permits the major phase to remain disperse, up to 90%v for a hydrocarbon and up to 75%v for water. Higher viscosity makes it more likely that a major phase will stay dispersed.

Settling and Coalescence—The Other Side of the Equation

So-called primary dispersions have drops ≥ 100 μm in diameter. Equilibrium drop size emerges from a balance between breakup and coalescence taking place dynamically within the dispersion device by drop/drop impact. Dynamic coalescence is, to put it mildly, in correlative disarray. Primary dispersion drops are also coalesced in a subsequent phase separator by drainage and/or rupture of the film separating touching drops.

Secondary dispersions denote droplets ≤ 1 μm in diameter. They can arise from supersaturation effects, discussed later in this chapter, or as "daughter" droplets ejected during rapid coalescence of large drops or tearing of a larger drop out of a bulk phase. Coalescence of such fine drops requires special approaches, since very low Stokes settling velocities preclude gravity settling. Use of electric fields, chemical coagulants, centrifugation, and flow through porous media are among the usual recourses. The last of these exemplifies coalescence via collision with a third body. Jackson et al. (1976) have shown aerosol collection by passage through several 2.5-cm-deep beds of fluidized alumina granules. Another third-body option that I believe would be more widely used if there were a more definitive treatment in the literature is full spray patterns of scrubbing liquor to absorb finely dispersed drops of a phase-compatible mist.

While interception of dusts is a complex art that space does not permit me to treat here, it is pertinent to our discussion of third-body scrubbing to note that a venturi scrubber is a very simple and effective device, achieving removal of dust particles down to 0.1 μm from waste gases, using atomized droplets in the 25- to 250-μm size range, with rather low energy consumption.

Another version of the third-body approach is provision of extended surface preferentially wetted by the disperse phase in phase separators to aid coalescence of primary dispersion drops. Use of such internals is not without problems, however. Fouling of such surfaces, used to shrink the volume required for phase separation, with dirt present in most practical situations, can lead to inconveniences in opening and cleaning equipment that would not be necessary with a larger, less complex layout. Another potential weakness is secondary droplets formed during rapid coalescence of primary drops at the wetted surface. The extensive experimental and theoretical work of Das and Hartland (1990) on the film drainage and rupture effects that dominate rate in "quiet" coalescence is recommended reading for troubleshooters who are deeply involved in coalescence problems.

Where coalescence is unwelcome because it reduces the interfacial area needed to complete needed mass transfer, but additional energy dissipation is not appropriate, there are several options. One is the use of nonwetted contact surfaces; another delays impact of the disperse phase upon a coalescing wall, as in careful centerline axial injection of the disperse phase in a pipe.

Stokes' Law (SL) and Its Byways

The familiar relation for the relative velocity of a spherical particle in a continuously surrounding fluid phase has served me well in innumerable troubleshooting situations, either to infer what the dispersed phase particle size must be or, given a particle size, the implication for slip velocity. However, the troubleshooter should be aware that isolated and spherical bubbles or droplets are an idealization. In general, both gas bubbles and liquid drops are distorted from their spherical shape, reaching mushroomlike, spherical-cap shapes when interfacial tension is low or the bubble exceeds a certain size (1.5 mm in the air/water system). Such deformed bubbles rise in zigzag fashion rather than by the simple verticality assumed in SL.

Even in the near-spherical situation, a low-viscosity dispersed phase has internal recirculation and oscillations that (particularly for liquids) reduce its rise velocity vs. the surrounding fluid relative to what it would be for a solid sphere (Hughes and Gilliland 1952). Further, even a solid sphere has a smaller rise velocity than the simple SL treatment calculates.[23] These various SL deviations are recapitulated in dimensionless form in Fig. 6-4.

Another complicating factor is that bubbles often travel in swarms. Gal-Or and Waslo (1968) analyzed mathematically the rise velocity of ensembles of drops or bubbles of similar size. Where the volume fraction of the dispersed phase is appreciable and the viscosity of the continuous phase is substantially greater than that of the dispersed phase, their results indicate a substantial reduction in rise velocity for the ensemble.

Example: Fluid Mechanics in an Extractive Upflow Reaction

As an example of the complexities of a seemingly simple system, consider an upflow open-pipe contactor I worked on, involving ex-

---

[23] This is another example of careful experiments opening up a subject that was supposedly settled by theory. It seems that a hard sphere carries an effective skin of "borrowed mass" from the surrounding fluid that increases effective drag by 50% or more.

Fluid Mechanics

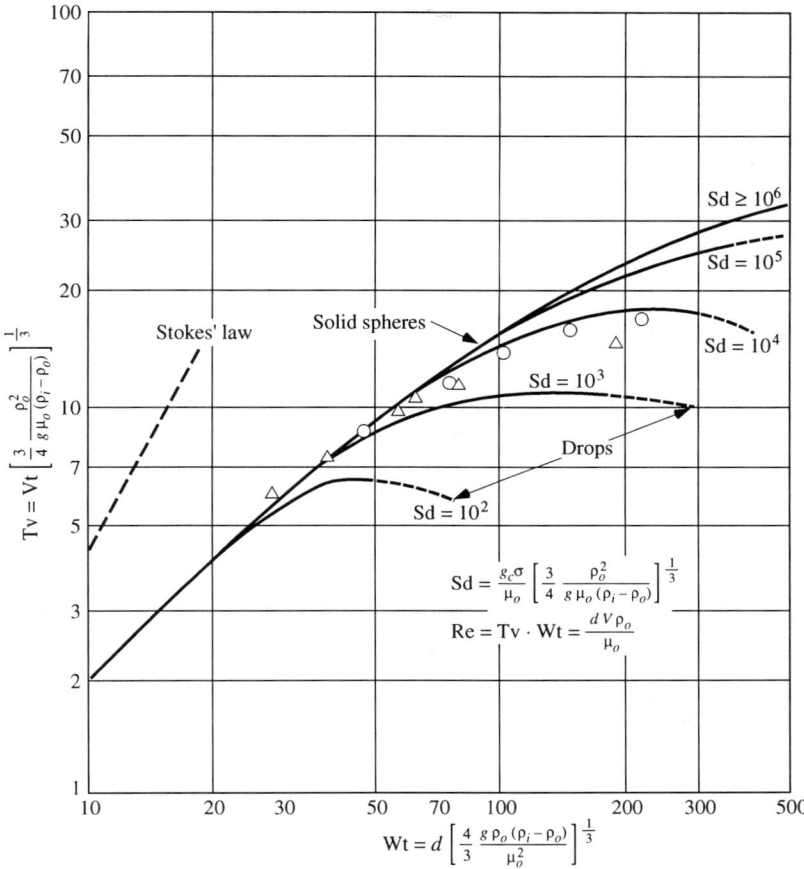

Fig. 6-4 Terminal Vertical Drop Velocity as Corrections to Stokes' Law (Dimensionless) (From Hughes and Gilliland 1952)

traction of a light liquid phase, initially dispersed into a heavier, aqueous phase via an entry nozzle at the base. The co-current flow is modified by substantial "slip" of the heavy continuous phase with respect to the lighter phase; both the limited residence time of the light phase and extraction from it depend upon the droplet size that defines the interfacial area and (via droplet rise velocity) its slip with respect to the heavy phase. The initial droplet swarm is modified by coalescence commencing somewhat above the nozzle injection at a distance that depends upon slip velocities. In the actual reactor problem from which this example is drawn, an empirical approach was taken whereby the preferential holdup of the heavy phase was inferred from the average density of the two-phase mixture in the tank, as measured by a differential pressure (dp) cell across the height of the tank.

Quasistable Dispersion

Viscosity Effects in Mixtures of Small and Macro-molecules

In multi-phase mixtures that do not separate over extended periods, a number of questions arise on transport properties and other matters.

We expect the viscosity of simple fluids to differ functionally from the effective viscosity of systems containing smaller (e.g., solvent) molecules among multimolecule aggregates such as polymers. Similarly, diffusion within a polymeric system is likely to have different dependencies on apparent viscosity for smaller molecules and for polymer molecules than the roughly inverse[24] relation with viscosity we expect in simple fluids.

Rag

This is one of those terms, like *gunk* and *polymer*, that have been coined to name real effects but that act to limit our thinking rather than to reveal the phenomenon to us. *Rag* denotes a cloudy pseudo–third layer that intervenes between two reasonably clear major liquid phases and hinders their disengagement. Several distinct mechanisms can create a rag.

For example, rag can develop from an amphoteric element like Fe or Al that precipitates as, say, a hydrous oxide in the course of processing. This hydrous oxide may constitute an inorganic polymer or a network of some rigidity and structure, preferentially wetted by one of the major phases but invading the other phase due, for example, to density differences. A second mechanism creating a rag is a tight surfactant-driven emulsion of one major phase in the other,

---

[24] In fact, the dependence is somewhat weaker, $\approx \alpha\ \mu^{-0.6}$.

with a high effective *Einstein equation* viscosity[25] that slows assimilation of the external phase into its proper clarified major phase.

A third mechanism that often figures in rags is a *Pickering* emulsion of the major liquid phases stabilized by fine solids that are partially wettable by both major phases and whose particle size is much smaller than the droplets of the emulsion internal phase. Total coating of the L/L interface by such solids enhances the rigidity and effective viscosity of the rag layer. The thermodynamics favoring the formation of such tight three-phase rags has been reviewed by Levine et al. (1989).

Rag is best addressed as a chemistry problem rather than a fluid mechanics problem, seeking ways to prevent its initial precipitation or to destabilize it chemically to recover the phase occluded in it. A pH adjustment is often surprisingly effective.

Microemulsions (MEs)

Most lattices, rags, and troublesome emulsions are not stable but have a viscosity/rigidity that delays indefinitely the thermodynamically favored coalescence. In MEs and microdispersions (MDs) we meet systems that are inherently stable. A ME will have a drop size of about 0.1 μm, obtained with sufficient surfactant[26] to bring the oil/water (O/W) interfacial tension down to 10^{-2} dyne/cm, much smaller than in an ordinary emulsion. A MD is distinguished from a ME by the stronger effect of the packing factor and by geometry in its formation and a lower domain size, 10 nm.

Such pseudophases are now often used in applications such as dry cleaning and have received much recent attention because of their significance in enhanced oil recovery schemes. From the standpoint of troubleshooting, the important features of MEs and MDs are their transparency and their solubilization effect on the volatility and other properties of the internal phase. We are also likely to en-

---

[25] A latex system wherein the internal phase constitutes a high-volume fraction of the dispersion has a viscosity manyfold higher than that of the bulk external phase. (Up to 93 %v is theoretically possible with optimal packing of spherical particles of varying size.) Einstein derived the hard-sphere equation that describes this effect.

[26] The dispersion entropy term overcomes the small but positive interfacial tensions of such systems. All of the considerable (about 2%v) surfactant population is in a monolayer at the O/W interface, which defines the arithmetic of drop size. A cosurfactant that distributes well between the O/W phases (such as a lower alcohol) may be necessary for ME stability. A lower alcohol alone may suffice to form an O/W ME. Water-in-oil MEs with an appropriate detergent have also been reported.

counter them frequently in new high-tech systems (e.g., in extraction of heavy metal ions from water by organic acids dissolved in the interior of a micellar system).

Heat Transfer and Phase Change

Evaporation of a Liquid Sprayed into a Gas Stream

In troubleshooting we may occasionally be required to design or to validate the performance of a liquid spray injection. Such a system may be for the purpose of rapid dispersion of a gas-phase inhibitor, a postreactor quench, humidification, fuel injection, and so on. This area is suitable to fairly exact modeling, permitting integral calculations over the trajectory life of the spray. The droplet goes to its effective wet-bulb temperature; coalescence and further breakup after the atomization are disregarded. The drag coefficient is well known as a function of droplet Re number, or the spray may be fine enough that slip between the liquid and gas phases is negligible.

Boiling Anomalies

Reboilers are a common source of process problems and provide a good transition to the complexities of gas-liquid fluid mechanics. The two areas referenced here are instabilities in vertical thermosiphon reboilers and reduced flux from film boiling. Certain instabilities are implicit in two-phase flow[27] and are simply brought to full realization in a heat-transfer environment.

Thermosiphon Instabilities

Thermosiphon instabilities, well known since the 1960s, may now be forgotten by the new generation of technical people. They rest on the nonlinear change of pressure drop and heat transfer as liquid converts to vapor in passing through a tubular heater. Two considerations dominate:

1. Choosing a percentage vaporization that maximizes the U value. This is basically a design issue, but it becomes a troubleshooting

---

[27] Wallis and Heasley (1961) give an interesting mathematical analysis of three distinct oscillatory modes that are possible in a two-phase, natural-circulation loop system.

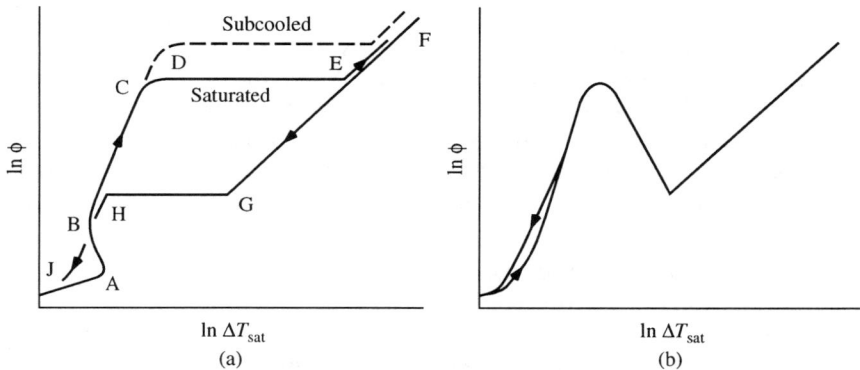

Boiling curve, heat flux versus wall superheat: (a) heat flux controlled; (b) wall temperature controlled.

Fig. 6-5 Metastable Transition from Nucleate to Film Boiling (From Kenning 1977)

constraint issue when process evolution leads to deviation from the 25–40% vaporization/pass that is near-optimum for most systems. In fouling-prone systems, lower percentage vaporizations must be used or forced-circulation systems adopted, which suppress boiling within the exchanger.

2. An unstable boiling cycle, wherein the vapor "lift" grows until flow choking results from increased heat transfer and percentage vaporization. Throttling the vapor return or (preferably) the liquid feed is the usual method for breaking this cycle.

Film Boiling

Film boiling refers to a sharp decrease in boiling flux through a metastable transition region as the wall/fluid temperature differential is increased and nucleate vaporization is superseded[28] by a blanketing film of vapor over the entire hot surface, as shown schematically in Fig. 6-5. This is not just a textbook curiosity. My experience suggests that it plays a role in:

---

[28] A recapitulation of the metastable transition to film boiling by van Ouwerkerk (1972) makes it clear why hysteresis will occur in passing from one regime to another. The nature of the transition depends strongly upon whether a wall temperature or a wall flux (e.g., electrical) is the boundary condition.

1. Relatively clean systems where an overly conservative choice of temperature differential ΔT puts the system near the critical heat flux for the transition to film boiling. Water-dominated data and scaled-surface data have led to expectations of 35°–50° C for the critical ΔT where nucleate boiling ceases. For relatively clean hydrocarbon systems at elevated pressures, however, the extinction of nucleate boiling can begin at a differential as low as 7° C (Kenning 1977).

2. Fostering the very fouling cycle that the designer feared when he provided an overly conservative ΔT between heating medium and process fluid. This can push the system into a film boiling mode where high metal skin temperature promotes degradation reactions that cause fouling deposits to accumulate. I have seen cases where a reduction in heating medium temperature (steam pressure) actually extended the time between reboiler cleanings.

G/L Flow in Pipes

We now pass into the more complex domain of coexisting G/L flow. Perhaps the first thing the technologist seeking to improve her skills in this area should do is to become familiar with the various characteristic flow regimes, mapped in terms of the superficial velocities of the respective phases. There are many published versions. Let us begin with the less complex case of horizontal flow. I have combined features of Sternling's treatment (1965) with Dukler's mapping (Taitel and Dukler 1976) in Fig. 6-6.

The chief purpose here is not to use quantitative computer programs to estimate pressure drop[29] but rather to master the qualitative aspects of the different flow regimes. In the following discussion, we will walk through the map as part of achieving physical identification with the texture of the problem. We are concerned with irregularities and inhomogeneities that stem from stratification by gravity, annular stratification, axial segregation, atomization into bubbles and droplets, and accelerational pressure drop. The process troubles that may arise from these effects include pulsing of pressure and/or mass flow, unwanted carryover of liquid components,

---

[29] See Cindric et al. (1987), based on activity of the AIChE Design Institute for Multiphase Processing (DIMP) in bringing some correlative order and doing new experimental work in this field.

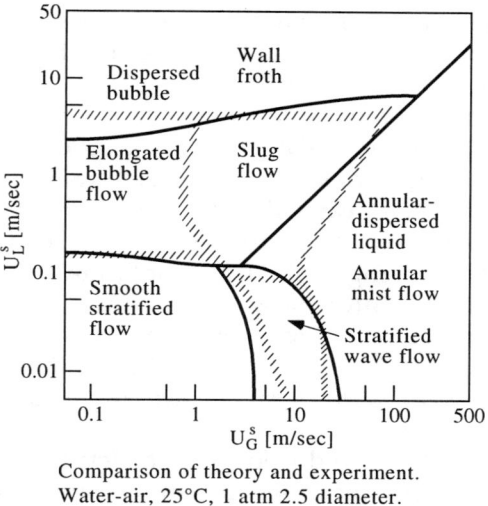

Comparison of theory and experiment.
Water-air, 25°C, 1 atm 2.5 diameter.
Experiment:—; theory: //// Mandhane et al.
(1974) regime descriptions.

Fig. 6-6 Air-Water Flow Regimes in Horizontal Pipe (From Taitel 1976)

mechanical failure, erosion, corrosion, deposits, and reduced heat transfer.

Horizontal Cocurrent Flow

Fig. 6-6 defines flow regimes for the air/water (A/W) system at 1 atm in 2.5-cm tubes. The theoretically derived boundaries between regions (Mandhane et al. 1974) are seen to be remarkably close to experimental observations.

Let us start with gas flows containing only token amounts of liquid. For relatively low superficial gas velocity (the lower left corner of Fig. 6-6), flow is smooth but vertically stratified, with the gas above. Wave formation in a still-stratified system begins as gas velocity increases further. Annular flow, with a liquid wall film surrounding a gas core containing some disperse liquid droplets, commences as the superficial velocity of the gas increases further and (as gravity forces become relatively less important) liquid is thrown to the walls of the pipe as a natural segregation of the shared-flow system. The basic concept here is that wall drag will cause slippage of the higher-density phase with respect to the gas flow. The liquid film is always rippled but is stable, while the ripples remain small in amplitude.

As gas velocity becomes large, we reach the mist regimes, shown at the lower right corner of Fig. 6-6. This is an especially important region because the generation of fine-mist entrainment is central to many problems. What creates the mist (and here viscosity and surface tension must surely be factors) is wave formation at the G/L interface, finally reaching such large amplitudes that the crests are torn off as mist and enter the gas core. This is akin to the "sea-spray" effect, in which curling over of an ocean wave crest permits small droplets to detach from the main body of liquid and enter the core of gas flow. For water systems, we are speaking of drops in the 50-μm size range.

At somewhat higher liquid velocities, the left-to-right transition in Fig. 6-6 as gas velocity increases encounters not wave phenomena but *slugging* of the liquid phase. Here liquid bridges the pipe cross section and begins to travel at about the velocity of the gas. Intermittent elongated gas bubbles punctuate the slugs of liquid (hence the alternative term *intermittent flow*). The rapid and somewhat random alternations of the two phases in slug flow can lead to vibrational/structural problems or to higher-frequency pressure pulsations or heat transfer anomalies. At still higher liquid flows, the elongated bubbles of gas become small bubbles in a *dispersed-flow* regime. As gas flow increases, the bubbles become a *froth*.

The boundaries between stratified, slugging (intermittent-flow), and froth (dispersed-flow) regimes as liquid flow increases are affected both by pipe size and by fluid properties like surface tension and gas density. Increasing pipe size shifts the transition between regimes to higher liquid velocities. Baker (1954) suggested groupings for adjusting regime boundaries for other fluid properties. For example, as surface tension decreases from the relatively high value of water, there is a leftward shift of region boundaries such that annular flow occurs at lower gas velocity in a hydrocarbon, higher-pressure system.

Froth flow can be considered benign for most process considerations; however, the *choked* regime that enters (upper right) at high velocities of both phases is a troublemaker. Both moment-by-moment variation in phase proportions and pressure drop rise rapidly as the phases compete for passage through a limited cross section. Timely intervention to provide coarse phase separation, followed by separate conduits for the two phases, should be considered when such a regime is possible.

Example: Pulsing Flow in Reactor Transfer Lines

In one startup of a train of bubble reactors, fluid mechanics was reasonably well behaved in the large-diameter G/L upflow reactors themselves but strong pulsations were experienced in the lines between reactors until the separate-conduit solution was adopted.

Vertical Two-phase Flow

For vertical two-phase flow, there are four possibilities involving upflow and downflow, cocurrent and countercurrent cases, of which two have major process relevance and a third has minor interest. A fusion of cocurrent upflow and downflow work by Dukler (Taitel et al. 1980, 1982) with Sternling's (1965) mapping of the four subcases is shown as Fig. 6-8.

Cocurrent Upflow

The first difference we note from horizontal flow is, logically, the absence of the stratified flow regime at low liquid rates. Let us first consider qualitatively the four major regimes, shown in Fig. 6-7. At low gas flow and low to medium liquid rates, we have bubble flow, with relatively coarse bubbles well dispersed in a continuous liquid stream. In large-diameter versions of this regime, characteristic of many sparged reactors, it is necessary to understand the extent of liquid circulation patterns induced by gas bubble plumes rising from a sparger injection and, stemming from this, slippage of the gas phase with respect to the liquid and effects on G/L mass-transfer contacting.

As gas flow increases, at low to medium liquid flow, bubbles coalesce to form first free-rising, spherically capped voids (see the earlier discussion of deviations from spherical-bubble SL behavior). The method of initial dispersion of gas via a sparger may be critical to deferring the onset of such coalescence. Ultimately, bullet-shaped gas *pistons* form that intermittently fill the whole cross section of the pipe, usually when the gas volume fraction approaches 0.3. The length of the intervening liquid *slugs* in stable slugging will be about

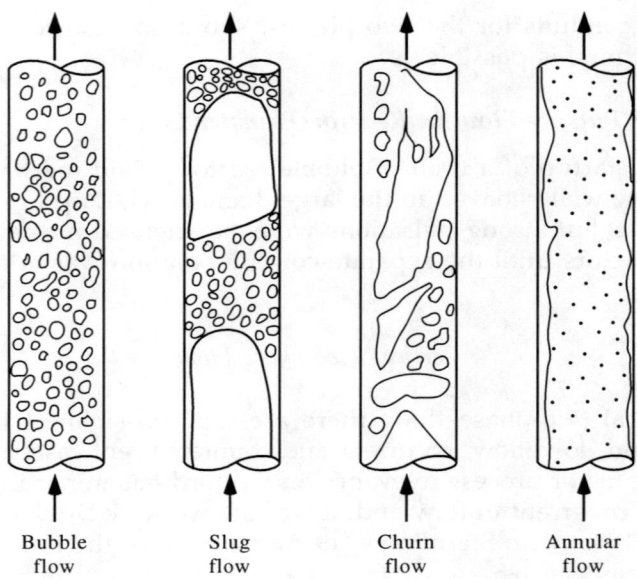

Fig. 6-7 Major Flow Patterns in Vertical G/L Flow (From Taitel 1980)

16 pipe diameters. It is not a regime that we normally choose to be in.

At still higher gas flows, these pistons of gas become distorted as the continuity of the intervening liquid is destroyed; an oscillatory, frothy *churn* flow takes over. Dukler showed that churn flow is an entrance phenomenon, organizing itself into slug flow farther down the pipe (Taitel et al. 1980). The dashed lines within the churn regime in the upper right quadrant of Fig. 6-8 are L/D parameters for this transition. However, at gas superficial velocities of 6 m/sec, this "entrance" zone may last for 500 pipe diameters and constitute the whole pipe length!

At gas rates above 15 m/sec[30] a more organized *annular* flow resultts, with gas flow in the core. The similarity of this regime to that for horizontal flow is not surprising, considering that forces other than gravity are dominant. A wavy liquid film flows upward at the wall, with sheared-off droplets entering the gas core as entrainment.

---

[30] This is for A/W at 1 atm; the transition gas velocity V_{SG} can drop to 1.5 m/sec in high-pressure hydrocarbon systems. The Froude number, $V_{SG}\sqrt{\rho_g/g_c D\Delta\rho}$, can unify the effect of pipe diameter D and fluid properties on several of the regime boundaries.

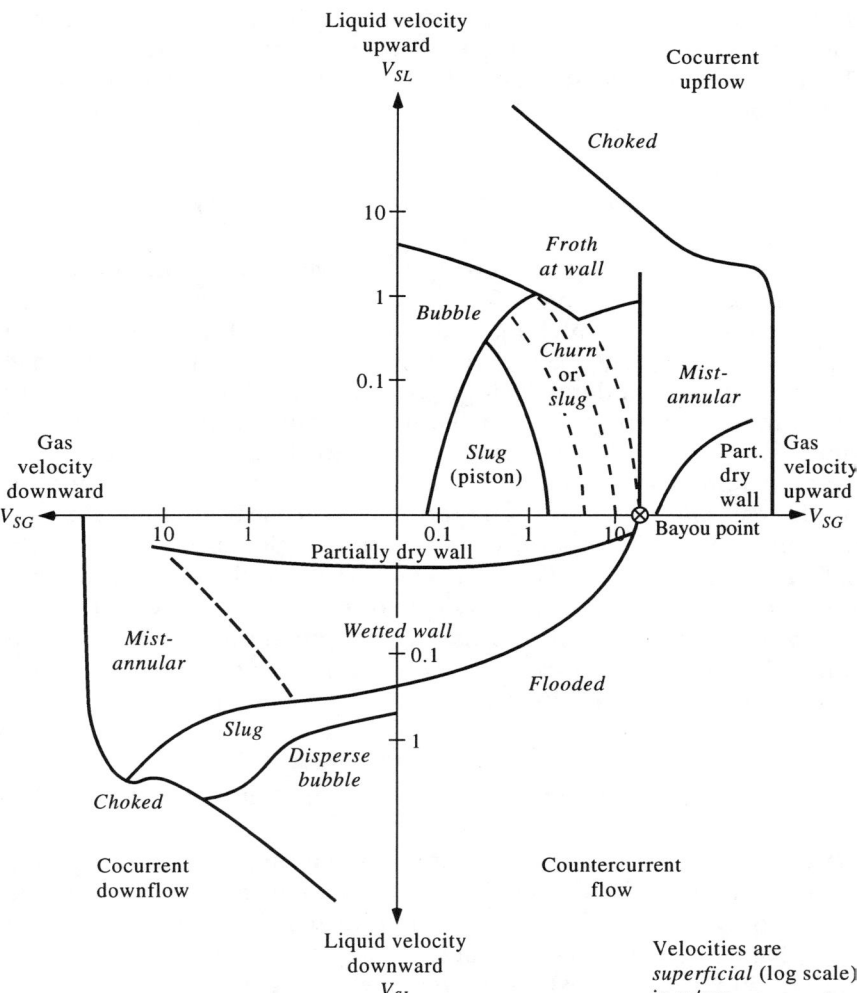

Fig. 6-8 Four-quadrant Map of Vertical G/L Flow Regimes

At high liquid flows and medium gas flows, the gas will tend to flow in a *disperse* bubble regime, similar to the horizontal case and to the low-gas-velocity upflow regime, except that at medium gas velocities, flow near the wall becomes a froth, as a sort of friction-drop-reducing adjustment.[31] In most of the regimes, slip of one phase with respect to the other is the rule rather than the exception. Choked flow, at very high velocities of both phases, is similar to the horizontal case.

Countercurrent Flow with Liquid Downflow

While our first thought in viewing the lower right quadrant of Fig. 6-8 may be of distillation systems, the insights here apply to open pipe issues, such as catching and run-back of entrainment and the capacity of gas-sparged countercurrent reactors. At vanishingly small liquid flow and moderately high gas flow, we find the *bayou point* phenomenon. Here a drop of liquid, introduced at the wall, will stand still. This point is closely related to flooding in wetted-wall columns (and also to the onset of annular flow in cocurrent upflow). As the gas rate falls and the liquid rate increases, we follow a continuous flooding boundary, to the right of which liquid will be partially entrained upward. For relatively light liquid flows and gas flows below the bayou point, the liquid may run down the wall in rivulets, with partially dry areas, having potentially harmful effects on contacting or wall heat transfer. The boundary of this partially dry-wall regime depends strongly on wetting behavior.

Cocurrent Downflow (lower left quadrant of Fig. 6-8)

Why would we configure a system this way rather than using horizontal cocurrent flow? Sometimes it is the shortest run connecting two upflow vessels. Sometimes it minimizes slip or backmixing of one phase in a G/L reactor system. Also, with gravity working for us, we minimize the extent of the slugging regimes, which are so prevalent in horizontal and vertical upflow situations.[32] In their place,

---

[31] Wicks (1991) has used this "anthropomorphic" way of looking at two-phase flow to derive a wide variety of phase stratification profiles based on a principle of minimum energy dissipation.

[32] The data in small pipes of Taitel et al. (1982) gives some indication of narrowing of the small slug flow regime as pipe size increases.

we see a continuation of the wetted-wall and rivulet flow phenomena shown in the lower right quadrant of Fig. 6-8. These can be construed as a variant of annular flow, which grows complex only when we enter a mist-annular regime at high gas velocity. With the exception of the minimal slug regime, cocurrent downflow is, in most respects, nearly a mirror image of cocurrent upflow. As liquid velocity increases, the narrow slug flow regime gives way to a dispersed-bubble regime and finally to a choked regime at very high liquid or gas velocities. If the mass transfer or reactor situation precludes the use of packing, a cocurrent downflow regime such as wetted-wall or bubble flow may be practical for our needs.

Countercurrent Flow with Liquid Upflow (upper left quadrant of Fig. 6-8)

This is of interest mostly to firemen using hoses in confined spaces.

Systems with Mass Transfer Gradients

Where chemical equilibrium exists between phases, the interface generally will be well behaved (except for the generation of annular flow mist discussed above). In nonequilibrium, a number of troublesome situations can occur. One such situation exists when local supersaturation evokes fine nuclei of a new phase.

Fog Formation in Condensing

Most fledgling engineers are somewhat aware that noncondensible gases can substantially decrease condensing heat transfer. The rationale for this effect is that noncondensibles lower the temperature at which condensation takes place. Schrodt (1972) has furnished analytical solutions for both heat and mass flux in these cases. Guidelines for a mechanical layout to facilitate venting of inerts are also available (Standiford 1979).

Much less well known is the common formation under these circumstances of fog, which has a number of potentially serious effects. Steinmeyer (1972) shows how the lesser molecular diffusivity of the higher-mole-weight condensibles relative to the noncondensibles gives a mass diffusivity significantly less than its thermal diffusivity to the vapor-gas mixture. This imbalance causes supersaturation of the condensible vapor before it reaches the cold wall. However, whether nucleation of a liquid phase actually occurs depends very much upon the presence of foreign bodies to serve as nuclei. Adiabatic expansion of a vapor and a noncondensable gas can also create fog, both because of the factors just cited and the speed at which the vapor crosses the dewpoint. The mathematics that define the boundaries of such regions has been given by Toor (1971).

Loss of material or the emissions that result from entrainment due to fogging in a partial condenser may be only part of the prospective PTS problems. In systems where there are two liquid phases in the condensate, we may also end up with a tight emulsion to decant when one of the condensibles supersaturates readily and the other serves as a source of nuclei.

We skirt crystallization, a complex field that deserves its own treatment. However, it is important for the troubleshooter to sense situations where supersaturation is likely to exist and the resulting nucleation is likely to outpace growth of the new phase, since this means fine particles and TROUBLE. Nucleation of a microphase need not always be a catastrophe, of course. It is used to advantage in, for example, the ceramics industry, where very fine crystallites of a second compound serve as a "nucleating dust" in cooling a melt to ensure the required fine-grained structure.

Spontaneous Emulsification at Interfaces

We should pay early attention to whether mass transfer is accompanying phase contact, including its direction. Rapid transfer in one direction can cause spontaneous emulsification at an L/L interface, while the same system, but with transport reversed, can show accelerated coalescence and inferior mass transfer. Spontaneous interface eruptions tend to occur when the solute is leaving a low surface-tension (against air) phase and entering a high surface-tension phase (e.g., water). (See the next chapter.)

Electrical Effects

We are generally accustomed to thinking of colloidal suspensions as stabilized by mutually repelling charges. A number of treating systems, as in water clarification, depend upon destabilizing such suspensions by adding a critical amount of a polyelectrolyte or other additive that neutralizes the electric charge. However, the effect is much broader than one might think; for example, the quality of fluidization of a solids/gas fluid bed is critically affected by the electrical charging of the particles. Even where we have droplets of a polar phase, not specifically charged but dispersed in a nonpolar liquid, dc and ac electric fields can be useful, via inductive effects, in both separation and intensified contact of the two phases (Clayfield et al. 1988).

In a different context, Hill has noted that interception of particles in aerosol filtration through fibers cannot be properly described without referencing electrical charge effects governing deposition as dendrite chains rather than as simple blanketing of the fiber surface (Nielsen and Hill, 1980). Boundary layer effects involved in static electricity generation are discussed in Chap. 16.

The Unsteady State

The impression may have been given, in the discussion of gas-liquid flow above, that pulsations are always harmful. However, in some cases, deliberate pulsation of flow can be very useful. Its only solidly established use is in clearing pipes partially fouled by wall deposits. But the characteristic effect of thinning the wall boundary layer is potentially useful in assisting heat transfer, flattening radial variation of axial velocity, and so on, and should be kept in mind.

Fluid mechanics is a very large field and a happy hunting ground for lifetime specialists. I hope the once-over-lightly treatment presented in this chapter has dispelled some of its mystique. In the next chapter, we turn our attention to fractionation, where fluid mechanics is a crucial element but where the prime objective is mass transfer between phases.

References

Acrivos, A., et al. *Chem. Eng. Sci.* 10 112–24 (1959).

Baker, O. *Oil & Gas J.* 52 (30) 185 (1954).

Calderbank, P.H. *Trans. Instn. Chem. Engrs.* 36 443–63 (1958); 37 178–85 (1959); (with M.B. Moo-Young) *Chem. Eng. Sci.* 16 39–54 (1961).

Cholette, A. and Cloutier, L. *Can. J. Chem. Eng.* 37 105–12 (1959).

Cindric, D.T., et al. *Chem. Eng. Prog.* 83(3) 51–6 (1987).
Clayfield, E.J., et al. Paper 44h AIChE annual meeting (Dec. 2, 1988).
Collins, S.B., and Knudsen, J.G. *AIChe J.* 16 1072–80 (1970).
Cooney, D.O. *AIChE J.* 20 1010–11 (1974).
Das, P.K., and Hartland, S. *Chem. Eng. Comm.* 92 169–81 (1990).
Fox, E.A., and Gex, V.E. *AIChE J.* 2 539 (1956).
Gal-Or, B., and Waslo, S. *Chem. Eng. Sci.* 23 1431 (1968).
Gumbel, E.J., *Statistics of Extremes* Columbia Univ. Press (1958).
Gunn, D.J., et al. *Chem. Eng. Sci.* 42 2163–71 (1987).
Hooper, W.B. CHEMTECH 20 (1) 54–7 (1990).
Hughes, R.R., and Gilliland, E.R. *Chem. Eng. Prog.* 48 497–504 (1952).
Jackson, M.L., et al. *IEC Proc. Des. Dev.* 15 266–72 (1976).
Karabelas, A.J. *AIChE J.* 24 170–80 (1978).
Kenning, D.B.R. "Pool Boiling" in *Two-phase Flow and Heat Transfer* Chap. 7, ed. D. Butterworth and G.F. Hewitt, Oxford Univ. Press (1977).
Kister, H.Z. Paper 62b AIChE annual meeting (Nov. 22, 1991).
Levine, S., et al. *Colloids and Surfaces* 38 325–64 (1989).
Levins, D.M., and Glastonbury, J.R. *Trans. Instn. Chem. Engrs.* 50 32–41, 132–46 (1972).
Mandhane, J.M., et al. *Int. J. Multiphase Flow* 1 537–53 (1974).
Mugele, R.A., and Evans, H.D. *Ind. Eng. Chem.* 43 1317–24 (1951).
Nielsen, K.A., and Hill, J.C. *AIChE J.* 26 678 (1980).
Schmidt, E.F. *Chem. Ing. Tech.* 39 781–9 (1967).
Schrodt, J.T. IEC *Proc. Des. Dev.* 11 20 (1972).
Standiford, F.C. *Chem. Eng. Prog.* 75(7) 59–62 (1979).
Steinmeyer, D.E. *Chem. Eng. Prog.* 68(7) 64–8 (1972).
Sternling, C.V. Institute lecture on two-phase flow at AIChE annual meeting, Philadelphia (Dec. 6, 1965).
Taitel, Y., Bornea, D., and Dukler, A.E. *AIChE J.* 26 345–54 (1980).
Taitel, Y., and Dukler, A.E. *AIChE J.* 22 47–55 (1976).
Taitel, Y., et al. *Chem. Eng. Sci.* 37 741–44 (1982).
Toor, H.L. *AIChE J.* 17 5–14 (1971).
van Ouwerkerk, H.J. *Int. J. Heat Mass Transfer* 15 25–34 (1972).
Veeraraghavan, S., and Silveston, P.L. *Can. J. Chem. Eng.* 49 346–53 (1971).
Villermeaux, J. *Chem. Eng. Technol.* 11 276–87 (1988).
Wallis, G.B., and Heasley, J.H. *Tr. ASME J. Ht. Transfer* 83(8) 363–9 (1961).
Weisman, J., and Efferding, L.E. *AIChE J.* 6 419 (1960).
Wicks, M. III. Paper 20lc AIChE annual meeting (Nov. 21, 1991).
Zlokarnik, M. *Intl. Chem. Eng.* 27(1) 1–9 (1987).
Zwietering, T.N. *Chem. Eng. Sci.* 8 244–53 (1958).

7

Classic Problem Elements: Distillation and Other Fractionation Systems

Thinking is more interesting than knowing but less interesting than looking.
J.W. von Goethe

Distillation—an Art in Transition

In the mid-1970s, during the first energy crisis, there was a feeling that distillation was passé. The process industry realized that the thermodynamic efficiency of this unit operation was typically very low, on the order of 10% or even less. But membrane processes, pressure swing adsorptions, and the like have been slow to displace distillation. In part this is due simply to easing real energy costs, in part to adaptation of distillation: heat-integrated coupling of columns, more widespread use of side draws, and reoptimization of the balance between plates (capital) and reflux (operating cost). In many processes, the energy usage per unit of product in distillation has been cut in half by adroit process engineering. Distillation is convenient and adaptable.

A few observations on problems and opportunities in L/L extraction and fixed-bed sorption are included at the end of this chapter as valid alternates to distillation in many applications.

In this chapter, we explore phenomenological quirks that can derail simple distillation and variants such as extractive and azeotropic distillation. We apply some of the fluid mechanics discussed in Chap. 6 to characterize flow regimes in distillation, invoke surface chemistry effects, and anticipate some of the nonideal phase equilibrium effects to be discussed in Chap. 9. Work of the industry research consortium Fractionation Research, Inc. (FRI), some ingenious laboratory experiments, and inferential diagnostic devices developed in recent years have provided new insights into distillation phenomena.

The bubble-cap trays that dominated the process industry for many years and the naive computational framework used to design columns in that era illustrate the imbalance created when a real grasp of the phenomena addressed is lacking and a design craft is unable to invoke all the options potentially available to it. The rediscovery of the utility of sieve trays, the invention of valve trays, structured packing, distributed downcomers, and a host of new and useful technology flowered once we rejected the "domination" of entrenched ideas and hardware. On the other hand, some of the new equipment is useful only in particular operating regimes. Forgetting the trade-offs implicit in its use is a source of many operating problems.

The distillation problems I have seen go beyond the mechanical arrangement of column internals; they intersect with the other classic sources of problems—control, fluid mechanics, peculiar phase equilibria, and trace chemistry—which are treated in other chapters. Some of these composite problems are included anecdotally in this chapter.

Just as distillation problems interact with these other classic problem elements to determine system performance, so we as troubleshooters must master the separate elements within distillation, like hydraulic loading, geometry options, and separation efficiency, before making practical decisions for improvement. The following sections provide this background, before getting into less well-known phenomena, applications, and some PTS examples.

Factors Limiting Column Capacity

Flow Regimes in Distillation

The *spray* regime represents one extreme of V/L interaction on a tray, with the vapor continuous and the liquid broken up into small drops. It occurs in vacuum columns or other applications with a high volumetric vapor load and a small liquid load; intertray upward entrainment of liquid drops is an important consideration and may decrease tray efficiency significantly well before classic flooding occurs. At the other extreme is the *emulsion* regime, in which high volumetric liquid flows disperse the vapor in a continuous liquid phase and bubble size is governed more by liquid than by vapor flow. Extractive distillation (ED) systems are often in this regime; foaming systems have maximum impact on capacity here, usually via downcomer backup. There is even a situation, near critical conditions at high pressure, in which downward vapor entrainment can lower tray efficiency (Hoek and Zuiderweg 1982).

Intermediate between the above extremes is the *froth* regime, comprising local regions where either liquid or vapor is momentarily the continuous phase. This transition takes the form of jetting in certain column types.

The Onset of Loading

Many troubleshooting problems occur during peaks in the business cycle, when columns are pushed to their throughput limits and planned new capacity is not yet in place. Since maximum value attaches to incremental production at such times, skills in squeezing more out of an existing column will be quite valuable. The effect of the flow ratio parameter on the maximum permissible vapor load parameter[33] is readily available in the literature. Our concerns here are more qualitative features: *in what form* the throughput limitation will show up and *which factors* decrease or increase tray efficiency from normal expectations. A phenomenological understanding of terms like *loading* and *flooding* will also help us wade through a sea of discordant data and jargon to get what we need for PTS purposes.

It is important to distinguish among entrainment, classic liquid backup, or simple pressure drop increase as the constraining effect. For example, in a packed column, loading commences when the surface of the liquid film flowing down over the packing reaches zero velocity because of the opposing drag of the vapor flow. This occurs at perhaps 70% of the vapor load for true flooding, but it can be the beginning of significant liquid entrainment. True flooding is

---

[33] The volumetric flows in these two parameters are weighted by the square root of density; momentum is the real figure of merit.

the point at which the *average* downward velocity of liquid in the film on the packing reaches zero. The onset of loading seems to be an abrupt transition only because entrainment is a powerful function (ca. seventh-power) of the vapor load.

What are the corresponding transitions in a tray column? We can identify the flood point as massive liquid backup on the tray, whether it is a downcomer or dual-flow type. The analogy to the load point is more elusive here.

Capacity Limited by Entrainment

One of the more striking examples of "less is more" is counterintuitive removal of trays to increase column capacity. In a tray column, tray spacing relative to the froth height on trays strongly affects the chance that a spray droplet will reach the tray above, rather than cease rising because of coalescing with another droplet or hitting the wall of the column to resume downflow. The increase in entrainment W as tray spacing is decreased from, say, 12 to 5 cm is well documented (Stichlmair 1978).

Normal entrainment in a column is much greater than is usually appreciated. It is common to have 0.1 kg of entrainment going up the column per kilogram of vapor. The effect of this on fractionation is minor until entrainment becomes a major fraction of liquid downflow (Colburn 1936). At low mass ratios of L/V (or in vacuum columns), entrainment can markedly reduce efficiency without creating classic loading symptoms like liquid backup in the downcomers. Even at higher L/V values, an increase of about 20% in vapor can triple entrainment relative to downflow, bringing loading and/or poor efficiency.[34]

Efficiency aside, liquid entrainment may pose quality or reliability problems in columns, even when it is only a minor fraction of the

---

[34] In one correlation (Kister and Haas 1987), entrainment grows inversely with the fourth power of tray spacing S. In another correlation, entrainment W is an inverse exponential function of S, that is, log W varies linearly with $68/S$ (S is in cm). The composite of these two approaches is that increasing tray spacing from 45 to 68 cm reduces W fourfold. If this decrease is from 60% of liquid downflow to only 15%, it can make a profound improvement in tray efficiency, enough to compensate for the loss of one-third of the trays.

Puppich and Goedecke (1987) showed that entrainment grows more powerfully for low-surface-tension systems (about the seventh power of vapor flow) than for high-surface-tension systems, (about the fifth power). A recent study (Koziok and Mackowiak 1990) claims to have included surface tension, tray spacing, and all key parameters for the first time in a correlation.

liquid downflow, if we are dealing with a corrosive liquid phase, such as an acid, or if a low-volatility trace impurity is thereby contaminating the overhead. In fact, one of the easiest ways to measure the level of entrainment is when a high-boiling color body or other easily detected trace impurity is used as a tracer.

Entrainment generated by normal distillation contact of the two phases has drops generally larger than 100 μm in diameter; a number of devices for intercepting such entrainment coming overhead off the top tray are available. Even within the column, such entrainment may be captured by tray liquid action and may then have to reinitiate itself in order to continue upward. But, as noted in Chap. 6, mechanisms such as rapid flashing or fog formation can generate droplets under 5 μm, when disentrainment is much more difficult. Normal tray action may be quite ineffective in capturing such entrainment. Results with single sieve trays show 2–3 μm to be the breakpoint between respectable and minimal capture by a phase-compatible scrubbing liquid.

Capacity Limited by Downcomer Backup

A larger column diameter is the (expensive) ultimate fallback solution for debottlenecking. However, in a tray column, the division between active "pattern" area and downcomer cross section is often negotiable. Particularly in foaming or high-liquid flow situations, we will look to downcomer enlargement. Increased tray spacing can also address flooding by providing more downcomer residence time for clarifying downflow. Aeration of liquid in a downcomer detracts from capacity both by occupying extra volume and by the high pressure drop of the two-phase mixture flowing out of the bottom of the downcomer. However, assuming that vapor downflow does not harm separation, widening the downcomer bottom seal may make a contribution in uprating foaming systems where full clarification in downcomers is unattainable.

Antifoams are a noncapital expedient for skirting entrainment or flooding limits and may supply an interim solution. The likely associated loss of contacting efficiency and possible contamination effects should be kept in mind, however (see below).

Limiting Pressure Drop

It is good practice to monitor differential pressure (dp) across column sections that are near loading. Sensitivity in this method is provided by the upward break in the dp–throughput relation as loading is reached. This stems from the fact that dp is being expended not merely on building liquid inventory on each tray but also on an accelerational drop when masses of liquid comparable to the upward flow of vapor are entrained from tray to tray. Close attention should be given by the process technologist to the hardware of dp measurement. Such installations can be unreliable because of the variable fluids content of the seal legs, failure to measure across the most sensitive section of the column, and other factors. As with process analyzers, it should be recognized at the outset that attention to small installation details may be as meaningful for ultimate success as the broad concept of the instrument.

This discussion implies that dp is being used as a monitoring/control tool for loading. However, there are occasions when pressure drop increases themselves limit throughput. Examples include situations when temperature differentials for reboiling are adversely curtailed or (less often) when α value shifts adversely in a tight separation. Here, a more open tray pattern area, even at the expense of tray efficiency, may provide acceptable debottlenecking.

Guidelines for the Choice of Trays and Alternate Internals

The intent of this section is not to teach tray design but to sensitize the technologist to special strengths and weaknesses of the various tray types. This information is useful for making appropriate selections in the design stage or for choosing the best available substitute in a retrofit. The two situations are, of course, inherently different: In the original design, there is considerable maneuvering room but a much more simplistic understanding of the system and its texture; in the retrofit, maneuvering room is much narrower but the grasp of the terrain more sure.

The relentless rise in construction costs and historically shrinking profit margins in the chemical industry in the decades of its greatest rate of expansion in a sense forced the addition of new options for column internals. Irrelevant detailing was pruned, but designs were tailored more closely to the character of the system served. A technical and economic issue was the kinetic energy wastage in the flow reversal of the gas phase inherent in a traditional bubble cap. Today bubble caps are specified only when even slight weepage of liquid between trays is unacceptable.

Dual-flow Tray

In *dual-flow trays* the critical physical aspect is the jostling of the liquid and vapor phases as they compete for passage through the same apertures (holes or slots) in opposite directions. There is an inherent instability here, with a rapid alternation of slot occupancy by the counterflowing phases. This unsteady-state situation detracts from the efficiency of contacting in dual-flow trays relative to most downcomer types. The efficiency trade-off may still be worthwhile to gain the cheapness and relatively low pressure drop of dual-flow trays. The pulsing action is an advantage in fouling services, and the simplicity of the tray lends itself to easy cleaning. A second issue with dual-flow trays is the transition from a froth to a spray regime (Takahashi et al. 1986).

A further disadvantage of dual-flow trays, decisive in many cases, is the poor turndown ratio, with liquid "dumping" through the holes when vapor velocity falls only a modest percentage below the maximum loading. This stems from the quadratic dependence of liquid holdup on gas velocity. Adding or removing blanking strips at annual turnarounds to re-rate the operating range, according to demand changes of the business cycle, is a reasonable way of coping. Hole size, its ratio to plate thickness, and the way holes are cut or punched in the tray all have a bearing on the fine points of performance. Vulnerability to vapor and liquid bypassing each other is shown in the sensitivity of dual-flow trays to imperfect leveling in installation[35] and to a fall-off in performance at large column diameters (vs. the gain in efficiency with downcomer trays from the path-length effect).

Valve Tray

The *valve tray*, in a variety of options, is the current staple of tray design. Its variable opening overcomes a major disadvantage of bubble caps and provides a broad operating range. However, it has more pressure drop than we may wish to accept in vacuum or other demanding services. Higher capacities are available through dumped or structured packing in debottlenecking situations.

---

[35] Some authors who dismiss leveling as unimportant are probably misled by the limited size of column diameters they investigated. Even in sieve trays with downcomers, a leveling study of 0.5- and 1.2-m diameter trays at a 3-degree incline gave a decrease of about 20% in mass-transfer performance (Kovshov et al. 1966).

Sieve Tray

Sieve trays are an inexpensive alternative to valve trays and have about the same efficiency. Use of downcomers compensates for a shortfall of dual-flow trays, but they still have turndown restrictions from dumping (remediable, as noted for dual-flow trays, by using temporary blanking strips). In fouling service, sieves are frequently substituted for valve trays. Here large holes are believed to reduce the chance that deposits will bridge across. The transition from an emulsion to a spray regime as vapor flow increases is an important issue in design and troubleshooting (Zuiderweg 1983).

Insightful tailored tray layouts such as the Linde sieve tray variant (Frank et al. 1969), which improves tray stability and bubbling action at minimal pressure drop (for vacuum service), continue to be introduced. Various distributed downcomer layouts can help high-liquid-flow systems by minimizing hydraulic gradient problems.

Dumped Packing

In an economically competitive situation, skilled vendors will often substitute *dumped packing* for trays to gain throughput or reduce pressure drop, as in ethylbenzene/styrene fractionators. Historic problems with flow maldistribution in columns without downcomers (dual-flow, dump-packed, etc.) are now corrected largely by entry-flow distributors and periodic redistribution along the length of the column to avoid the "collusion"-type bypassing of liquid and vapor countercurrent flows referred to in Chap. 6. A recent study defines experimentally the maldistribution of gas and liquid in columns with both dumped and structured packing (Stichlmair and Potthoff 1990).

Intalox packing provides much higher capacity than Raschig rings, with about the same efficiency. Other shape variants are also available. Billet (Billet and Mackowiak 1988a; Billet 1989) has been a prolific gatherer and correlator of hydraulics and efficiency data for various dumped and structured packings.

Structured Packing

Structured packing represents a notable attempt to take control of the V/L counterflow in a manner that minimizes pressure drop per theoretical stage relative to trays while avoiding some of the channeling and other problems of dumped packing. Its initial impetus came in heavy water separation in the nuclear power industry. Earlier versions were metallic gauzes, progressing to corrugated sheets with punched holes to reduce costs.

Well-designed structured packing sections can deliver a theoretical stage per 46 cm height (in nonstripping applications), twice as many stages/height as (e.g., trays at 61 cm spacing that have a 70% tray efficiency). Pressure drop can also be attractively low for vacuum separations, well under 1 cm of water differential per stage. However, some limitations must be kept in mind. Structured packing is expensive on a volume basis relative to random packing and even to trays (particularly when alloy is required). It has a definite role in upgrading separation performance in an existing fixed-height column. It may win on an initial design where close-boiling compounds and/or minimum pressure drop are involved, as in ethylbenzene/styrene fractionation, where it makes the difference between one and two columns needed to do the job.

There are limitations other than cost. One relates to possible rivulet flow of the descending liquid phase, reducing the area for mass transfer. Probably related to this rivulet flow, liquid residence time is 30–50% of that in a tray column, posing control and other problems. A minor second liquid phase in particular can plunge more readily than in a tray column. (Of course, the reduced pressure drop, valuable in vacuum service, stems in part from the lower liquid holdup.) Residence time may also have greater variance than in a tray column, leading to problems when, for example, a reaction must be accomplished in the column. To achieve the prospective low heights of a transfer unit HTUs requires considerable care in liquid distribution. Finally, in high liquid-load services, valve trays may prove more productive.

Tray Efficiency

In Chap. 1 we alluded to the distortions introduced by using the theoretical stage rather than the dynamic concept of transfer units to describe mass transfer for computational purposes. Before we even consider tray efficiency prediction for the major components, we should be aware of the mathematical distortions caused by the theoretical stage method relative to trace impurities that are much lighter or much heavier than the key components. As seemingly innocent a procedure as carrying out a column calculation on the basis of 30 theoretical stages rather than 50 trays of 60% efficiency can overpredict rejection of minor components critical to our performance goals.

Incomplete contacting of vapor with liquid on a tray causes rejection of heavies in a rectifying section to be less when calculated with (N/E) real trays of fractional efficiency E than when calculated using N theoretical plates. Similarly, stripping rejection of very light components will be less with a real tray calculation. The difference in rejection can be an order of magnitude. Such errors are in addition to the reduction of tray efficiency by physical properties like lower diffusivity of a heavy component.

Absolute prediction of tray efficiency is still chancy, but there is now a fair body of data relating vapor-side and liquid-side transport (as given by N_G and N_L) to flow parameters; we can make selective configuration changes in the internals of an already existing column with more confidence. Proper attention to how the two transport resistances are combined is often more important for a troubleshooter than getting N_L and N_G just right. (See below.)

In packed columns, wetting of the packing (not just as delivered but as conditioned in the process) by the liquid is an additional strong consideration affecting efficiency.[36] For a pure component, the surface tensions of the G/L and S/L interfaces define the tendency to flow in rivulets rather than sheets down that surface. However, in a mass transfer system, concentration gradients can strongly affect this bunching/spreading tendency, and thereby contacting efficiency, in both packed and tray columns. (See the discussion of the Marangoni effect below.) Shi and Mersmann (1985) have separated the area term from the mass-transfer coefficient k_L in terms of the response to system parameters. With residence time (see below) added, N_L is effectively defined.

Liquid- and Gas-side Resistances in Distillation on Trays

The use of tray efficiencies derived from tests in total-reflux and other experimental systems in which vapor-side resistance to mass

---

[36] If one is driven to a polyvinylidenedifluoride (PVDF) or polypropylene packing by other considerations in an aqueous system, it is not necessary to settle for poor wetting. There are techniques for hydrophilizing such surfaces (Hüttinger and Rudi 1983). By the same token, one should not take it for granted that an initially water-wetted packing like ceramic will not become hydrophobic from process deposits over time.

transfer is controlling can overstate tray efficiency when applied to stripping conditions. The number of transfer units per tray in the individual phases, N_L and N_G, give more information. (The notation used in this section is as follows. Subscripts: G, gas phase; L, liquid phase; OG, overall, based on gas-phase units, combining resistances of both phases; MV, PV; Murphree or point efficiency. E is the fractional efficiency; m is the local value, at that compositional point, of the equilibrium y/x ratio for the component in question; G, L are gas and liquid flow rates in molar units.) N_L is usually between 1 and 2, while N_G is slightly smaller, ranging from 0.5 to 1. The variation of N_L and N_G with flows and physical properties is discussed by Zuiderweg (1983) and Lockett (1980). A stripping factor S ($\equiv mG/L$) far from unity will often determine which phase dominates the overall N_{OG}, given by a sum of resistances:

$$\frac{1}{N_{OG}} = \frac{1}{N_G} + \frac{S}{N_L} \qquad (7\text{-}1)$$

This equates to a point vapor tray efficiency $E_{PV} = 1 - \exp(-N_{OG})$.

This yields an integral Murphree efficiency for well-mixed vapor and liquid in plug flow across the tray by

$$S \cdot E_{MV} = \exp(S \cdot E_{PV}) - 1 \qquad (7\text{-}2)$$

However, for a dual-flow or other tray without a crossflow gradient,

$$E_{MV} = E_{PV} \qquad (7\text{-}2a)$$

The following table illustrates the effect of stripping factor variation on local tray efficiency for a cross-flow tray with fairly normal hydraulics and $N_G \approx 1$, $N_L \approx 1.5$.

S	N_{OG}	E_{PV}	E_{MV}
0.3	0.833	0.565	0.616
1	0.60	0.45	0.57
5	0.185	0.169	0.264
20	0.0698	0.067	0.14

(The notation used in this section is as follows. Subscripts: G, gas phase, Lm liquid phase; OG, overall, based on gas-phase units, combining resistances of both phases, MV, PV; Murphree or point efficiency. E is the fractional efficiency; m is the local value, at that compositional point, of the equilibrium y/x ratio for the component in question, G, L are gas and liquid flow rates in molar units.)

Increasing the stripping factor S weights N_L downward, making liquid-side resistance dominant and E_{MV} anomalously low. This occurs because the tray efficiency E_{MV} is oriented to the vapor-phase driving force by referencing a vapor composition that is in equilibrium with bulk liquid composition. Its dropoff at high mG/L is, in a sense, a mathematical fudge factor. Conversely, in a rectification situation (mG/L < 1), gas-side resistance tends to dominate, with tray efficiency being normal or better than average. The O'Connell plot relating tray efficiency decline to increases in relative volatility α and liquid viscosity μ rests on the above situation, in addition to the inverse relationship between diffusivity and viscosity (O'Connell 1946). In this section, the subscripts are as above for a trayed column.

Liquid- and Gas-side Resistances in Packed Distillation Columns

Where we spoke of N as the number of transfer units per physical tray, in a packed column we speak of NTU, the number of transfer units in a given height of packed column. The inverse—height divided by NTU—is HTU or singly H for brevity in the equations. If an equilibrium stage concept is used instead of the dynamic concept of transfer unit, HTU or H becomes HETP, the height of an equilibrium theoretical plate (or stage).

A parallel treatment for combining phase resistances in packed columns uses HTUs in place of reciprocal NTUs:

$$H_{OG} = H_G + S \cdot H_L \tag{7-3}$$

Thus, even if the hydraulics that give N_G and N_L stay constant, the overall HTU = H_{OG}, based on gas-phase driving forces, grows linearly with the stripping factor S when the latter exceeds unity and becomes proportional to S as liquid-side resistance comes to dominate transport. If one insists on using an equilibrium stage concept for a differential device like a packed column, HETP grows as the log of S as S becomes $\gg 1$:

$$\frac{\text{HETP}}{\text{HTU}} = \frac{\ln S}{S - 1} \tag{7-4}$$

These relations are supported by the results of Hanson et al. (1978) in steam stripping of sparingly soluble organics from water when the stripping performance did not improve with steam increases beyond a certain point, not from flooding but from the decrease in NTU_{OG} offsetting the effect of increasing S. The effect of large deviations of S from unity were also shown by Koshy and Rukovena

(1982). By contrast, the work of Bravo et al. (1982, 1985, 1988, 1990) on structured packing is inherently biased toward gas-side control, both by taking data at total reflux and by focusing on vacuum systems, which tend to reduce liquid film thickness.

Residence Time Considerations

A rough estimate of residence time per tray is frequently needed in troubleshooting. This can be 2–10 sec on a sieve tray with downcomer, somewhat less for dual-flow trays, and as much as 20–30 sec for bubble caps. Of course, in bubble caps, it can be argued, as with a packed column, that there is a semistagnant component of the residence time that is not fully "in play."[37] It should also be recognized that residence times will decrease at higher liquid flows, since holdup does not increase in proportion to flow. The significant parameter is liquid residence time per liquid-side transfer unit N_L if we are concerned with some side-reaction event relative to stripping out of a component. Residence time is, of course, not independent of mass-transfer performance; Billet (1988b) has nicely cross-related these factors for packings.

In a specific problem situation, it may be worthwhile to establish residence time by using a tracer test. Use of pressure drop as equivalent to liquid holdup for defining residence time is a good approximation only for a dual-flow tray with low entrainment.

Complications from the Presence of a Second Liquid Phase

In recent years, I have been encountering (or recognizing) more often the effect of a second liquid phase in detracting from fractionation performance, either directly by allowing a component to shortcut undesirably to the bottom of the column or indirectly by affecting residence time or tray efficiency. New computer programs have been provided for calculating tray-by-tray equilibria when two liquid phases and one vapor phase are present. But the algebraic convergence problem, which is important to computer scientists, is not the real problem.

---

[37] Compare the classic discrimination of static and dynamic holdup components in irrigated packing (Shulman et al. 1955).

Fallacy of V/L/L Equilibration in a Column

The central issue is the failure of the minor liquid phase to equilibrate with the major liquid phase or the vapor phase. A rough calculation from a notional Sherwood number of 2–5 for external transport from drops of the minor liquid phase reveals that a minor phase can equilibrate with the major phase in the residence time available on a typical tray only if it is well dispersed as fine droplets under 0.5 mm in diameter.[38] In addition to the low mass transfer between the liquid phases, equilibration is further slowed by the low concentration driving force through the major liquid phase before the minor liquid-phase component(s) can reach the V/L interface and volatilize.[38]

Plunging Effects of a Minor Liquid Phase

Consistent with the effects discussed above, a minor liquid phase may descend much further in a column than equilibrium calculations would lead us to expect. Plunging of a minor water phase[39] in a hydrocarbon like liquid may be the first time a novice encounters two-liquid-phase phenomena in columns. Water thus reaching column bottoms can lead to interesting complications.

Nucleation of a Trace Water Phase in a Column

Physical sources of a trace water phase include incomplete phase separation, solubility changes from temperature shifts, and volatilization of a water-solubilizing component initially present, such as an alcohol. There are also chemical sources. In situ nucleation of a water phase can occur by acid-base neutralization reactions as an addition of NaOH to counter HCl acidity in distillation of organic chlorides. HOH formed in the neutralization on the trays of the column may not seem great in quantitative terms, but solubility in the

---

[38] Only when the minor phase spreads spontaneously at the V/L interface, as in antifoam applications, is it reasonable to expect good mass-transfer contact with the major phases. The unlikelihood of a minor water phase existing at the interface between the vapor phase and the major liquid phase is established by surface tension energetics (see the discussion of Ross defoaming below).

[39] Only recently have I noted the publication of experimental work in this area (Weiland 1991).

organic liquid phase can be quite low; in addition, the equimolar salt formed can substantially reduce the volatility of water from the resulting aqueous saline phase. When water formed this way reaches a steel reboiler, the Fe engendered by corrosion may create additional HCl by catalyzed dehydrochlorination, necessitating additional NaOH dosing and a resulting vicious circle of self-perpetuating trouble, as described in Chap. 3.

Superheat Explosions

Perhaps the most dramatic example of plunging of a water phase is the explosive delayed vaporization of very fine droplets of water.[40] The not uncommon dislodging of whole sets of trays from pressure surges in commercial columns may stem from phenomena similar to this explosive vaporization. The prerequisite fine droplet formation is much more likely in a supersaturation context from minor chemical reactions or abrupt temperature shifts. This is another instance where we must consider superheating and other supersaturation phenomena as distinct possibilities, rather than assuming that equilibration is the norm.

Complications from Buildups in an Entrapment Zone: Phaseout

The nucleation of a second phase within a distillation column can be due to an entrapment zone or "bulge" related to abrupt shifts in volatility of a minor component; these shifts all caused by its non-ideal phase equilibrium behavior in relation to the compositional profile of the major components. An example is the appearance of an ice or hydrate phase between the ethylene-rich and ethane-rich zones of a C_2 splitter column in a modern steam cracking olefins plant. This and other striking examples are detailed in Chap. 9.

Material Balancing of Bulges

The fractionation vectors causing entrapment of a minor component can lead to its buildup in the bulge zone to concentrations well above

---

[40] These superheat explosions are not confined to water. In fact, many industrial accidents involve light hydrocarbon liquids spilled into bodies of water. The common element seems to be vaporization delayed by a liquid's superheating under surface tension forces (Reid 1976), with very rapid flashing at a temperature $T \leq 0.88 T_c$ (critical temperature in absolute units) (Porteus and Blander 1975).

its presence in the feed. Material balance will eventually be reasserted when the trace component's buildup permits it to work its way upward or downward out of the column, against adverse absorption or stripping factors,[41] at a rate matching its arrival in the column feed. Meanwhile, however, the buildup may have undesirable side effects.

Side-draw Arrangements

A side-draw purge at the center of the bulge zone can further limit the magnitude of the buildup. Let us note here briefly some of the physical arrangements for withdrawing a purge flow from such a bulge to limit its buildup. A minor draw will probably tap into a downcomer. The tray should be liquid-tight, without weep holes, and completely drainable via the draw line. It should not be possible to unseal the downcomer feeding the draw tray. It is preferable to avoid a vertical draw line, but if one must be used, a good vortex breaker should be installed at the tee and the draw line sized for a low linear velocity.

Since only a limited number of draw positions can reasonably be provided, delicate control may be needed to hold the column profile so that the bulge concentration peaks at the draw tray position.

Control-related Problems in Distillation Columns

Distillation control schemes may be crudely divided into material-balance (MB) splits and quality control (QC) schemes (see Chap. 8). Frequently, a sensible choice between these two is betrayed by a faulty implementation. For an MB split, an example involves a top draw on flow control where a flow cycle in reflux is induced by poor level control instrumentation of the accumulator. For a QC scheme, examples of faulty implementation are the use of a process gas chromatograph (GC) whose sequence cycle is too slow for the dynamics or the use of tray temperature as an index of composition when temperature sensitivity to compositional shifts is either too great or too slight.

---

[41] I generally use $KV/L \approx 1$ as a condition to define roughly the center of a bulge, but good calculational aids are needed for calculating the magnitude of the bulge. One approach is to establish the column heat and mass traffic for major components only and then "piggyback" onto this the trace entrapment.

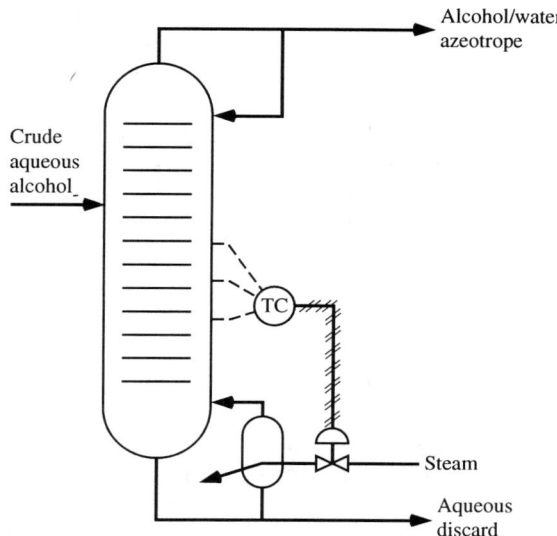

Fig. 7-1 Multipoint Temperature Control of Alcohol/Water Stripper

Example: Tray Temperature Sensing as a Surrogate for Composition

The practice of using multipoint temperature control in stripping of alcohols from aqueous bottoms streams illustrates the sensitivity problem when temperature is used as a measure of composition (see Fig. 7-1). The nonideal solution behavior of alcohols in water is pronounced, leading to a temperature profile that is flat near the column bottom (see Figs. 9-1 and 9-2) but inflects sharply farther up, through a "knee," to a final flat profile as the azeotropic composition is approached. Choosing a control tray in this knee is intended to provide sensitivity of response.

This expedient creates two new control difficulties in exchange for the one it eliminates. The lesser difficulty is that moving sensing up the column moves it dynamically farther from the manipulated variable: reboiler steam. The more serious problem is that the temperature knee shifts abruptly up or down the column in response to extremely fine moves in stripping heat balance. A multipoint averaging of several sensors along the knee is an attempt to dampen this sensitivity.

A better solution begins with moving the temperature sensor much lower in the stripper and addressing the complications attendant on this approach. For example, the obstacle to using a low-sensitivity temperature point is "noise" (rather than intrinsic difficulty of the measurement), which is caused largely by pressure fluctuations. Since we need temperature only as a surrogate for composition, the noise problem can be minimized by pressure-compensating the temperature measurement. A second element, reducing control loop sensitivity for good tunability, is achieved by manipulating a secondary, trimming, valve rather than the main reboiler steam valve from the temperature control (TC) signal (see the discussion of control loop gain in Chap. 8).

Control Near Loading

While the standard answer to this problem is to avoid this region, control here may be a necessity for economic reasons discussed earlier. A split-range scheme, where dp becomes the dominant controlled variable when the column nears loading, has much to recommend it. Depending upon the characteristics of the particular fractionation, a switch to reflux or reboiler heat (or even feed rate) reduction, keyed to dp measurement, takes over from the normal control scheme whenever dp exceeds a set level. While this may cause some reduction in the quality of separation, it is nothing compared to the difficulty of recovering control of a column once it has become fully loaded.

Level Control

Level control misconceptions cause many problems in distillation, particularly in sequences of columns with minimal intercolumn inventories. Overtight control of bottom sump or overhead accumulator level can produce large swings in feed to the next column, with disturbances amplifying disastrously through several columns. One must recognize that in 80% of applications a level controller is really a flow smoother (see Chap. 8). The widest possible deviations in level should be accommodated by the control scheme to insulate flow to the next column from any disturbance in the prior column. In a two-column sequence where overhead takeoff feeds the next column, this involves putting top draw on feed control (FC) and tuning the accumulator level control (LC) that sets reflux to the first column for minimal oscillation.

Complex Phase Equilibria

General-purpose distillation design programs do not work very well in special situations such as extractive distillation (ED) and azeotropic distillation (AD). Even when special programs overcome the mathematical difficulties (some of which mirror sensitivities of the real system), a few problem-specific computational runs do not give one the needed level of insight. With AD, I prefer the graphical overview provided by the residuals method of Doherty and coworkers (Doherty et al. 1989; Pham and Doherty 1990).

In both AD and ED, we are trying to use nonideal behavior of a major component occasioned by the presence of an added third component to achieve separations that would otherwise be impossible. We can view AD as the case where the added component travels up the column with the component whose volatility it enhances (hence entraining agent) and ED as the case where it travels down the column.

Azeotropic Distillation (AD)

Chap. 9 provides some guidelines on the choice of an entraining agent based on nonideal solution behavior. Perhaps the most rudimentary example of AD is the use of the binary heteroazeotrope to reject traces of water from a hydrocarbon such as benzene. Enrichment of water takes place as it goes up the column, but only in the overhead condenser does a second liquid phase appear, providing an aqueous discard layer and an upper layer that is refluxed to the column. The bottom product is bone-dry benzene.

Where there is substantial miscibility, as in an alcohol/water binary, a third component is added to provide heterogeneity and thereby also to boost the volatility of water. Drying of aqueous ethanol (EtOH) using benzene or pentane as the azeo agent is a common example. The trace of this agent in the overhead aqueous layer discard is subsequently recovered and recycled, since the intent is to keep it locked within the system. The azeo agent must be light enough so that none of it leaves in the dried bottom (EtOH).

In an elaboration of this process, a crude, wet lower (C_2 to C_4) alcohol feed emerges as a pure, nearly dry bottoms product while despatching overhead both water and hydrocarbon impurities (see the introduction to this system in Chap. 5). This process depends on maintaining a water-rich zone in the upper stripping section (where the volatility of both the azeo agent and the hydrocarbon impurities is enhanced relative to the alcohol) and above this a zone enriched in the azeo agent that sends water overhead with a minimum amount of alcohol. Even after the pattern of nonideal phase behavior underlying this two-tiered rejection is understood, the technologist is faced with the critical problem of controlling the column inventories of both water and azeo agent.

Too dry a stripping section will reverse the hydrocarbon/alcohol separation. Too wet a section will give excessive water in the alcohol bottoms product. Degrees of freedom for control in such an AD may be fewer than one would normally expect (Ravaglio and Doherty 1988). While the water profile may be well controlled by the stripping section's temperature, wide swings in the amount of azeo agent in the zone above may be little reflected in temperature changes there. One form of inventory control for the azeo agent is a process GC on a key tray sample. Another is in/out mass balance control of the azeo agent, with "drift" corrected by periodic tray analyses.

Example: Azeotropic Drying: Low Tray Efficiency at High KV/L

Azeotropic drying of hydrocarbons, described above, although competitive at today's energy prices, has extremely low tray efficiencies, 6–20% being commonly reported. Why is this? As shown above, the low efficiency is an artifact of the way we calculate tray efficiencies, using vapor-phase compositional driving force. A high stripping factor, KV/L, for the trace water component, because of its high γ value (low solubility in hydrocarbon), has the effect of magnifying the liquid-side resistance to mass transfer. Only trace components whose K value is high will exhibit this effect; major components tend to have KV/L near unity, which is why the effect strikes us as abnormal.

One of the deficiencies in azeotropic drying practice is the refluxing of a water-saturated upper layer (UL) from the overhead accumulator. A better procedure would be to pass this stream through a mole sieve or other adsorbent drying bed before returning it as column reflux, thereby reducing the boil-up required for the distillation. It might be uneconomical to use such a drying bed to do the whole job of taking water from, say, 40 ppm down to 5 ppm in the hydrocarbon stream. But once we have boosted the water content up to saturation, say 200 ppm, by distillation, the driving force for a drying bed is at hand and should be used.

There is nothing wrong with a mixed strategy, here and elsewhere, in process improvement. The same approach takes advantage of the concentration buildup of a trapped component in a distillation column to purge it, as in the C_2 splitter example presented in Chap. 9.

Extractive Distillation (ED)

This unit operation came into its own in World War II with the need to separate close-boiling or azeotroping pairs of compounds, as in the separation of butadiene from butene-2 required by a crash program to develop synthetic rubber in the United States. After the war, the search for ED environments with maximum separating power led to the use of polarity indices such as the solubility parameter for screening potential ED solvents. A later stage brought functional group contribution methods of correlation for nonideality. There was also much more boldness in choosing novel compounds, whose cost or stability problems were outweighed by their greater separating power. I participated in the substitution of acetonitrile for acetone in C_4 separations, which led to a worldwide burst of development involving acetonitrile and more exotic third-generation solvents.

Table 7-1 shows experimental rating of solvents (Gerster et al. 1960) for a pentane/pentene separation by infinite-dilution activity coefficients γ. We scan the data not only for selectivity (conveyed by the $\Delta \log \gamma$ entry), but also for a good carrying capacity of the solvent (low value of γ^∞). One might end up choosing, for example, NMP over other more selective solvents, on the basis of high C_5 solubility.

Even when there is less to work with than the unsaturation difference discussed above in distinguishing two molecules, enhancements in relative volatility can often be achieved that are sufficient to offset the complications of ED relative to conventional distillation. For example, n-heptane/methylcyclohexane, with an α value of only 1.07, can be brought up to an α value of 1.5 by appropriate solvent and temperature selection.

Issues in ED Practice (Fig. 7-2)

A number of practical issues arise in designing and operating an integrated ED system in addition to the primary separation sought.

Table 7-1
Ranking of Solvents for ED Separation of Alkanes/Alkenes by Non-ideal Solution Behavior

Solvent	Log γ^{∞} n-Pentane	Log γ^{∞} Pentene-1	Δ Log γ^{∞}
Acetone	0.723	0.500	0.223
Acetonitrile	1.310	0.975	0.335
Dimethylformamide	1.148	0.856	0.292
Dimethylsulfolane	1.100	0.810	0.290
Dimethylcyanamide	1.129	0.837	0.292
Furfural	1.219	0.947	0.272
Propylene carbonate	1.529	1.207	0.322
N-Methylpyrrolidone (NMP)	1.031	0.739	0.292
Methanol	1.431	1.248	0.183
Nitromethane	1.728	1.332	0.396

The recovery of the heavier key component (e.g., butadiene) from the extractive solvent leaving the bottom of the primary (ED) column (2-butenes as the light key having gone overhead in the ED column) poses a number of problems. The solvent must be sufficient heavier than the heavy key so that stripping can readily free the recycle solvent of the heavy key. Otherwise the recycle solvent, entering high in the ED column, will recontaminate the light key overhead product with heavy key. On the other hand, too high-boiling a solvent will create temperatures in the bottom of the solvent stripping column that cause degradative reactions.

Minor components that are heavier than the heavy key but lighter than the solvent will accumulate as bulges in the solvent recovery stripper and require a purge. Processing of a purged vinylacetylene (VA) bulge in a Union Carbide ED system for butadiene is believed to have led to a 1969 Texas City explosion, although it took accidental total reflux operation to build up the minor impurity VA to a dangerous level (see Chap. 16).

In general, the solvent will be light enough to be a significant impurity in the light key overhead in the primary (ED) column. Top knockback sections (above the solvent entry tray) used to minimize this are kept short, because the separation between the keys provided by the ED may reverse, once the solvent becomes a minor component. Foaming (see below) is a likely complications within and just below the knockback zone. Even with the knockback, sub-

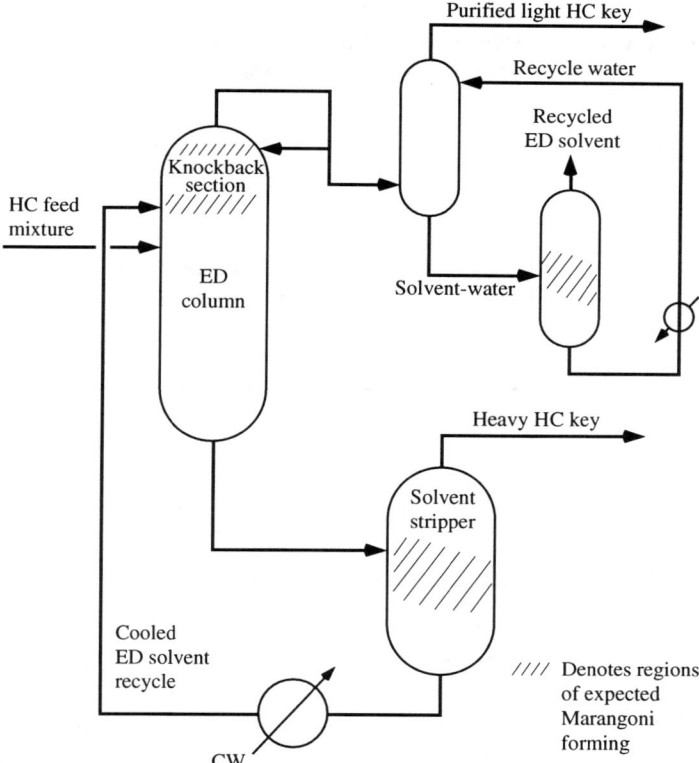

Fig. 7-2 Integrated ED System Layout denotes regions of expected Marangoni foaming

sequent separation steps are required to remove traces of ED solvent from the light key.

Table 7-2 lists some compounds used for ED in industry today.

Solubility Constraint

The effectiveness of separation is lost if the compounds being separated exceed their solubility in the ED solvent. This was the cause of poor performance in an ED system I was involved with, where limited miscibility with water of various low-polarity trace impuri-

Table 7-2
Some Commercially Useful Extractive Distillation Solvents and Their Applications

Water Ethylene glycol Phenol	Nonpolar/polar compounds
Acetone Acetonitrile Dimethylformamide Furfural N-methylpyrrolidone Methylsulfolane β-Propiolactone	Hydrocarbons by degree of unsaturation

ties made aqueous ED a good choice for purifying a product that was freely miscible with water.

Example: Failure of Aqueous ED Caused by Phaseout

In general, solubility in water drops faster than vapor pressure as the carbon number of hydrocarbonlike substances increases. Hence (γp^0) increases for heavier impurities. There had therefore been an expectation that all such impurities (e.g. thioalkanes, heavy olefins and oxygenated hydrocarbons) would distill overhead, while the major product, diluted with water, passed downward in the ED column. When this column became a focus of concern and was closely monitored, however, rejection of many key impurities was well below theoretical expectations. It turned out that a small but variable amount of heavier hydrocarbons, soluble in the crude feed, phased out upon water dilution in the ED column.

These compounds themselves were of no concern with respect to product quality, being readily rejected in final overheading of the dry product. However, a number of critical lighter impurities partitioned preferentially into the relatively small second layer created by these heavy hydrocarbons, where they exerted normal rather than a water-enhanced volatility. It might be thought that if, say, 20% of an impurity initially phased out into this skim layer of hydrocarbon, this would constitute at worst a 20% reduction in its relative volatility in ED—not too serious in view of the high α values in ED of the critical impurities. However (as noted earlier), the assumption that the minor liquid phase will stay in equilibrium with the major

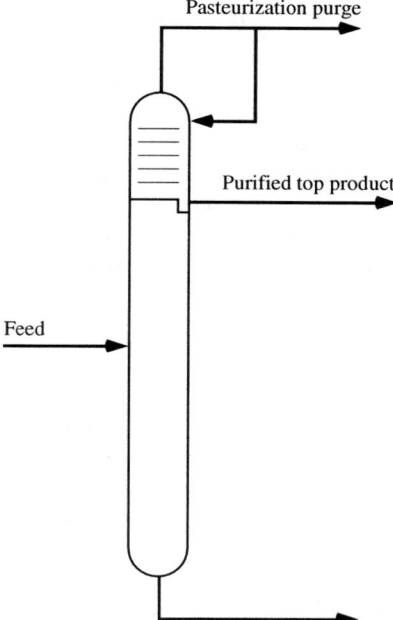

Fig. 7-3 Top Pasteurizing Column Section for Light-end Purging

aqueous phase is not correct. Thus, an initial 20% phaseout may mean that rejection in ED of that impurity may not exceed 80%.

One solution to this phaseout problem is an L/L extraction *preceding* the extractive distillation (Litzen and Bolger 1988) to reduce the population of heavy hydrocarbonlike impurities, so that the remainder stay within their solubility limit in the ED column.

"Pasteurization" (see Fig. 7-3)

I have not determined the origin of this nickname, but the connotation of light sterilization of milk is quasiappropriate. Instead of germs, however, we are eliminating impurities significantly lighter or heavier than the key components, without the expense of a separate column. For a light impurity the main product becomes a side draw, with the pasteurization section above it operating at essentially total reflux. A small liquid or vapor top purge stream enriched in the light component being rejected is withdrawn for treating or disposition. There is a mirror-image version, where the main product is removed as a vapor side draw and a heavy reject purge is taken at the bottom of the column.

Pasteurization sections are useful retrofits provided when an additional function in rejecting a trace impurity is forced upon an existing distillation column. (A wholly new plant is more likely to use a separate column.) Adding height to a column to provide pasteurization trays without detracting from existing functions can be a quick and inexpensive expedient.

Example: Rejection of Trace Methanol from Acetone

Acetone made by any of several oxidative processes will contain a few hundred parts per million of methanol (MeOH). For certain end uses, this MeOH may be objectionable. Let us consider a pasteurizing section, with acetone product taken as a side draw from the rectifying section (with appropriate changes in internals), as a potential solution. The best enrichment we can hope for in distillation at one atmosphere is the MeOH-acetone azeotrope, in which MeOH is 12%w. (Pressure distillation might improve this value.) If the top purge were burned as fuel, near-attainment of the binary azeotrope would imply discard of only about 0.2% of the acetone product if most of the MeOH were in the azeotrope purge stream. But this scenario is too optimistic. With only a modest number of pasteurization trays, enrichment will be substantially less.

The relative volatility in the pasteurization section is readily estimated from activity coefficients at infinite dilution (see Chap. 9). In this case, we can get a value of 2.3 for γ^∞ from a modern source (Hirata 1975), which, with relative vapor pressures of 0.8 near 60° C, gives $\alpha^\infty \approx 1.8$. The six theoretical trays provided by an eight-tray pasteurization section can enrich MeOH 34-fold at total reflux. If the MeOH content in the main product stream of acetone were reduced from 200 to 40 ppm, then total reflux pasteurizing would raise the MeOH in the discard stream to only 1400 ppm. The desired 80% reduction in impurity level would thus (by material balance) require purging about 11% of the product stream. This is uneconomical and also violates the premise of total reflux. Nonetheless, the calculation is useful in suggesting other options, such as performing an 11% purge earlier in the process, when other factors may help attenuate MeOH, or seeking a separate end use for high-MeOH acetone.

Distillation & Fractionation Systems

A better estimate of what pasteurization can do is obtained from an infinite-tray pinch calculation for the smallest concentration of the impurity that can be attained in a (top) side draw:

$$X_{DO}/X_{DM} = K + (K - 1)R \qquad (7\text{-}5)$$

where X_{DM} is the concentration of light impurity in the side-draw; K is the equilibrium y/x at the draw tray ($\approx \alpha^\infty$); X_{DO} is the overhead concentration of impurity; and R is the reflux/top draw ratio without pasteurization.

With $K \approx 1.8$ and $R = 3$, eq. 7-5 indicates that at best a fourfold reduction of impurity can be achieved by pasteurization. To get a 10-fold reduction at this reflux ratio, we need $K \approx 3.2$, typical of the volatility required for effective use of the pasteurization approach.

Mechanical/Layout Sources of Problems

In my experience, a frequent error in distillation troubleshooting is that mechanically oriented people fail to give adequate credence to process/conceptual problems as causes. However, for balance, let us examine some areas where a mechanical focus pays off.

Reboiler Circuit Problems

A wide range of problems from erratic boil-up can result from superficial configuration errors in hydraulic elements, either in the original design or from hastily engineered debottlenecking retrofits. Two elementary errors to avoid are the following:

1. Depending on symmetry alone to share the load equally between dual kettle reboilers. It is best to have shared liquid entry and vapor return lines, with the second reboiler acting as a "satellite" to the first.

2. A sump arrangement at the bottom of a column with a thermosiphon reboiler whereby all the return liquid from the reboiler mixes with the liquid off the bottom tray. This is to be avoided not merely because it reduces the mass transfer driving force in boiling, but also because accumulation of heavies may interfere with hydraulic action at startup or after some period of operation. A similar problem can stall a kettle reboiler system. The usual preference is division of the bottom sump so

that an elongated downcomer from the bottom tray goes straight to the bottom head (but with a dirt purge hole). A useful elaboration of this system stabilizes the head feeding a thermosiphon.

Thermal and Side-reaction Related Problems in Columns

Problems from Excessive Skin Temperatures

Inadvertently designing into the film-boiling regime through the use of an excessive temperature differential (ΔT) can limit or destabilize boil-up. I have seen provision of an excessive steam supply pressure to give extra (ΔT) for a fouling cycle become a self-fulfilling prophecy, with film boiling reducing heat flux below that available from nucleate boiling and promoting surface fouling. Excessive ΔT can occur inadvertently, as in the use of an elevated steam condensate return header that raises the steam condensing temperature. As noted in Chap. 6, the 'critical ΔT' for entering the film-boiling regime can be as low as 7°C, depending upon the fluid, the cleanliness of the system, and the convection conditions.

Inorganic Fouling

The presence of inorganic impurities with inverse solubility dependence, precipitating from an organic stream as temperatures rise going down a column, is more common than one might believe. Another mechanism is solubilization of salts by a minor water component, which distills overhead upon entering the column stripping section. Understanding this phenomenon is not exactly the key to a solution, but it at least precludes wasted effort. Root-cause solutions like upstream deionization are recommended for early consideration.

Fouling from High-molecular-weight Organics

Such fouling can limit distillation performance either by coating the reboiler heat transfer surface or by blocking free area of the trays themselves to a degree that reduces capacity. Where this results from an in situ side reaction, inhibitors or other additives may be useful. However, such an approach should be adopted only after a mechanism (e.g., ionic or free-radical polymerization) of fouling is established that makes the additive plausible. If solids are already present in the feed to the column, it may be worthwhile to try dispersant additives that minimize adhesion to the surfaces of the column internals.

Opportunity for Solvent Cleaning of Organic Deposits

Operating/maintenance supervisors tend to stay with mechanical methods (brushing or hydrojetting, for example) for removing deposits. However, post facto cleaning with solvents is often an attractive alternative. Of course, once polymeric deposits have become cross-linked, shutdown for a formidable mechanical cleaning job may be unavoidable. However, laboratory investigation can sometimes find an optimal solvent and circulation temperature to provide in situ clean-out of deposits without a lengthy shutdown. Using steam and hot water for decontamination may hard-cure polymers in place and preclude more sophisticated approaches. Such procedures may also make it impossible to obtain a representative polymer sample for testing a solvent approach.

Baking Deposits Off

In the spirit of lateral thinking, I must mention a well-established "reversal" approach to this problem: Gummy polymeric deposits are deliberately cooked until they are reduced to a cokelike scale, susceptible to subsequent removal by hydrojets.

Self-aggravating Formation of Light Ends in Vacuum Systems

Reboiler decomposition reactions may be catalyzed by particular metal surfaces. This factor should be considered, along with corrosion, in material selection when reboiler bundles are replaced. In Chap. 3 we discussed reboiler dehydrochlorination of organic chlorides in the context of corrosion/neutralization. In distillation of a high-boiling organic chloride, such a reaction can cause trouble simply by generating HCl as a fixed gas that loads the vacuum system. The problem is self-aggravating since an increase in pressure raises reboiler temperature, causing a further increase in HCl formation. Even water, from chemical dehydration of a heat-sensitive compound, can produce a similar effect. Heavy end formation, proceeding from the olefinic species formed in the above situations, often accompanies light end formation, with fouling and other consequences.

Separations Thwarted by Aliasing Reactions

Sometimes a side reaction that adversely affects the quality of one of the distillation products occurs on the trays. *Aliasing* is the term applied to one such situation. It connotes the reversible conversion of an impurity into a species that is lighter or heavier than the parent compound, enabling it to escape the rejection expected in a given distillation column. An example from my experience is chloral, CCl_3CHO, avoiding rejection by back-and-forth masquerading as its hydrate. Many aldehydes alias in this way or via trimerization. Entrapment of crotonaldehyde on the EtOH-rich side of the ethanol-water azeotrope at 1 atm was attributed by Ikari et al. (1986) to acetal formation with EtOH.

Reactions of Minor Components Trapped in Bulges

Another quality-impairing situation occurs when a slow side reaction becomes significant even in the limited time on a distillation tray because it involves a trapped component whose residence time greatly exceeds that of the major components. An example of this occurred in stripping light organic compounds from a sulfuric acid catalyst system, where heavier unsaturated organic impurities were trapped in the middle of the stripper. Here oxidative action of the acid, quite slow at the temperatures involved, turned out to be significant for the trapped compounds.

Distillation with Reaction

Frequently, by design or by process evolution, a distillation column is expected to accomplish a chemical conversion in addition to separations. Many examples involve adding inorganic reagents to drive a chemical reaction to completion, perhaps taking advantage of a volatility difference between reactant and product. The ability to shift chemical equilibria to advantage has already established distillation-cum-reaction in the synthesis of esters such as methyl acetates

(Agreda et al. 1990). This may become a major processing category, with development of catalysts whose activity and structure fit a distillation environment (see Chaps. 11 and 15).

Product volatilization away from the reagent on the trays, either to avoid undesirable sequential reactions or to overcome an equilibrium limitation, has been a dominant consideration in many systems I have worked with. I used to favor dual-flow trays in such applications, partly because of potential fouling by inorganic reagents and their saline products but mostly to maximize stripping relative to liquid residence time. The assumption was that downcomers did not provide useful residence time, since there was no L/V mass transfer concurrent with the reaction. Recently I have become convinced that downcomers, by stabilizing tray action, may enhance the mass transfer effectiveness of a tray more than in proportion to the added liquid residence time, and my preference has switched to sieve trays with downcomers.

Foaming Systems: The Zuiderweg Criterion

Several generic kinds of foam can occur in process systems. Long-lived foams, due to viscous effects or colloidal solids that impede drainage of the film between gas bubbles (an essential step in clarification), are not our concern here. In most distillation troubleshooting, we are dealing with a milder kind of foaming system. Ross called it *champagne* foaming (as distinct from beer foaming), a modest degree of surface activity that hinders vapor bubble coalescence and leads to a need for extra time in a downcomer for clarification of the froth entering it, perhaps two to five times normal (e.g. 7 vs 2 sec). Such foaming, linked to the magnitude of surface tension gradients by Zuiderweg and Harmens (1958), impairs column capacity while enhancing contacting efficiency. It is one of two important subsets of surface tension effects discovered by Marangoni.[42]

Zuiderweg's work did not generate as much follow-up work as it merited because it was thought initially that the effect was important only in laboratory-scale tray columns. Subsequent work indi-

---

[42] In the first Marangoni effect, dealt with here, surface tension forces stabilize a phase interface. The second effect refers to spontaneous convection cells, from surface tension forces related to mass transport, that can grossly disrupt a phase interface.

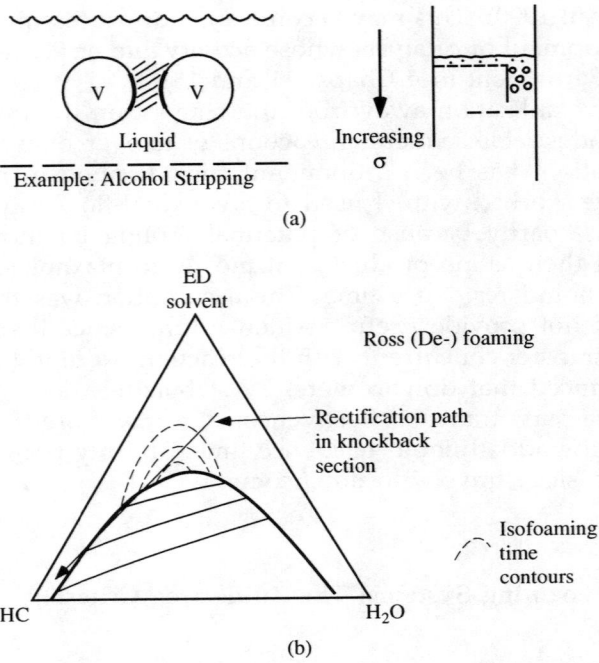

Fig. 7-4 (a) Marangoni Foaming by Inhibition of Bubble Coalescence; (b) Foaming/defoaming Regions in an ED Knockback Section

cated a meaningful differential between so-called positive and negative systems even on the commercial scale in most columns.

Nature of a Zuiderweg-positive System

A Zuiderweg-positive system is one in which σ increases ≥ 0.5 dyne/cm per tray in descending the column. The hindered coalescence of vapor bubbles in a continuous liquid, when they jostle against one another, stems from enhanced surface tension at the thinnest point of the liquid film separating them (since V/L mass transfer is most complete there) (see Fig. 7-4a). In a negative system, the film is weakened at the point of impact and coalescence is facilitated. Thus, in an absorber/scrubber where transfer into the liquid from the vapor lowers interfacial tension (a negative system), we should expect sub-par contacting efficiency. In the converse stripping operation, where loss of a volatile component increases surface tension, we should achieve superior stage efficiency[43] but at reduced capacity.

The Zuiderweg criterion for a trayed column assumes a dispersed vapor phase in a continuous liquid phase, that is, a foam regime. For a spray regime on a tray, where entrainment rather than downcomer backup is the capacity constraint, the effects are in principle reversed because the curvature of the liquid phase is inverted.

Mass transfer effectiveness ratios approaching 1.5:1 of positive vs. negative systems may apply. The offset between negative and positive systems also holds in packed columns but the mechanism is slightly different, resting upon better wetting for positive systems and less rivulet bunching.[44] The addition of surfactants to negative systems can make up part of this difference in both packed (Francis and Berg 1967) and tray (Brumbaugh and Berg 1973) columns.

Identifying Zuiderweg-positive Systems

Table 7-3 lists a number of binary systems that meet the Zuiderweg-positive criterion in at least part of the column. Since the criterion is framed in terms of $\Delta\sigma/\Delta N$, the change in surface tension per tray, the identity $(\Delta\sigma/\Delta N)/(\Delta\sigma/\Delta x) = (\Delta x/\Delta N)$ was used to identify the compositional gradient, mole fraction/actual tray, for a positive system, based on handbook binary σ data to generate $\Delta\sigma/\Delta x$.

One reason why Zuiderweg-positive systems are common is the self-association of polar, higher-surface-tension compounds like water and many ED solvents, which are higher-boiling by virtue of these very properties and tend to be the distillation bottoms product. Concentration gradients, however, may confine foaming to a limited zone, such as the solvent entry tray and the knockback section just above it in ED of alkanes from olefins (the third item in Table 7-3).

But in stripping the ED solvent to recover olefin from the "fat" solvent from the bottom of the extractive column, we can expect a positive Zuiderweg criterion throughout most of the column (see Fig 7-2). And even relatively polar compounds like alcohols and nitriles have enough surface tension differential to be strong foamers in being stripped from water. With a compound like hydrogen peroxide, with

---
[43] But recall that high stripping factor (KV/L) will lower tray efficiency.
[44] Mass transfer gradients aside, a packing with low contact angle and reasonable spreading of liquid should always be chosen.

Table 7-3
Binary Distillation Systems with Positive Zuiderweg Coefficients (Units in dyne/cm and mole fraction)

Components Light/Heavy	$\Delta\sigma/\Delta x_2$		Critical $\Delta x_2/\Delta N$	
	Top	Bottom	Top	Bottom
Paraffin/olefin	4	4	0.12	0.12
Olefin/aromatic	4	4	0.12	0.12
Paraffin/ketone	6		0.08	
Ketone/phenol	16	6	0.03	0.08
Aromatic Hydrocarbon/phenol	7	18	0.07	0.03
Alcohol/water	5	2000	0.10	0.0002
Nitrile/water	8	600	0.06	0.0008
H_2O/H_2O_2	7	6		0.08

a higher surface tension than water, even a system in which water is the lighter component can be a foamer.

Remedial Action for Zuiderweg-positive Systems

In the more common situation of a Zuiderweg-positive anticoalescence system, our strategy will usually be to sustain the increased contacting efficiency without accepting a capacity penalty. This tends to rule out antifoam additives except as an interim expedient. A first recourse might be an enlarged but sloped downcomer to add clarification time for the relatively short-lived foam while subtracting only moderately from pattern area. An entry area as large as 35% of the column cross section can be considered, tapering to about 10% at the bottom, for a reduction of perhaps 40% in pattern area from a standard design.

We do not deliberately design so as to obtain a positive or negative Marangoni system; this is set by broader process considerations. But by understanding the phenomenon, we can selectively modify only those trays most involved, for example, trays in the middle of a column recovering the $MeCN/H_2O$ azeotrope for reuse from aqueous washes in an ED systems using acetonitrile (MeCN) as the ED solvent (see seventh entry in Table 7-3).

Ross Defoaming Criteria

The work of Sydney Ross at Rensselaer Polytechnic Institute has provided a real breakthrough in troubleshooting surface-active systems. Years ago, it was commonly believed that the onset of a second liquid phase in distillation promoted foaming. Ross and Nishioka (1975) showed that it was actually in a particular region *just outside* a two-liquid-phase region (see Fig. 7-4b) that the maximum foam lifetime usually occurred in a distillation system.

Equally important was Ross's finding that appearance of the second phase will cause defoaming only if it spreads spontaneously at the main V/L interface. Young's rule for the free energy of spreading at an interface provides a simple test of this criterion if the minor phase can be characterized. For example, a nucleated minor aqueous phase will not spread over a hydrocarbonlike major liquid phase, since the surface tension of water would be higher than the hydrocarbon surface it displaced. Silicones are effective antifoam agents in aqueous systems both because their low surface tension facilitates such spreading and because of the low dosage required by their limited solubility. In hydrocarbon systems, their elevated solubility makes them ineffective (they might even be pro-foamers).

Interplay of the Ross and Zuiderweg Criteria

Example: Foaming in an Extractive Distillation Knockback Section

Many of the issues discussed above came together for me in an early problem of coping with a capacity-limited ED column. (The following discussion expands on a troubleshooting example given in Chap. 5.) I started with the assumption that foaming in downcomers of the solvent entry tray (see Fig. 7-2) and the trays immediately below was responsible for limiting capacity, from the appearance of a second liquid phase. Transitory incomplete mixing of hydrocarbon-rich reflux from the knockback section above into the entering solvent was the implicit (incorrect) scenario invoked. Much study was given to ternary diagrams to determine how to stay out of the two-phase region.

A bit later, with a grasp of the Zuiderweg criterion, surface tension measurements simulating the phase relationships in the plant column were made in pressurized laboratory apparatus. These showed that the entire knockback section amply met the Zuiderweg criterion for downcomer foaming. However, the middle of the knockback section was in a two-liquid-phase region and was excluded from foaming by the Ross criterion (see Figs. 7-2 and 7-4b). The sharp increase in surface tension just above and on the solvent entry tray, where the major liquid load begins, was now understood as the point of maximum foaming tendency.

An interim solution used a silicone antifoam agent whose action met the Ross defoaming criterion. Later, the critical trays were modified to handle the greater volumetric flow of unclarified liquid; antifoam usage was discontinued. Bolles (1967) has described a parallel experience with a furfural ED problem.

Example: Deepened Understanding of Antifoam Usage

Early in my career, we routinely used a whale-derived higher alcohol mixture (Ocenol) as an antifoam in several aqueous distillations. When the product was withdrawn for ecological reasons, we initially had no basis for finding an acceptable substitute. The Ross criteria, with handbook information on the water solubilities and melting points of the higher alcohols, came to our rescue. We learned that antifoam action required that the solubility limit be exceeded, and that spreading be thermodynamically favored and rapid.

Ocenol was primarily a C_{17} unsaturated alcohol. Economics dictated finding a substitute with an effective dosage of <20 ppmw. Ineffective initial attempts with C_{12} and C_{14} saturated alcohols were attributed to their excessive solubility. We also came to appreciate that unsaturation had been useful in providing not only a lower melting point but probably also less viscosity than the corresponding saturated alcohol at column temperature. The Ross criterion of spreading at the V/L emulsion interface helped us to understand a further requirement: We needed an excess over simple solubility to furnish a film equivalent to at least two monolayers of thickness at the V/L interface.

Invoking our Avogadro number/molecular surface units conversion kit (see Chap. 5) and assuming a foam of 30%v vapor in 1-mm-diameter bubbles, we estimated that about 8 ppmw excess over simple solubility would make an antifoam work in the Ross sense. Finally, it became clear that the dosage of a long-chain alcohol must be adjusted upward, both for elevated column temperatures and for the presence of lower-molecular-weight organics that increased the solubility of the antifoam. A mixture of higher alcohols that satisfied all these requirements was ultimately found.

This discussion of an important phenomenon closes with a caution on possible downstream complications from antifoam additives. I have seen several cases where supposedly safe silicone (polysiloxane) compounds used as antifoams dissociated under process conditions into fragments, which traveled to remote parts of the process and poisoned catalysts or created other unacceptable effects. Hence it is not just "less is more" aesthetics that suggests the need to work around a foaming situation before resorting to antifoam dosing.

Classic Errors in Configuration and Operation

Although this book claims to be a first in the overall treatment of process troubleshooting, distillation per se has benefited in recent years from publications (Kister and Hower 1986, 1990; Lieberman 1985) describing vividly a variety of errors made in this field.

Energy Efficiency

Heat pumps and coupling of the heat loads of two columns are frequently used techniques for improving energy efficiency in distillation, nearly doubling the efficiency of a number of processes. However, we start from quite a low efficiency in a Carnot cycle/ thermodynamic *availability* sense. An availability diagram is a valuable tool in sensitizing us to the enthalpic, work, and entropic elements in process thermodynamic efficiency (Fitzmorris and Mah 1980; Kotas 1986).

Zuiderweg (1975) notes that Marangoni effects have important consequences in heat transfer as well as mass transfer. For example, in a positive system we can increase both boiling heat flux and the transition flux at which nucleate passes into film boiling. Negative systems can increase condensation rates both by dropwise condensation and by rivulet flow of condensate (Ford and Missen 1968).

Heat Transfer on Distillation Trays

Occasionally, we need to know how effectively distillation trays can transfer heat, rather than mass, as in injection of a superheated vapor into a column. Sensible heat transfer to the V/L interface detracts from mass transfer (Ruckenstein 1970). Thus conventional trays are likely to give only 25–40% efficiency of enthalpy transfer in a quench situation. However, in humidification (all heat transfer to supply evaporative latent heat), heat transfer can somewhat outpace mass transfer (Azizov 1972).

Liquid/liquid Extraction

Extraction can be less energy-intensive than distillation.[45] It merits serious consideration in contexts where the separation factor β is much more favorable than the prospective relative volatility α via distillation. This can result from nonideal behavior in a second liquid phase or from canceling out an adverse vapor pressure ratio. The physical chemistry factors in choosing an extractive solvent have been reviewed (Hampe 1986). There are also many cases involving high-boiling components where distillation would cause degradation or other complications.

Thus, one chooses sulfolane extraction for removing benzene from close-boiling aliphatic hydrocarbons and water-immiscible solvents for extracting trace phenol from water[46] over distillation.

Design Issues

The process engineering of L/L extraction is still in its formative stage.[47] There are only a few (mostly German) articles in the literature that give a reasoned overview (Blass et al. 1986). The three main issues of phase distribution equilibrium, mass transport fundamentals, and equipment options interact with each other.

Our first judgment may be whether internal agitation and/or a tray configuration, rather than packing, is needed. The choice rests on both the number of separation stages needed and the difficulty

---

[45] Note that environmental considerations can cut into energy savings. Where an aqueous stream used to extract impurities could formerly be discarded, water reuse processing may now be required.

[46] In the latter case, the logic employed carries us on to fixed-bed adsorption as the ultimately preferred alternative.

[47] The field has been handicapped, relative to distillation, by the absence of a consortium like FRI, which is able to work in large-scale equipment. Much of the good-quality data come from small-scale university work. Thus stacked Bialecki rings may be recommended for high performance (Billet 1988b), which are hardly practical for large-diameter and tall columns.

of dispersion. Properties such as phase equilibrium, viscosities, densities, and interfacial tension (σ) are relevant.

For example, dispersion difficulty, plus the need for a large number of stages, might dictate the use of a centrifugal contactor, whereas mixer-settlers might be indicated if fewer stages were required. Packed towers, with relatively low power input, work best with systems with σ values of 10 dyne/cm or less; if more than five transfer units are needed, the progression would be to a perforated tray tower and then to some sort of agitated staged device.[48]

Packed Columns

Despite their limitations, packed towers are frequently used for their simplicity and ready rearrangement, particularly when the extraction is a peripheral cleanup operation and is not central to the process. Our treatment here will deal with some judgment elements that go into configuring or reconfiguring such units.

Predicting hydrodynamic capacity a priori is still chancy. As troubleshooters, however, we will likely know whether we are flooding; the literature can suggest repacking options that may provide some incremental capacity. Capacity and mass transfer goals will often be in conflict. Improved extraction efficiency is the troubleshooting goal that demands the most intelligent handling.

Mass Transfer Resistances of the Separate Phases

Once we have settled on a particular extractor type, we must try to define separately, as in distillation, the resistance to mass number of the respectives phases and how they are combined via the phase distribution coefficient m (Leibson and Beckmann 1953) (see eq. 7-6). We use m to denote the phrase equilibrium ratio here, in place of the K used earlier on distillation. Only if m varies strongly with composition will detailed attention to phase equilibrium data be necessary.

In a sieve-plate column, much of the mass transfer occurs during drop formation and close tray spacing makes sense. However, in packed towers, there is little surface stagnation. Here the internally

---

[48] Horvath and Hartland (1985, Fig. 3) give capacity/HTU trade-offs in common stirred-column devices. From Blass's review (1988b, Fig. 14), sieve-plate columns may be the best choice for high mass-transfer performance while avoiding axial mixing that cuts into staging.

circulating large-drop film model of Handlos and Baron (1957) gives the basis for estimating disperse-phase transport, with circulation perhaps promoted by passage over the packing. For the continuous phase, a penetration model is more appropriate.

Packed Columns

Despite their limitations, packed towers are frequently used for their simplicitiy and ready rearrangement, particularly when the extraction is a peripheral cleanup operation and is not central to th process. Our treatment here will deal with some judgment elements that go into configuring or reconfiguring such units.

Predicting hydrodynamic capacity a priori is still chancy. As troubleshooters, however, we will likely know whether we are flooding; the literature can suggest repacking options that may provide some incremental capacity. Capacity and mass transfer goals will often be in conflict. Improved extraction efficiency is the troubleshooting goal that demands the most intelligent handling.

Key Parameters in L/L Extraction in Packed Towers

The key selection parameters in configuring a system are which phase to make continuous, the type of internals, and a packing material *not* wetted by the dispersed phase.

Nonwetted Packing

Unlike the other criteria, all writers are in agreement on this one. A dispersed phase that wets the packing[49] will coalesce very rapidly. The packing is there primarily as a network to reduce axial mixing! In an agitated column, baffles serve this function. Packed columns resist coalescence of the dispersed phase better than open spray columns of comparable flows.

Choice of Internals

Substituting smaller-size packing will generally reduce HTU, with some loss in throughput capability. The newer packings, such as

---

[49] Even in a sieve-plate column, holes are punched to minimize droplet wetting/spreading. For a case where we do want a minor phase to spread on a packing, see the discussion below on laminar contactors.

Intalox saddles and Pall rings, serve more to increase capacity than to improve mass transfer. This suggests that a combined switch in size and type might achieve some gain on both fronts.

Choice of the Dispersed Phase

A difference from distillation is the freedom to make the denser phase the dispersed one, although extreme flow proportions may dictate this choice. If a very high degree of raffinate purification is required, that phase will probably be made the dispersed one, since axial mixing, which is the main threat to the separation, occurs primarily via the continuous phase. Beyond this, the basis for choice is still controversial. I suggest choosing as the dispersed phase the one that minimizes HTU.

Combining Phase Resistances (See Earlier Parallel Treatment for Distillation)

The droplet phase, constrained to a lesser turbulence, usually has the lower Sherwood number (Blass 1986). Typical mass transfer coefficients for the disperse and continuous phases might be 50 and 300 kg/m²/hr, respectively, assuming that molecular diffusivities of 10^{-5} cm²/sec apply in case.[50] (This difference may be rationalized on the basis of a fuller realization of turbulence in the exterior phase.) If the raffinate phase is a light hydrocarbon with a higher diffusivity than the aqueous extract phase, and if, in addition, phase equilibrium favors the raffinate ($m \gg 1$), then overall HTU can be minimized by dispersing the raffinate phase:

$$\frac{1}{K_{od} \cdot d_p} = \frac{1}{D_d \cdot Sh_d} + \frac{m}{D_c \cdot Sh_c} \qquad (7\text{-}6)$$

where c, continuous phase; d, dispersed phase; p, particle (or droplet); o, overall or combined; d_p, drop diameter; D, diffusivity; m, raffinate/extract distribution constant; Sh, dimensionless Sherwood number = kd/D (d is drop diameter); k, mass transfer coefficient; K_o, overall mass transfer coefficient in units of cm/s.

---

[50] For comparison, a rough generalization from mixing-impeller work is that the composite coefficient (with the area term) $k_d a$ for the disperse phase will be about one-third that of the continuous phase, and this differential grows with power input/volume.

On the other hand, distribution strongly favoring the aqueous phase would suggest that the aqueous extract be made the disperse phase.

Interfacial Tension Effects

Interfacial tension σ_i affects performance in several ways: via the Marangoni effect, which modifies mass transfer rates, and via the facility of coalescence of the disperse phase.[51] Mass transfer that increases σ_i in the direction of continuous phase flow, as in distillation, hinders coalescence and enhances mass transfer. Thus, if the higher surface tension phase (e.g., water) is continuous and we are extracting a surface-active compound (e.g., alcohol, organic acid) into that phase from a less polar phase, a negative Zuiderweg index[52] results, connoting reduced mass transfer. In this case, reversing the phases would be indicated if consistent with the HTU minimization/phase distribution strategy described above. If extraction were into the organic phase, the Zuiderweg analysis would suggest keeping water as the continuous phase.

Role of Surface-active Impurities

We are often tempted to add surfactants to enhance mass transfer. The evidence is that either enhancement or retardation can result (Mudge and Heideger 1970). Low-mole-weight surfactants that are fluid seem to enhance, while longer molecules that stiffen the interface block transport. That interfacial viscosity and elasticity are key parameters in coalescence is known; however, there are few data to try to apply.

---

[51] The second Marangoni effect of spontaneous convection cells disrupting the interface can also figure in L/L extraction (as in drop formation at a nozzle) when mass transfer strongly lowers σ_i. Such cells can even cause emulsification. Recent work (Slin'ko et al. 1988; Vasquez et al. 1990) attempts to relate interfacial turbulence to a "Marangoni number." Another driving force may also create emulsification in ternary L/L extraction systems when the diffusion path creates zones of supersaturation (Ruschak and Miller 1972).

[52] Here we have made two assumptions: that a given concentration of a surface-active substance reduces σ of a high-σ phase more than that of a low-σ phase; and that σ_i will parallel the difference between the surface tensions in air of the respective liquid phases. There are better options than this crude Antonov's rule for exact prediction of σ_i (Fowkes 1963, 1964; Nakahari and Arai 1989).

Revisited Example: Sprucing Up a Packed L/L Extractor

In Chap. 5 we introduced an example involving misemphases and omissions in upgrading a L/L extractor due to insufficient attention to the context of the existing process. Let us review the decision elements here against the design criteria listed above.

In substituting a superior packing shape, ceramic material was retained, with water still as continuous phase, in view of its 4/1 volumetric predominance over the hydrocarbon (HC) stream being washed. The new inlet dispersion plate for HC was specified as stainless steel, a dubious choice, since this metal tends to be preferentially wetted by HC-type phases. Extraction of alcohol (ROH) into a continuous water phase is detrimental to enhanced separation in that it makes this a Zuiderweg-negative system. The molar distribution of ROH favors the HC phase by about 2/1. With about a 36/1 molar ratio (V_c/V_d) of water/HC flows, the analog of a distillation stripping factor, mV_d/V_c, will be about 0.055, placing mass-transfer resistance decisively in the internal (HC) phase. This fails to maximize K_{od} (minimize H_{od}). These two factors argue for a serious attempt to make water the dispersed phase[53] by using the minimum practical water flow, possibly adding incremental water at the end of the tower if a pinch results.

Laminar Contactors

Where cocurrent contacting suffices, and where a minor phase that contains a generous stoichiometric excess of the treating agent (e.g., NaOH) is to remove from a major (e.g., hydrocarbon) stream a trace (e.g., acidic) impurity by extraction/neutralization, the contacting problem essentially focuses on the *aspect ratio*. This involves arranging for the major phase to have an extended interface with the minor, treating phase. This is commonly achieved by massive energy inputs in either a pipeline contactor, with periodic orifice-plate redispersion of the treating phase, or a mixer/settler system, wherein the minor phase is recirculated from the settler to increase it to a significant volume fraction of the two-phase system in the mixer section.

A more elegant system, promulgated by Merichem with its Fiber-Film devices (Norris 1975), uses preferential wetting of a bundle of

---
[53] Such considerations explain why, surprisingly at first sight, water is often made the dispersed phase. Flowing from this choice would be a switch to a packing such as polypropylene.

very fine fibers (glass or stainless steel, according to the system's chemistry and wetting characteristics) by the minor phase to provide the extensive interface needed. The wetting phase also moves much more slowly axially along the fiber bundle than the major phase treated. This results in both lower energy input and a less onerous phase separation afterward. This unit operation deserves wider application than it has received.

Fixed-bed Sorption

I use the term *sorption* as a generic decriptor because *adsorption* is frequently used imprecisely to characterize separations that go well beyond laydown of a surface layer on an extended surface of activated carbon, alumina, and so on. Because nonlinearity is intrinsic to adsorption most of our (limited) discussion is deferred to Chap. 9. Suffice it here to point out that regenerability is the crux of most sorption operations and deserves early testing before investing much effort in the fine points of the adsorption isotherm or breakthrough loadings. Only if an impressive adsorption equilibrium and a small amount of impurity to be removed make it economical to discard the bed after a single use (including environmental disposal cost) do we have a real bargain.

Example: Bulk Sorption Defeats Resin Bed Regenerability

Recently, I had advocated using a polystyrene resin bead bed as a generic purifying device for removing moderately heavy chloroorganic impurities from an aqueous stream. The weight loadings and distributions obtained in batch tests were impressive. I assumed that regeneration with moderate-temperature steam would give removal of the sorbates adequate for a workable cyclic process, but laboratory tests soon showed that slow *kinetics* of desorption made the proposed process infeasible. The very favorable weight loading upon sorption meant that the organics were *plasticizing* the resin beads, a bulk phase penetration that forecast slow removal of the sorbate because of the poor diffusion through the unplasticized, sorbate-depleted outer zone of resin bead in the regeneration part of the cycle. Too good a sorption equilibrium spelled trouble.

Sometimes we can reverse an unfavorable desorption equilibrium by "going the long way around." An example is the regeneration of a chelating resin used to pick up divalent copper from effluents.

Displacement by H^+ in a direct regeneration is inefficient, but a two-step regeneration, in which NH_4OH is first used to complex off the Cu, succeeds (Sengupta and Zhu 1991).

Having dealt in this chapter and the previous one with problems that stem largely from fluid mechanics gone awry, we turn in the next chapter to the more elusive management of the opportunities and pitfalls of instrumentation and control.

References

Agreda, V.H. et al. *Chem. Eng. Prog. 86*(2) 40–6 (1990).

Azizov, A.G. *Intl. Chem. Eng. 12* 249–51 (1972).

Billet, R. *Chem. Eng. Technol. 11* 139–48 (1988b).

Billet, R. *Packed Column Analysis and Design* Ruhr-University Bochum (1989).

Billet, R., and Mackowiak, J. *Chem. Eng. Technol. 11* 213–27 (1988a).

Blass, E., et al. *Ger. Chem. Eng. 9* 222–38 (1986).

Bolles, W.L. *Chem. Eng. Prog. 63*(9) 48 (1967).

Bravo, J.L. et al. *IEC Funds. 21* 162 (1982); *Hydrocarbon Proc. 64*(1) 91–5 (1985); *I. Chem. E Symp. 104* A183–A201(1988); *Chem. Eng. Prog. 86*(1) 19–29 (1990).

Brumbaugh, K.M., and Berg, J.C. *AIChE J. 19* 1078–80 (1973).

Colburn, A.P. *Ind. Eng. Chem. 28* 526 (1936).

Doherty, M.F., et al. *AIChE J. 35* 1585–91, 1592–1601 (1989).

Fitzmorris, R.E., and Mah, R.S.H. *AIChE J. 26* 265–73 (1980).

Ford, J.D., and Missen, R.W. *Can. J. Chem. Eng. 46* 309 (1968).

Fowkes, F.M. *J. Phys. Chem. 67* 2538–41 (1963); *Ind. Eng. Chem. 56*(12) 40–52 (1964).

Francis, R.C., and Berg, J.C. *Chem. Eng. Sci. 23* 685 (1967).

Frank, J.C., et al. *Chem. Eng. Prog. 65*(2) 79–86 (1969).

Gerster, J.A., et al. *J. Chem. Eng. Data 5* 423–9 (1960).

Hampe, M.J. *Ger. Chem. Eng. 9* 251–63 (1986).

Handlos, A.E., and Baron, T. *AIChE J. 3* 127–36 (1957).

Hanson, D.N., et al. *IEC Funds. 73*(3) 170–4 (1978).

Hirata, M., et al. *Computer-aided Data Book of Vapor/Liquid Equilibria* Elsevier (1975).

Hoek, P.J., and Zuiderweg, F.J. *AIChE J. 28* 535–41 (1982).

Horvath, M., and Hartland, S. *IEC Proc. Des. Dev. 24* 1220–5 (1985).

Hüttinger, K.J., and Rudi, H. *Chem. Ing. Tech. 55* 867–9 (1983).

Ikari, A., et al. *IEC Proc. Des. Dev. 25* 859–62 (1986).

Kister, H.Z., and Haas, J.R. *I Chem. E. Symp. 104* A483–94 (1987).

Kister, H.Z., and Hower, T.C., Jr. Paper 86c *AIChE Miami Beach meeting* (Nov. 1986); Paper 129f *AIChE Chicago meeting* (Nov. 1990).

Koshy, T.D., and Rukovena, F. Paper 128c, AIChE annual meeting (Nov. 19, 1982).
Kotas, T.J. *Chem. Eng. Res. Dev. 64* 212–29 (1986).
Kovshov, A.N., et al. *Khim. i Neft. Mashinostroenie* 1966(6) 7–9.
Koziok, A., and Mackowiak, J. *Chem. Eng. Proc. 27* 145–53 (1990).
Leibson, I., and Beckmann, R.B. *Chem. Eng. Prog. 49*(8) 405–16 (1953).
Lieberman, N.P. *Troubleshooting Processing Operations* Penwall (1985).
Litzen, D.B., and Bolger, S.R. U.S. Patent 4,762,616 (to Shell Co.) (June 22, 1988).
Lockett, M.J. *Trans. Instn. Chem. Engrs. 59* 26 (1980).
Mudge, L.K., and Heideger, W.J. *AIChE J. 16* 602–8 (1970).
Nakahari, S., and Arai, Y. *J. Chem. Eng. Japan 22* 315–17 (1989).
Norris, B.E. *Hydrocarbon Processing 54*(9) 127–8 (1975).
O'Connell, H.E. *Trans. AIChE 42* 741 (1946).
Pham, H.N., and Doherty, M.F. *Chem. Eng. Sci. 45* 1823, 1837, 1845 (1990).
Porteus, W., and Blander, M. *AIChE J. 21* 560–6 (1975).
Puppich, P., and Goedecke, R. *Chem. Eng. Technol. 10* 224–30 (1987).
Reid, R.C. *Amer. Scientist 64* 146–56 (1976).
Ross, S., and Nishioka, G. *J. Phys. Chem. 79* 1561–5 (1975).
Rovaglio, M., and Doherty, M.F., Paper 172i AIChE Washington meeting (Dec. 1988).
Ruckenstein, E. *AIChE J. 16* 144–6 (1970).
Ruschak, K.J., and Miller, C.A. *IEC Funds. 11* 534–40 (1972).
Sengupta, A.K., and Zhu, Y. Paper 70d AIChE annual meeting (Nov. 21, 1991).
Shi, M.G., and Mersmann, A. *Ger. Chem. Eng. 8* 87–96 (1985).
Shinskey, F.G. *Distillation Control* McGraw-Hill 1977.
Shulman, H.L., et al. *AIChE J. 1* 247–64 (1955).
Slin'ko, M.G., et al. *Dokl. Akad. Nauk 303* 429–32 (1988).
Stichlmair, J. *Fundamentals of Gas/Liquid Contact in Trayed Columns* (in German) Verlag Chemie-GmbH (1978).
Stichlmair, J., and Potthoff, R. Paper 199b AIChE annual meeting (Nov. 1990).
Takahashi, T., et al. *J. Chem. Eng. Japan 19* 339–41 (1986).
Vazquez, G., et al. *Intl. Chem. Eng. 30* 228–35 (1990).
Weiland, R.H., and Merrell, P. Paper 61a AIChE annual meeting (Nov. 21, 1991).
Zuiderweg, F.J. Paper 19e AIChE Houston meeting (Mar. 1975).
Zuiderweg, F.J. *Chem. Ing. Tech. 55* 459–66 (1983).
Zuiderweg, F.J., and Harmens, A. *Chem. Eng. Sci. 9* 89 (1958).

8

Classic Problem Elements: Control and Process Dynamics

If a person offends you and you are in doubt as to whether it was intentional, do not resort to extreme measures; simply watch your chance and hit him with a brick.
                                                                    Mark Twain

The intent of this chapter is to convey not expertise, but a sense of what is worthwhile and necessary in process control and an awareness of the many frills and irrelevancies that choke the field and make us lose sight of simple truths and strengths. I attempt to convey to the technologist who has no real background in control enough understanding that he or she can communicate comfortably with instrumentation or computer control specialists in configuring or reconfiguring control schemes.

The Mission of Control

Quality and Control

These two subjects are inevitably linked. However, the quality control of an earlier era, with sampling, inspection, control charts, and feedback corrections of a process, has in recent years been superseded by a broader concept of what quality means and how to achieve it and profit from it. I am not referring to the faddist management preoccupations with the "quality process" typified by the Baldridge Awards and endless meetings and slogans. Rather, we are beginning to appreciate (belatedly) the waste-cutting and insightful approaches of quality consultants such as W. Edwards Deming and Joseph Juran who preach elimination of slogans, quotas, and numerical targets in favor of truly understanding and stabilizing our existing processes. Such understanding and control are an essential preliminary to breaking through to new levels of performance. It is in this sense of process comprehension that I suggest the reader approach control.

Control Without Dynamic Buffers

Control is more important today than ever. At the beginning of my career, a wise man in distillation design, Dan Sarno, told me that he always slipped enough overdesign into his columns to have more trays than were required for the separation in a steady-state sense. While Dan was far from unique among designers in such conservatism, he was different in being driven by process dynamics concerns. His intent was to accommodate dynamic "wobble" of the composition profile within the column, keeping the end concentrations within specification.

Today competitive pressures have eliminated intercolumn surge and many natural dynamic buffers; the control system *must* work. By "work," we do not and cannot mean absence of any deviation from the desired state or path of the process. Even sustained cycling, as practiced in some flight systems, may be an acceptable control solution if self-adaptive features keep its amplitude small. The payoff of sound control, then, is partly in making good on the gamble that management has taken by capital conservation in a tight design that eliminates surge.

Control as Part of Optimization

The other major element is the ability to operate reliably in economically attractive modes, without fear that process dynamics will carry the process into regions with high penalties in terms of downtime, corrosion, or tube failure. Although control is inevitably linked to optimization of process operation as a goal, and to computers, in this chapter I emphasize control at the individual loop level. Control is not Cinderella, to be reduced to a scrubwoman for her glamorous stepsister, the computer.

Professional Divisions and Digressions

Computers Are Not Control

The strong, pervasive presence of process computer control (PCC) over the last two decades might seem to have made control a robust part of process operating technology. Not so. Never has "the better is the enemy of the good" been more of a truism. The all too human tendency to vault over basic needs to address "higher" concerns must date back to the Egyptian priestly elite building the pyramids while 10% of the population was suffering from the dread "river disease" schistosomiasis, or the United States building a "supercollider" while our bridges crumble for lack of maintenance.

What these similes are intended to convey is that performance achieved in single-loop control and reliable process analyzers is lagging well behind the superstructure of computer control and the mathematics of process optimization. There has been a real divergence of what should be closely related efforts, with computer control experts pursuing global tasks of computer architecture and language while neglecting the more mundane "infrastructure" tasks. In my opinion, building the control infrastructure should precede the pursuit of higher goals such as optimization or advanced control. In a few cases, computer control can override a wobbly infrastructure and make a contribution. More often it constitutes a cosmetic front, hiding from management the seriousness of problems at the grass roots.

Computers per se are cheap, but process interfaces and elaborate systems architecture with unnecessary features are not. The result of rushing past the basics has been to price control computer systems beyond the reach of many medium-scale process units. The technologists on such units typically need to master two or three dominant single-loop control problems as a prerequisite to carrying out their responsibilities for the unit, which may later lay a basis for process computer control (PCC).

The Benefits of PCC

In citing excesses of PCC, there is no intent to deny the real benefits that accrue from well-chosen applications. Batch processes, enforcement of constraints on reactors or large distillation columns to minimize energy cost or maximize throughput, and control of parallel modules like furnaces or compressors are positive examples. Features like fine time-base resolution for following off-normal events and graphic nested displays of the process flow diagram offer the operator and the technologist information well beyond that provided by the control room panels of yesteryear.

Control based on a calculated variable is much easier than with the older hardware elements. But beware of using computers for constrained optimization via process models that have not been thoroughly validated in a differential sense. I have seen elaborate computer models that were so bad that operators turned computer control off because the moves made by the optimizer were contrary to common sense; in some cases the sign, and not merely the magnitude of the step suggested by the optimizer, was wrong.

A Needed Convergence: Technologists and Instrumentation Specialists

There is a certain language/concept gulf, which predates PCC, between technologists and the instrumentation specialists who should be partners in their efforts. The former may be struggling with a dynamics problem via endless controller retuning when a slight process change would be the effective solution. Or an authoritative colleague is needed to tell him that half of the pressure drop across a heat exchange system should be reserved for the control valve. Yet the instrumentation peer to provide this information is not there; she has been physically preempted or the dialog has been preempted by a computer team preoccupied with "higher-level" problems.

Plant technologists bear significant responsibility for the lack of dialogue, even in a no-computer mode, by arrogating to themselves decisions on control configuration that should be a matter for interdisciplinary consultation. Instrument engineers are often handed nearly insoluble problems by having the most amenable control handle (e.g., feed rate to a reactor) ruled off-limits by a process engineer or operations manager who lacks a real grasp of control fundamentals. The corollary of this situation occurs when a severe startup control problem is experienced that never surfaced at the design stage because of the timidity and lack of experience of the instrumentation people in addressing process issues. It is preferable for each discipline to do some second-guessing in full consultation with the professional for whom that skill is basic. Redundancy pays!

Commonsense Control Basics for the Technologist

The All-important Sensor

Integrity of the measurement, choice of the manipulated variable, and the algorithm used to link them are the basics of control. The first is most important and is all too often not properly executed.

Example: A Process Analyzer Lacks an Engineered Sample Loop

The front-line people have correctly decided that a process analyzer, rather than the indirect control via temperature they have been using, will solve their control problem. Unfortunately, it will be installed in a process context where an inadequate sampling loop will defeat the overall scheme, for lack of an experienced crew of sampling-loop specialists. Proper configuration of a sampling system can be as demanding as process design. It cannot be given to a hardware specialist working in isolation.

Control Reasoning Without the Experts

In what follows we will assume that the technologist, in sorting out the conceptual side of a control problem, is largely on his own or must know enough to hold up his side of a discussion with an instrumentation peer. What tools and insights must he use? Some useful rule-of-thumb approaches will be described first, followed by a sampling of problems specific to certain types of units and concluding with a brief look at the some new frontiers in process dynamics.

Why Feedback Control Is Still the Main Scene

Feedback correction of entering disturbances is second best but robust. It covers a multitude of circumstances and does not require in-depth process knowledge. It is reminiscent of what has been said about democracy: "the worst conceivable form of government except for all the alternatives." The analogy to democracy vs. benevolent despotism in political systems is not an idle one. Those who think that the rich texture of a real (process or human) system can be rendered by a limited set of differential equations tend to emphasize feedforward (read autocratic) control. The response is faster, more exact—if only the real system weren't so ornery!

My approach is to try for minimum "overshoot" correction via a feedback control loop, with secondary attention to speed of response. The reader is advised to become familiar with the stability analysis of a simple loop involving transport delay (or "dead time") θ, first-order lag T, and process or instrumentation "gain" K from a basic text (Buckley 1964, 1985; Harriott 1964). In my experience, the lay technologist readily grasps the idea of instrument amplification and can appreciate the significance of dead time from the common shower bath temperature control problem. But chemical engineers, familiar with dimensionless groupings like the Re number, made up of products and quotients of dimensional variables whose units cancel out, seem to have trouble appreciating the similar combining of process and control elements to obtain a dimensionless loop gain, K_L. Three loop gain examples presented in this chapter illustrate derivation of the maximum controller gain achievable from a stability limit for K_L.

Example: Process Sensitivity in Tray Temperature Control

In Chap. 7 we alluded to the control sensitivity of a stripper column recovering an alcohol-water azeotrope, with a water bottoms discard. The feedback control scheme (see Fig. 8-1a) attempts to control the mass split between overhead alcohol (azeotrope) and water bottoms discard by adjusting boil-up on the basis of a temperature control (TC) point. Control is jittery; the temperature break jumps a few trays up or down with minute changes in reboiler heat input. Rudimentary static reasoning was to locate the TC sensor at a sharp break in the column temperature profile (see Fig. 9-2). Now dynamics teaches us that measuring composition or its surrogate close to an end of the column can improve control by reducing dead time and process gain, provided that sensitivity of measurement can be maintained.

Dead time and first-order lags in the response to a steam flow change arise from the mass of metal and liquid in the reboiler, the column below the control tray, and the thermowell encasing the

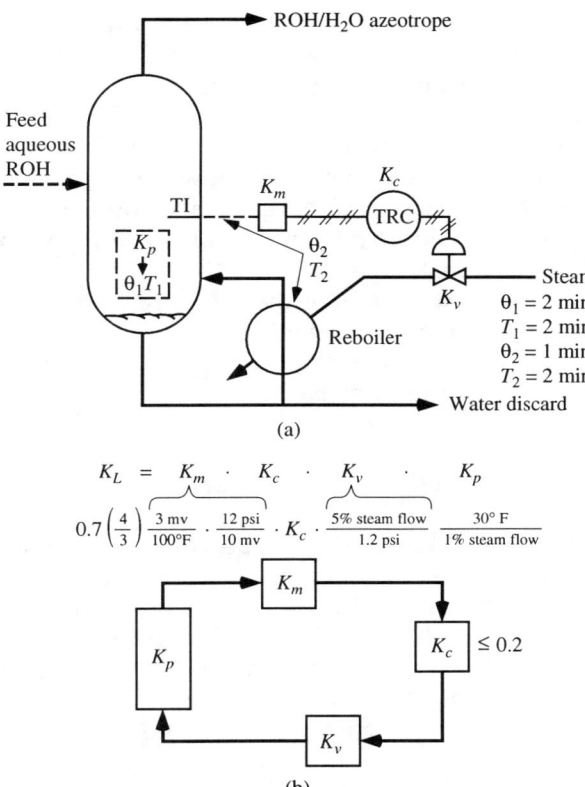

Fig. 8-1 (a) Feedback Control Scheme for Alcohol Stripping from Water (b) Limitation of Controller Gain by Intrinsic Sensitivity and Dead Time in the Control Loop

control tray thermocouple. The ratio of total time θ to total lags T in the circuit is critical to how much gain can be sustained in the overall loop before instability results.

Loop Gain Rule of Thumb

I have found very useful an approximate solution for maximum overall loop gain K_L as directly proportional to T/θ. This proportionality is the reason pure reset action is theoretically the best controller in a dead-time-dominated process loop (Buckley 1964). The proportionality constant for a well-tuned loop (providing an oscillatory but well-damped response to a step disturbance) is 0.4 for proportional-only control but increases to 0.7 for a two-mode proportional plus integral (PI) controller (where reset time will be $\approx T$).

This relationship is useful in several ways to improve feedback loop controllability. We can increase allowable K_L by reducing the dead time (see the discussion of pH control below); we can decrease process gain or sensitivity to permit the controller gain K_c to increase inside the fixed product, $K_L = K_m \cdot K_c \cdot K_v \cdot K_p$, where the subscripts refer, respectively, to the gains of the measured variable signal, the controller, the manipulated variable (e.g., a valve[54]), and the process. If neither of these is feasible, we can add first-order lag to the system (via a blending tank or electronics).

Reconfigurations to Improve Stripper Control

For example, in our alcohol stripper, process gain would be the ultimate tray temperature response to a change in reboiler steam flow. A K_p of 20 means that a 1% shift in the steam rate will cause a 20°F shift (relative to a 100°F control span) in the temperature control point. A steam valve characteristic, such that a 1% drop in input signal causes a 2% drop in steam flow, constitutes a loop element with K_v of 2. Measurement gain would be the thermocouple signal shift in millivolts (mv) per temperature shift on the control tray. Controller gain would be valve movement per mv error signal. There can be other unit conversions in terms of analog/digital, voltage/current, or other hardware in the loop. However, the final K_L value is dimensionless and can be used, with due caution for the linearizations involved, to evaluate stability.

Fig. 8-1a shows the individual dynamic elements of the alcohol/water stripper control loop. The loop gain of 0.93 (see Fig. 8-1b) means that stable controller gain is initially quite limited, ≤ 0.2. Once we have determined how individual process and instrument elements contribute to loop gain and dead-time θ, however, several potential improvement options suggest themselves. Use of a second, trimming steam valve for control corresponds to reducing K_v. Using an average of three tray temperatures serves to dampen the process sensitivity in degrees Fahrenheit per steam change. Moving

---

[54] There is a wide variety of valve characteristics; we will use local linearization of flow relative to valve stem movement; linearity is, of course, not generally true of the full valve stroke.

the temperature control tray closer to the bottom of the column helps increase permissible K_c in two ways, by reducing both θ and K_p.

A Place for Feedforward (FF) Control

We should not rely totally on feedback (FB) control if that means not using all we know about the system. In many cases, a combination of FF and FB elements is desirable (facilitated by a control computer). For example, if feed rate disturbance is frequent, we adjust reboiler steam in a fixed ratio to feed rate, with a suitable dynamic delay. If the concentration of the reactive component in a reactor feed decreases, we make an anticipatory change in the flow of the reagent with which the feed is reacting. If ambient temperature shifts in a system with significant heat losses, an anticipatory adjustment is warranted—for example, reducing reflux flow to a column to offset greater subcooling of reflux. The key consideration is which input—flow, concentration, or heat—causes the most frequent or most serious upsets. FF approaches need greater precision in defining dynamics and nonlinearities than do FB loops.

Example: Need for Feedback Corrector of Azeo Agent Inventory

An example of a pure FF approach is an azeotropic distillation in which the top draw of the column is adjusted by calculation to make the net withdrawal of the entraining agent exactly equal to its entry in the column feed. The intent here is to maintain a constant inventory of the entraining agent in the section of the column where it works to adjust key volatilities. However, over time this inventory will drift significantly up or down due to analytical or computational errors. There must be a FB element in the critical section of the column, using temperature or (more likely) a process analyzer, to sense drift in azeo agent inventory and make adjustments.

Coping with Dead-Time Domination

It is sometimes feasible to reduce the effect of dead time θ by physically reconfiguring a system, as in the use of static mixer elements in pipeline blending, or by adding first-order lag, as discussed above. The Smith predictor (Smith 1959) also deserves more use. It furnishes an approximate compensator (dead-time compensator,

DTC)[55] for dead-time that permits feedback PI control with adequate gain. Its underutilization may stem from the poor results achieved when dead-time is inexactly characterized or from inappropriate application of conventional tuning techniques to the three-parameter control system.

Sampled-data Control

The delay involved in many instruments measuring composition poses a special case of dead time. The case considered here is a quality control instrument, such as a process GC, whose output changes only every τ minutes, delayed by sample preparation or sample processing time. Even with shortcuts such as single-peak analysis, it is difficult to get τ for such a GC below 2–5 min. Mosler et al. (1967) studied the design issues for a continuous PI controller dealing with sampled data; process dynamics was approximated by a first-order lag T and a dead-time θ. Achievable controller gain was shown to be roughly the same as it would be if the sampling delay τ were absent, unless $\tau \gg \theta$.

My conclusion from the above and other information in the literature is that a sampling delay will complicate matters significantly only if the process contains very little dead time, as in some column-end distillation quality control loops. I suggest use of an old rule of thumb that takes a sample-and-hold element as added dead time equal to half of the sampling period and then reasons from the analogous continuous system. If control results with sampling delay are unacceptable, we can try a continuous analyzer (e.g., a spectrophotometer or a single-peak mass spectrometer in lieu of a chromatograph).

Other Control Variants

Cascade Control (CC)

CC has often been urged as the solution to a problem when I suspected (sometimes mistakenly) that the problem went deeper. Let

---

[55] The DTC introduces a term $[1 - \exp(-\theta s)]$ into the denominator of the Laplace transfer function, which is not invertible into the time domain. The Smith predictor (in a real-time domain) is usually realized by taking the first few terms of an infinite series (Padé) representation of the above transcendental function and inverting them piecemeal.

me share what I have learned about the appropriate use of cascade systems. The two-loop system confers advantages for handling load (e.g., utilities) changes by working on an "interior" variable before the process output shows any deviation.

Example: Cascade Control of Reboil Heat

Suppose that a key process variable, such as boil-up in a distillation column, is disturbed by fluctuations in the supply pressure of steam providing heat. In single-loop control, the disturbance would have to work its way through the metal of the reboiler and the additional thermal lags inside the column before a deviation in the sensed variable—tray temperature in the stripping section—caused movement in the steam supply valve. In the CC 2-loop version, the output of the tray temperature controller (TC) becomes the setpoint against which actual steam chest pressure can be compared in a secondary or inner pressure controller (PC) loop. Only the output of the PC acts directly upon the process via a reboiler steam flow valve. The master controller is concerned with the primary process objective of holding the cutpoint inside the column. However, at steady state, the TC output has "found" the right chest pressure setpoint by FB action; when the load change causes chest pressure to dip, corrective action starts almost immediately.

CC is effective when the inner loop is faster and is affected first by the disturbances. Flow controllers or valve positioners are frequently put in the inner loop. One would not choose CC to protect against a compositional upset in a column feed, for example, since the response lags of such a disturbance are long.

Fuzzy Control

My purpose here is to demystify terminology that serves to wall off laypeople from an area of potential usefulness to them. Precisely because *fuzzy control* became a buzzword for theorists, ordinary practitioners may shy away from it. Yet it is used to advantage on the Japanese bullet train between Osaka and Tokyo. Stripped of jargon and hype, *fuzzy control* means, in a practical situation, a cautious approach that avoids one-dimensional stereotyping of a process or disturbance, has fall-back options, and tolerates some ambiguity. Such a scheme can stem from an appreciation that nonlinearities change the nature as well as the magnitude of the appropriate control response.

As a nonexpert, I believe that we are using fuzzy control when we feel a need to define the control algorithm conversationally. "Hit the brakes softly but repeatedly," to avoid skidding an automobile on a barely wet road, is in a real sense a fuzzy control algorithm. It circumvents the simplistic logic of a traditional power-braking system that applies maximum control effort because it isn't programmed to hedge with respect to brakes locking and skidding. Even something as simple as a wait-and-see algorithm for a sampled-data controller like a process GC can qualify as fuzzy control, if it reflects a caution to avoid overcorrecting on the basis of one abnormal output, until we know whether we are dealing with a brief spike or a sustained input disturbance.

"Advanced" Control

If any buzzword is abused in the control field, *"advanced" control* is it. The term has been used to cover everything beyond FB loops based upon the conventional three-mode proportional/integral/derivative action (PID) controller. Is use of a quadratic element in proportional control, so that large errors receive much greater control action than small errors, advanced control or common sense? Is it advanced control to spill load from a trim feedback controller (reboiler steam or reagent flow in pH control, for example) back to the main valve so that the trim valve stays in the center of its stroke? Many of these examples reflect simple ideas that were a bit awkward (but quite possible) to enforce with old-fashioned pneumatic controllers. How much less pretentious it would be to call these *nonlinear* and *self-centering control*, respectively!

Buzzwords tend to dull discrimination. Internal reflux compensation in distillation, termed *advanced control* by some, is valuable against condensing disturbances, but it makes things worse if feed flow is the primary disturbance. The level control algorithm discussed below deserves to be termed advanced control.

Simple and Advanced Algorithms for Level Control (LC)

Liquid LC is an example of a situation where the strategy of experienced operators using manual control can be embodied in a fuzzy algorithm, which can evolve to something more formal mathematically as we gain experience with it.

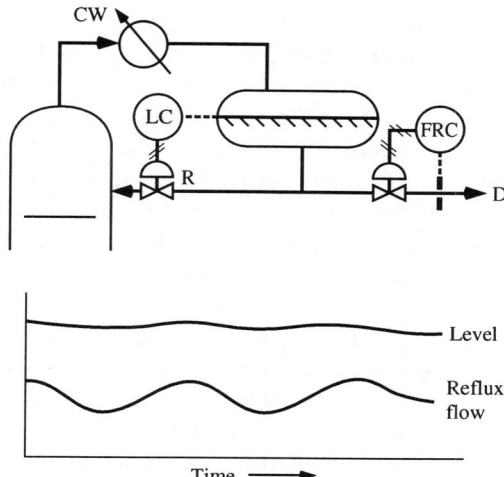

Fig. 8-2 Oscillatory Flow from Improperly Configured Level Control

Difficulties with LC begin with its name. Most LCs are in fact flow smoothers, and the volume controlled was provided largely to furnish surge capacity. An elastic, shock-absorbing response of level to flow disturbances is what is needed. But if an operator is told that the instrument is a LC (rather than a flow smoother) and is given a display of level but not of flow on her panel, she will tighten the reset (integral) action of her PI controller until a large outflow cycle results while the level is held near the setpoint (see Fig. 8-2). If the oscillation period corresponds to the repeats/minute controller setting of integral action, this is prima facie evidence of such overtuning. For a hardware resolution, one backs off reset action and adjusts proportional control to a more reasonable compromise between the two "missions," or invokes a so-called "inverse derivative" controller, or substitutes a proportional-only controller. However software-enforced fuzzy control (*my* jargon) can offer a more satisfying solution.

Example: A Software Fuzzy LC Algorithm

A colleague of mine, Chandler Barkelew, interviewed experienced operators to find out how they reacted manually to input flow disturbances in the LC of an important distilling unit. The answers indicated that very little action was taken as long as the level deviation was modest, and strong action was taken only when the level reached an extreme that threatened overflow or loss of pump suction. More to the point, the operator made adjustments based not just on the position of the level, but also on how fast it was changing and in which direction: little action if the level was moving rapidly toward the setpoint, strong intervention if the level was both far off the setpoint and moving further away. The result of these interviews was embodied in a nonlinear digital control algorithm, which proved much more robust for a variety of flow disturbances than traditional optimal control.

Barkelew drew a second insight from his LC work: With a wealth of evidence on its inadequacy, why are we so often saddled with the proportional/integral (PI) controller in such an application? The answer lies at least partly in mathematical idealism, which quashed common sense, turning the real problem into a "clothes horse" for a display of optimal control theory. If the time average of level deviation squared is used as the value function to be minimized in an optimal control approach, the linear PI controller emerges as optimal! But the use of level rather than flow deviations is absurd. The theoretician has walked around the real problem as surely as the operator who takes level control literally.

Barkelew suggests that many single-loop control problems merit a fresh look by a technical person operating the real system, making manual corrective moves, and observing responses until a control algorithm is formalized from experience. With the variety of software and firmware elements now available, implementing such a discovered algorithm will not be difficult.

It would be unfair to leave this discussion of LC problems without citing cases where faulty sensing, rather than the control concept, is the heart of a control problem. Noise and false readings seem to be particularly common in level measurement at the bottom of a distillation column. Turbulence and velocity effects, froth distortions, manometer-type oscillations, seal/purge problems on pressure taps, and other fluid-mechanical complications suggest the need to monitor signals closely before closing the level/flow control loop. False readings because of unrepresentative fluid legs or temperature differentials are common. There are also situations of restricted volume where tight level control *is* the mission.

Some Dynamic Freaks

Inverse Response

An early "wrong-way" response to a step disturbance, followed by a slow normal response is not uncommon. The following are some representative examples:

1. Distillation columns containing dual-flow trays, in which a boil-up increase causes an initial dumping of liquid inventory rather than the expected inventory buildup on the trays;
2. Boilers in which welling up of the liquid level from vapor nucleation in the bulk of the liquid after a pressure decrease may fool a liquid-level indicator into assuming that inventory is building;
3. Reactors, in which a step increase in feed temperature gives an initial inverse response on temperature differential measurement across an adiabatic reactor bed (i.e., a reduced temperature rise), used as a measure of conversion.

An inverse early response can confound simpleminded control schemes. In each example cited, we can find a physical basis for the short-term inverse response. For example, in the reactor, a traveling thermal wave involving the heat capacity of a catalyst must reach the end of the bed before a normal response can begin. But control against such effects, particularly when a significant delay is involved that acts like dead time, is difficult. From a control theory standpoint, the problem with an inverse response is that it destabilizes a feedback control loop by contributing 180 degrees of phase lag of the 360 degrees required for oscillatory behavior.

A "heavyweight" solution for the reactor problem is an approximate z-transform model (enforceable with a computer), used online as a reference compensator for generating a dynamically corrected temperature differential to feed a conventional PI feedback controller (Roffel and Chin 1987). Buckley et al. (1975) give a less esoteric solution to inverse response of column inventory to boil-up; they used reflux to control the bottoms level (normally an avoided method).

Hysteresis

Hysteresis effects, in which output changes asymmetrically as the input disturbance reverses direction, as in ignition/extinction of exothermic reactions, are also challenges to deal with.

Dynamic Characterization of Process Systems

Process Noise

Noise tends to imply white noise, spread evenly across the frequency spectrum. However, process noise is rarely white. Information on the noise pattern specific to a particular process is usually hard to come by. When it is available, formulating a robust, near-optimal control system is greatly facilitated. For example, suppose that we know that the feed rate to a column is a distorted sinusoid (originating in a mistuned upstream level controller not under our control). Our control strategy will be to try to attenuate disturbances in that frequency range. If we know, in another case, that a once- or twice-a-shift step change in a plant utility system from scheduled load changes is the chief disturbance, the speed of feedback response that we seek may be considerably reduced.

We should know something of process noise before attempting derivative control, attractive though it is in some applications, since this is inherently noise-amplifying.

Perturbation Tests

Perturbation tests to determine the open-loop dynamics of a process once provided a meeting point for theoreticians and control practitioners. The theoretical people recognized that they needed to learn from an operating plant. Sinusoidal perturbation with auto- and cross-correlation analysis of the output responses was often done and is still the preferred method for high-quality work. Repetitive perturbation over a large number of cycles can elicit the true process response from a signal submerged by noise, even at low levels of perturbation. The effort need not involve a long succession of individual frequencies to cover the desired frequency range; a signal that is a mixture of pure sinusoidal frequencies can be used all at once. Yet it seems that people now settle for step tests and occasionally single pulses.

This situation may testify more to the split discussed earlier, with theoreticians preferring models rather than the actual system and plant people being less willing to take their system off closed-loop control for an extended period of testing. A step test may suffice for a rough characterization by dead time plus one first-order lag. However, in the presence of noise, repetitive testing is required in any case. I have seen good results with a technique called *pseudo-random binary noise* (PRBN), which administers small perturbations in a sequence that approximates white noise, yet has an autocorrelation function equivalent to an ideal pulse input. The PRBN seems to extract a time constant for a second (minor) lag more efficiently than step or square-pulse testing; it is more vulnerable to drift or extraneous upsets during the logging period than sinusoidal testing.

With powerful computational resources and extended time, process identification today can be accomplished largely in the closed-loop mode. However, for troubleshooting (as distinct from design), only dead time need be estimated closely. In sorting out roughly a control problem on-line (with tuning to be done later online), I prefer perturbing the process (with FB loops left open) with a single square pulse, whose time base is a fraction of the expected value of the dominant first-order lag T and whose amplitude is sufficient to override process noise and to excite any inherent nonlinearities.

Deducing Process Parameters from Process Noise

In the 1960s a rash of electronic devices was developed to estimate online the transit time through a process, using native flow noise in existing inputs and output, via shortcut Fourier transforms of auto- and cross-correlation of functions. Beck and Plaskowski (1987) have reviewed the technique in the context of today's powerful programmable computers. Because of the high-frequency content of such noise, special-purpose devices must still be used to achieve the computation in real time. Any results based upon natural-variation noise must be accepted with great caution.

If we could develop techniques whereby variance as well as mean residence time were estimated online, there would be an enormous potential for monitoring flow anomalies in multiphase systems (see Chap. 13).

Some Common Measurement Problems

Composition Sensors

The online measurement that is lacking is often composition, measured on a time scale appropriate to the system dynamics. Temperature is still commonly used as a substitute for direct compositional information. Determined ingenuity must be exercised in providing true analytical information. For example, in azeotropic distillation (AD), years of desultory control effort have gone into attempts to make do with a tray temperature in the upper stripping section for control, rather than fight for a workable process analyzer that will unambiguously track the entrainer inventory there. Bozenhardt (1988) describes a typical AD system with alcohol, water, and an ether azeo agent in which, in addition to various kinds of control cleverness, a midcolumn process GC was an essential element in reliable control and optimization.

Suitability of relatively cheap and continuous spectroscopic or dispersive instruments based on infrared (IR), ultra-violet (UV), and refractive index, should be canvassed *before* a process GC is developed. Dielectric constant, and even density and viscosity, also merit consideration for measuring composition. There is no universal selection scheme to decide which device will work in a particular application. It surprised me when IR proved adequate to define residual propane in the high-purity propylene product from a C_3 splitter.

Level Sensors

For lack of a level sensor, we frequently cripple a control system when we ought to be fighting to get reliable level detection. One example concerns a chlorine vaporizer system generating superheated chlorine vapor to feed a delicate reactor system. Lacking the critical measure of liquid inventory, we made do with wobbly control, based upon vapor temperature and head pressure, to control steam flow and admission of liquid Cl_2 into the horizontal vaporizer.

Temperature Measurement

Temperature sensing is less of a problem, but there are three trouble spots. One is the dynamic insensitivity of thermocouples (and even error from axial conduction to the vessel wall) induced by thick-wall thermowells, installed for low maintenance, without weighting other factors. A second, related issue is flow distortion when a relatively thick thermocouple bundle is inserted in one of a number of parallel reactor tubes packed with catalyst. This may make the axial temperature profile of the sample tube unrepresentative of the other tubes. Third, the optical pyrometry involved in remotely assessing

Complexities of Controlling Distillation

Reverberation and Nonlinearity and Interacting Loops

In my experience, two key elements complicate distillation control. One is the nonlinearity of the compositional response to even modest disturbances, particularly in high-purity separations. The other is the "reverberation" effect of reflux and reboil, which gives longer time constants for compositional response than simple holdups and throughputs would indicate. Holdup in the overhead system may add additional significant lags.

There is a huge literature on distillation control; its volume and complexity tend to obscure the simple insights that a troubleshooter seeks. One gains by confining attention to a few sensible authors who highlight fundamentals. On interaction of loops, for example, I prefer to follow someone who suggests configurations that partition the top from the bottom control loops naturally (Rijnsdorp 1963, 1965) rather than someone who invokes dynamic matrix control and other nonrobust mathematical forms of decoupling. The benefits of the physical over the mathematical approach have been demonstrated by McDonald and McAvoy (1985).

Let us list some of the basic considerations, beginning with the environment in which the distillation column is set.

A Variety of Incoming Disturbances

Most common are feed disturbances: flow rate or composition or thermal condition. Disturbances from utilities such as steam supply pressure, cooling water supply temperature or pressure, or ambient temperature are the next most common. The choice of control scheme may depend upon the most likely kind of disturbance in the particular system addressed. One should have knowledge of such matters and weigh them beforehand. Composition and flow control will usually have primacy, with thermal disturbances being significant for their effect on these variables and not in an energy-saving context.

Dynamic Lags as Defined by Physical Parameters

In considering dynamic lags, there are three factors:

1. The longest lags are compositional responses delayed by the liquid holdup on trays. Wahl and Harriott (1970) have shown that this major time constant varies roughly as the square of the number of trays; it may be many hours in tall columns.
2. Hydraulic liquid flow responses to reflux or boil-up changes are faster. The hydraulic lag for a typical downcomer-type tray is 10–20 sec; the summation effect for a number of trays in series is a lag of which a major fraction will be dead time.
3. Vapor flow responds very rapidly to boil-up changes; this is why boil-up is often chosen over reflux as the manipulated variable.

Before considering its amplitude and time scale, one needs to know the *sign* of the dynamic response. For example, tray holdup will increase as liquid flow increases. However, early in my career, I learned that the response to a vapor flow increase differs in sign by tray type: Bubble-cap trays build holdup both immediately and later, while dual-flow trays tend initially to lose inventory. If this "inverse" response were sustained, one could address it with inverse control action. However the longer-term response is normal for both tray types (see the earlier discussion of inverse response).

Time Base of Disturbances

Good control people like my former colleague, Ben Stanton, view a control system as a filter. They seek to move the onset of loop oscillation to a higher frequency or, if that is not feasible, to move process disturbances to the low-frequency end of the spectrum (periods >100 min). High-frequency elements (pressure and flow control loops), with response times ≪1 min, do not matter. They are swallowed by the capacitance of a distillation system.

A feed composition variation, however, will generally be quite slow, involving as it does the long time constant of some upstream unit. FB quality control loops can be crafted to handle such a compositional disturbance, with instruments ranging from process GCs to newer devices like flow injection analyzers (FIAs). However, the nature of the disturbance affects what can be done. A compositional FB loop may make things *worse* if confronted by feed flow or temperature as the main upset, because these may be in a mid-frequency range (period of about 1 hr) where the control system adds enough phase shift to cause oscillation. Thus, good flow and thermal regulation is an almost essential preliminary to compositional control. An exception might be to go to FF compensation if we are sure of a dominating compositional cycle entering from, for example, a reactor upstream with rapidly drifting performance.

Column Control Scheme Options

The choice between so-called material-balance (MB) and so-called quality control (QC) for the basic column control scheme was once considered a fundamental divide in column control. Today it is more like a hierarchy, in which MB control is the basic minimum scheme for a column, with a single product stream dominating the quality issue. With modest separation demands, sufficient trays in place, and V/L traffic approximately right, only failure to enforce the appropriate top/bottom (or side-draw) flow splits can cause the column to go off specification on product quality or recovery.

Today the increase in product purity requirements makes dual QC loops or direct QC at one end of the column increasingly common.

MB Control

Let us recapitulate the basic forms of MB control in binary separations. All of them rest upon adjusting the split of feed (F) to top product (D), defined by D/F, rather than quality of separation, defined by reflux/feed ratio (R/F). The *direct* MB schemes move the draw flow at one end of the column to hold a quality sensor (usually a tray temperature) constant; the draw at the other end (usually chosen to be the larger flow) will be on LC to avoid an overdetermined system.

The *indirect* MB schemes manipulate reflux or boil-up, allowing both top and bottom product draws to be set by LC. Other things being equal, I prefer the direct schemes. Shinskey (Shinskey 1977) and others have shown that manipulating product draw to control purity avoids coupling of the column's material and heat balance envelopes.

After what has been said about LC, it will be appreciated that reflection is needed to decide which of the internal traffic streams, reflux or boil-up, to put on LC. The other will be put on FC, or on ratio to feed, if we expect feed rate to vary. Another basis for choice is which stream's flow variation is likely to upset a downstream column. If the temperature control tray is in the lower part of the column, there is a strong preference for boil-up as the manipulated variable. Reflux is chosen only if the TC point is well up in the column because of greater dead time in liquid flow manipulation. Beaverstock and Harriott (1973) have shown this nicely in a methanol-water column. Crisscross schemes, like bottoms-level setting reflux, are generally avoided.

If pressure is controlled by top cooling (e.g., by partial bypass of process vapor around the condenser) and the bottoms draw is on FC, then the bottoms level must be on heat source pressure control and the accumulator level on a loose top-draw control. Here we try for tight bottoms LC so that changes in bottom draw translate rapidly into boil-up changes. Conversely, if the top draw is flow controlled, the bottoms draw must be on bottoms LC and reflux on overhead LC. The former LC will be on loose control to avoid cycling of boil-up, while the latter LC will be tight so that top-draw changes are rapidly reflected in internal reflux changes.

QC Schemes

The term *QC scheme* should be reserved for systems where quality/recovery at one end of the column is not obtained at the expense of these parameters for the major component at the other end. That is, R/F is being adjusted primarily to change end-to-end fractionation, not just to trim the D/F split. In a column with an adequate number of trays, this should not be necessary; however, in a limited-capacity situation involving, in effect, a trade-off of throughput against quality of separation, it may be a valid control mode, provided that flooding constraints are honored.

McNeill and Sacks (1969) make an interesting claim: that scatter plots of tops vs. bottoms analytical data can reveal, by their shape, whether unsteadiness of the material balance split or a fundamental shortfall in fractionating power (denoted by a hyperbola shape) is at issue.

PC Schemes

PC is straightforward via three workable handles: the heating or cooling medium or light ends drawn from the condenser; it should act upon the one with the greatest variation. A small flow of vapor bypassing the condenser provides excellent PC if minimal subcooling of reflux is acceptable. Trim bypass flows around heat exchangers deserve first consideration in many TC schemes (as in control of column feed temperature) because they take the dynamics of heat exchanger metal out of the control loop.

A flooded condenser, where the accumulator is run liquid full and backs a liquid leg into the condenser, which may be located above or below the accumulator, is convenient and stable, albeit wasteful of utilities, somewhat sluggish, and leading to variable subcooling of reflux. Use of a "blanketing" fixed gas introduced into the accumulator vapor space gives smooth control when the condenser is above the accumulator, but this sacrifices driving force and gives variable reflux temperature. In subatmospheric control, fixed gas is drawn out by an eductor or other gas-moving device. Throttling the cooling water (CW) flow was frowned upon in the past because of fouling concerns on the CW side, but it may deserve reconsideration with controls that limit exit CW temperature.

Floating PC Option

Many columns have considerable self-regulation if the heating and cooling media are stable, since pressure rises increase overhead condensation and/or decrease boil-up. It is not self-evident that column pressure should be tightly fixed; this is often done only so that a tray temperature will correspond directly to composition. However, this regulates pressure only at one chosen end; loading changes can affect tray temperatures via small changes in differential pressure in a way that is significant for control. A variety of options is available for pressure/temperature combination as a surrogate for composition. The vapor pressure bulb sensor is an underused option; an online computer makes direct, linearized pressure compensation easy; differential temperature between two trays greatly reduces pressure sensitivity.

In using pressure as an optimizing variable, the lowest pressure, by riding against a condensing or shell vapor-handling constraint, is usually optimal. The earliest report of this was by a Shell group (Rijnsdorp 1967; Rijnsdorp and Maarleveld 1969). It has since been popularized and widely promulgated by Applied Automation and other companies. The logic whereby gains in relative volatility α can outweigh the vapor loading tendency as density decreases, is set forth in Chap. 9.

pH and Other Strongly Nonlinear Control Loops

The essence of pH control is addition of a small stream of acid or base to a larger process stream to bring its pH to a desired value (usually but not always near-neutral). The blending usually is done in a pipeline or stirred tank. High process gain aspects are a central problem in this application, aggravated by dynamic aspects related to dead time. The major response of a stirred tank to, for example, a dye injection pulse in the feed will be a perfect first-order lag with a time constant T approximating the nominal residence time. However, there will also be a pure time delay θ comprising 1–5% of T in the response; one can think of this as the time for a peripheral half-circumference circuit from inlet to outlet (Cholette and Cloutier 1959). In large tanks or where power input is under 10 hP/kgal, this dead-time proportion may be 10% or more.

A θ/T ratio <0.1 would normally allow a generous loop gain (see the previous discussion). However, in the context of the very high process gain of a pH system, low controller gain may give poor control.

Example: Dynamic Neutralization Loop

Imagine a titration curve (an experimental rather than a calculated one should be used whenever possible) for a system in which 34%w HCl is being used to adjust a process stream from pH 10.5 to pH 7.5. The slope of the titration curve will be much steeper at the operating point, per increment of acid, than the slope in the vicinity of the influent pH of 10.5, so that the average flow of acid (assumed to equal 50% of the span of flow control of reagent) achieves a pH shift of −3.0 units, while a further 1% increment in this flow will lower the pH another 1.0 unit. The process gain K_p, linearized at the operating point, is then a high value of 200 pH units per flow span.

If the quality of mixing in the tank gives a ratio of dead-time θ to first-order lag T of 0.04 (e.g., 6 sec and 2.5 min, respectively), then the rule of thumb cited earlier indicates that a loop gain of 18 is the maximum achievable. This means that, using a linear valve characteristic such that HCl reagent flow is proportional to valve signal, the pH controller can have a gain of only 0.36 if we are seeking a pH 5-to-9 control span. A loop block diagram is shown in Fig. 8-3, based on a pneumatic system with a signal span of 12 psi. Controllability would be much worse with purely inline mixing, with a

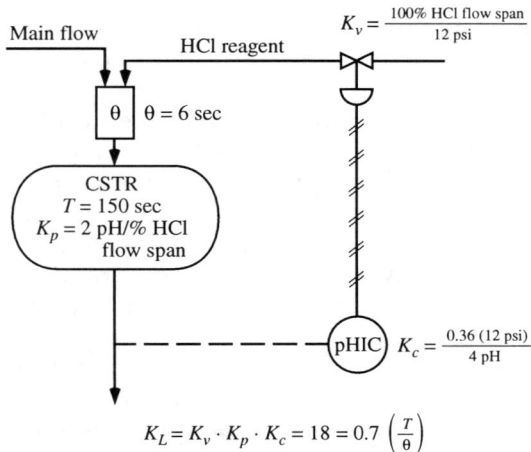

Effect of CSTR nonideality and pH sensitivity on permissible gain in a neutralization feedback loop

Fig. 8-3 Neutralization Control Loop Gain Block Diagram

pipeline delay comparable to θ of the tank but without the large first-order lag T of the tank. In recognition of this arithmetic, many pH control systems employ a two-valve system: A large valve carries the main reagent flow, and a small trim flow supplies the dynamic component. A quick-opening valve characteristic would give some of the same advantage.

pH Control of Effluents

pH control of effluent as a utility for a variety of uncoordinated units is a particular challenge. A large variety of upstream upsets is possible, and the blending tank volume is likely to be large enough to have a significant dead-time component.

Reactor Dynamics

A distinction must be made between fixed-bed catalytic reactors and open tubular or CSTR reactors. In the former, a major consideration will be the differing rates at which thermal and compositional disturbances propagate down the bed from an inlet disturbance. This can be a major factor even in an adsorption bed.

Wave phenomena can create very high localized hot spots in dynamic transitions such as burnoff of deposited carbon in a catalyst bed with steam and air. The potential for complexities, which is awesome in the steady state, grows even more formidable in a dynamic context. We can best obtain insight into such effects if we address dynamics in the time domain. Computational power is available for numerically solving partial differential equations describing dynamics of even a two- or three-spatial-dimension reactor. The alternative of taking it into the s- or z-transform domain, for combination with control loop elements for a controllability analysis, faces the difficulties of achieving local linearization in a distributed system. There are also problems in inverting a transform-domain solution back into real time.

Approximate Characterization of Reactor Dynamics

It is not surprising, then, to see open-loop time-domain responses to pulses or step inputs as the method of choice to characterize both the long-time-constant responses of the reactor bed in toto and (in a laboratory setting) finer-structured dynamic elements of the catalyst bed, like effective diffusivity and adsorption dynamics. For most reactor dynamic problems in the plant, we will turn open-loop time-domain responses to such perturbation inputs into rough two-time-constant (T and θ) approximations to dynamics. In an open tubular cracking reactor for olefins production, for example, Laspe (1956) found a dead time of 1 min and a first-order lag of 7 min by open-loop step temperature perturbation. In a tubular packed catalytic reactor, dead time and heat capacity of the bed will dominate the dynamic response.

An inverse initial response to, for example, a feed temperature excursion in an exothermic packed-bed reactor is quite common and may even remain at steady state. Conventional nonlinearity, where responses to large and small disturbances are not proportionate, is, of course, present, but it is not as critical to control as taking account of dynamic time constants and factors like inverse response.

Special Cautions on Reactor Dynamics

Feed/product heat exchange, a natural for energy conservation in a distillation system, must be used cautiously in a reactor system because of its potential for dynamic instability. Recycling of unreacted feed via a fractionation column can also increase the dynamic sensitivity and slow the response relative to a once-through reactor. Not surprisingly, exothermic reactions tend to bring more dynamic complications than endothermic ones. Liquid exothermic reactions over fixed-bed catalysts tend to be tamer than gases because thermal effects are somewhat damped out. However, in liquid polymerization reactors, unusual effects can arise from rapid changes in physical properties such as viscosity, which affect heat transfer, mixing, and inherent reaction kinetics dramatically.

Methanol and ammonia synthesis and methanation of carbon oxides are complex, widely used systems. Together with bulk and emulsion polymerization systems and more generic CSTR and tubular-flow reactors, they have received many stability/control/optimization theoretical analyses in the literature. For those for whom differential equations displace reality, there are fine adventures into Liapunov stability analysis and Pontryagin trajectory optimizations. For the rest of us, there are adventures of a different kind, emphasizing FB control based on rough modeling, but with an intimate grasp of the topography of the particular process.

Parametric (Thermal) Sensitivity

We need to be particularly wary of instabilities inherent in the physics and chemistry of even seemingly simple reactor systems. For example, in trickle-phase hydrotreating to remove olefinic hydrocarbon impurities, the liquid phase is a substantial thermal moderator and the heat release is seemingly limited by reactant availability. But once a flow deviation deprives a small "clump" of catalyst of liquid irrigation, temperatures may rise substantially in the resulting hot spot, initiating hydrocracking reactions of the already saturated light hydrocarbons, resulting in a thermal runaway. If such a hot spot occurs near the wall of the high-pressure reactor, a serious accident may ensue. Needless to say, monitoring/detection is as important here as design. A more common example of such parametric sensitivity is hot-spot runaway in wall-cooled, multitubular catalytic exothermic reactors (see Chap. 11), where heat generation outpaces removal.

CSTR Dynamics

Control of an exothermic CSTR received its first stability analysis in classic papers by Amundson (Aris and Amundson 1958; Bilous and

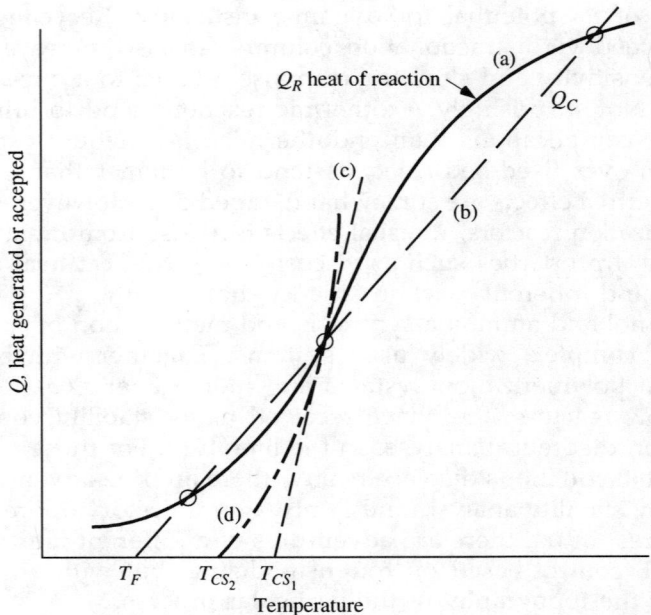

Fig. 8-4 Cold-shot Stabilization of Metastable Reactor Heat Balance

Amundson 1955). A popular rendition (see Fig. 8-4) is the S-curve Q_R of reaction heat release, intercepted at three points by the straight-line locus Q_c (b) of heat uptake, two of which are stable steady states (SS), with an intermediate metastable one. Westerterp et al. (1984) give a useful vector diagram illustrating dynamics near the various SS. Multiple phases in the CSTR (or enzyme/substrate interactions in bioreactors) increase the possibilities for multiple SS, and hence for complex dynamics.

Frequently in industrial practice, it is desirable to hold the middle or metastable SS, a state of "teetering" in nominal heat balance that will drift, if perturbed, toward another, inherently stable SS. The low-conversion or high-conversion stable SS are usually less desirable because of low productivity and low selectivity, respectively. A proper control system *can* hold the middle heat balance point. One crude method involves controlled cycling around the metastable point by a fast-cycle variation of the reactor feed rate. A smoother approach, involving flow control of a "cold-shot" recycle liquid injection, is shown in cooling locus (d) of Fig. 8-4.

Example: Nonlinear Q_c-T Relation Stabilizes a Middle SS

The S-curve Q_R (a), as a function of CSTR temperature, denotes the adiabatic possibilities; it is intersected at three points by the locus (b) of heat uptake, a straight line pivoting from the feed temperature, whose slope represents the heat capacity of the feed. The middle intersection is metastable; a minimum requirement for stability is that the slope of Q_c exceed that of Q_R at their intersection. One way to achieve this is by a large added flow of cold-shot recycle, entering at a temperature such that the combined reactor influent is at temperature T_{cs_1}, higher than before but still below the desired reactor temperature. Q_c (c) is now steeper and a (single) stable intersection is achieved.

The disadvantage of this expedient is the considerable increase in reactor volume required by dilution effects. By making cold-shot flow proportional to reactor temperature deviation, a FB control can be created wherein Q_c (d) is a *quadratic* function of temperature, with a steeply rising slope as the intersection with Q_R is achieved. A colder source temperature of the cold shot can be employed, with less dilution than in (c).

Cold-shot Cooling

Cold-shot cooling has dynamic advantages over jackets and even interior cooling coils in heat removal from CSTRs. Dynamics of the latter are slow, hindered by intervening metal and its heat capacity. In addition to the direct injection of coolant, reflux cooling is another common expedient for improved dynamics. By maintaining an inventory of cold reflux, the metal lag is avoided. In a polymerization, instability with respect to reaction runaway may result if a light reactant is used for reflux cooling; a better choice is a volatile diluent or a reaction product.

Control of Catalytic Cracking

One of the more controversial reactor control issues has been in fluidized catalytic cracking. There are two key handles: (1) the rate of air to the regenerator, where coke is burned from catalyst to sustain activity, and (2) the ratio of recycled regenerated catalyst (RC) to oil fed to the fluidized and/or riser reactors, where endothermic cracking occurs. Conventional strategy had been to set air feed to the regenerator to hold a given oxygen content in the flue gas from coke burnoff and to set the temperature of the fluidized reactor bed by adjusting the recycle of RC.

Gould et al. (1970) sparked the debate with an optimal control paper suggesting a rearrangement whereby regenerator temperature is set by air intake and flue gas oxygen content by RC recycling. I do not offer a conclusion on the issue, but recent developments seem to confirm the desired middle heat balance SS as metastable. Traditional control, in this view, had simply been sloshing around a metastable equilibrium point. A control scheme for holding this middle state cleanly has yet to be unveiled.

Limit Cycles

Many theoretical papers have been devoted to limit cycles, a sustained oscillatory behavior of a reaction system, sometimes even on an open-loop basis. Limit cycles are made possible by multiple SS possibilities, but a system will not necessarily gravitate to a lower or upper stable SS; there may be a sustained cyclical path approachable from one or both sides. A limit cycle in a one-phase reactor system involves both thermal and compositional variation.

A *phase plane* plot of concentration vs. temperature (with time as an implicit variable) organizes our thinking about limit cycles; such a plot will be an ellipsoid or other closed locus traversed by the reactor. A real 3-min limit cycle has been demonstrated in the laboratory (Fortuin et al. 1980) for the acid-catalyzed hydrolysis of glycidol in a CSTR with cooling coil. Such effects can exist in commercial reactors as well, usually with undesirable effects. After-running or "dieseling" of an automobile engine after the ignition has been shut off is a familiar example.

Configurational Changes to Improve Reactor Dynamics

Before resorting to exotic kinds of control or expensive developmental instruments, it is well to consider reconfiguring the existing loops or minor changes in process equipment or operating mode.

Example: Feed Rate Modulation to Hold Metastable Heat Balance

This gas-sparged reactor, the first stage of a reactor train, was to be held at a medium-conversion heat balance point, using both indirect coil cooling and cold-shot injection. For the original design conditions, the reactor SS may have been stable. However, in the course of pushing higher throughputs through the train, it became a metastable SS. The best retrofit control scheme for stabilization might have been quadratic heat removal via flow control of cold recycle, as discussed above. However, neither the support technologist nor the process designer was familiar with these matters. A crude but readily implementable alternative would have been to modulate the feed rate of the gaseous reactant, as discussed earlier. This was vetoed by an operations manager who considered reactor feed rate his domain, not available for control purposes. By default, a third-best alternative was adopted: a lower conversion with a large flow of hot recycle.

Add-on blending time can be used to improve the controllability of a reaction system dominated by dead time. This can be a real CSTR or, at much less cost, the electronic equivalent of a stirred tank.

Example: A Synthetic Blending Lag Improves Stoichiometric Reaction Control

A chemical conversion involving reaction with a basic reagent is carried out on the trays of a distillation column so that the volatile reaction product can be removed from the reaction liquid as formed, to avoid secondary reactions. The control problem is adjusting addition of base to the column feed to hold 10.5 pH in the process effluent from the bottom of the column. A control loop diagram is shown in Fig. 8-5.

Stoichiometric reactions like this one demand constant adjustments to counter drifts in the concentrations of either reagent or reactant in their respective (aqueous) streams. In this case, a minor side reaction also consumed variable amounts of base. A slight excess of base, monitored by effluent pH, was needed to complete reactant conversion in the residence time available in the column. The time on the trays comprised a dead time of about 4 min, between the point of adding reagent and the column bottoms. Only small blending lags were present in the control loop.

To avoid the poor control that had been experienced, we tried adding an electronic "lag" unit between the pH indicating/control unit and the controller of reagent addition to column feed. This unit furnished a first-order lag of 8 min, with a gain of 0.2. With a maximum loop gain K_L of 1.4 for stability with a PI controller (from

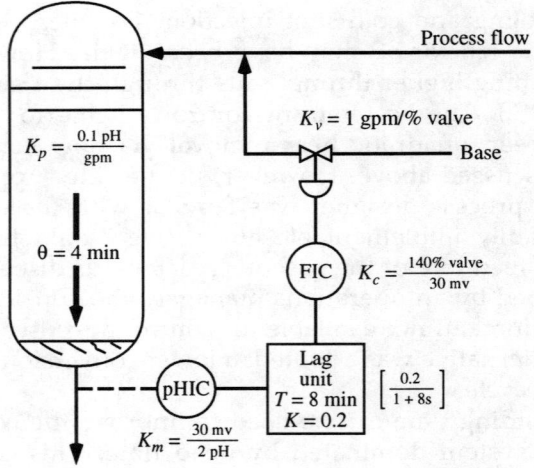

Fig. 8-5 A Synthetic Blending Lag to Improve Stoichiometric Control

$K_L \approx 0.7\ T/\theta$), we backed out the knowns in the control loop to derive permissible gain of the flow controller. The loop elements included a titration curve slope K_P of 0.1 pH unit per incremental gallons/min (gpm) of basic reagent, a pH controller with an input span of 2–12 pH, an output span of 30 mv, a gain of 5 (to give K_m), and a linearized valve characteristic K_v of 1 gpm per percentage valve movement. This yielded a permissible gain $K_c = 1.4$ for the flow controller, of reagent addition (i.e. 140% output valve movement per 30 mv of pH controller output), with a reset time of 8 min (to match the system lag). The price paid for stable control of the reconfigured control system was a 20- to 30-min settling time for a step disturbance. The only disadvantage of a synthetic relative to a real blending tank on the column bottoms was a lack of smoothing of effluent concentration variations, not a strong factor here.

Example: Reconfiguring pH Control By Lateral Thinking and Spaghetti

Suppose that we are obliged to resort to inline pH adjustment, which involves not only a dominating dead time, but a noisy signal as well. This example illustrates the generation of a novel solution to getting from our problem (dead-time dominance and very high process gain) to the goal (smooth control) by using lateral thinking

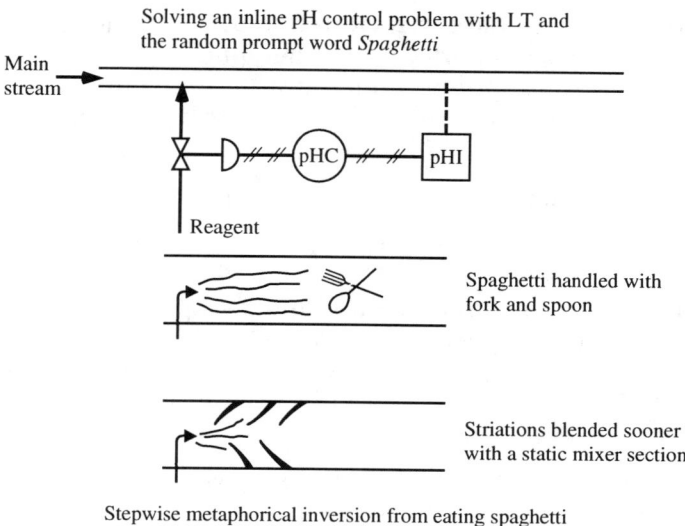

Fig. 8-6 LT: Spaghetti Metaphor for Improved Inline pH Control

(see Chaps. 2 and 5). A random word is taken from a dictionary: *spaghetti*, not connected to either the problem or the goal. We try by mind stretching to link this random word both to the problem and to the goal, to avoid the straight-line "rut" from problem to solution typical of vertical, project-type thinking. (Another solution-generating option is metaphorical conversion: see Chap. 5.)

The stepwise metaphorical inversion is illustrated in Fig. 8-6. My linkage of spaghetti to the *problem* was to identify strands of spaghetti with the striations of reagent immediately after injection, which must disappear by blending/reaction before we are ready to sense pH for feedback control. How do we make spaghetti disappear? If you eat it as I do, you use a fork and spoon to wind it into tractable masses for chewing. The fork/spoon imagery was then forcibly related to the *goal*, transmuting itself into the elements of a *static mixer* that can dramatically shorten the length of line required, hence line mixing delay relative to first-order lags in the system.

In most problems of mixing mobile fluids, I would advise the use of options other than a static mixer; for example, a length of open pipe can do the mixing with less overall pressure drop. Here, however, time or length, not energy expenditure, is the value function and makes the static mixer a preferred solution. While some striking successes have been reported with other quick-blending expedients, such as a centrifugal pump run backward, these effects are risky to scale up or down and have produced major disappointments.

Dynamics Issues in Other Unit Operations

Utilities

Boiler and furnace control are outside the scope of this book. However, one can learn a great deal of value to process systems by technology transfer from problems involving sensor performance, turbine control, opacity, and other effluent issues.

Centrifugal Compressor Control

A major issue in this field has been avoiding *surge* (backflow with destructive possibilities) at low throughputs. In default of a proper control system, excessive *kickback* (recycle to suction) to avoid surge will be used, with wastage of energy or capacity. A significant FB control component is required because the compressor performance curves (head-suction and flow-speed characteristics) are not as well known a priori as the vendors' models suggest. A flexible operation uses speed as a variable, further complicating the dynamics. Here it is wise to recognize the relatively narrow range of good operability and to size the compressor neither stingily nor overgenerously for the application.

Nisenfeld (1974) has chaired a useful symposium on the realities. *Reset windup* of controllers, which used to bedevil this field, has been largely eliminated. Digital computers have definitely improved control by displacing a large number of analog elements with limited response speed, accuracy, or reliability from the usual complex control scheme (including startup and emergency shutdown). Balancing parallel/series arrangements of compressors is particularly challenging in terms of surge, especially when one machine is being brought on- or offline (Nisenfeld 1976).

Heat Exchanger Dynamics

The distributed nature of tube-in-shell or double-pipe exchangers introduces resonance peaks in amplitude response at frequencies that correspond to fluid residence time and harmonics thereof. These resonances, which occur also in tubular reactors, can cause an oscillatory response to a step disturbance in feed rate when, for example, countercurrent exchange of heat is involved. In such a case, temperature control via a PID controller is unpromising. A small bypass flow around the exchanger is a preferred control solution.

Thermosiphon reboilers often used on distillation columns can become unstable from the vaporization/lift/circulation/increased-heat-transfer interactions. A throttling valve or orifice in the feed line to the vaporizer resists such positive feedback effects.

Unsteady-state Processing (USSP)

Process dynamics is not just a nuisance to be mastered. There is a potential opportunity for processes deliberately carried out in unsteady-state (USS) mode, with productivity or quality gains over their SS counterparts. For example, an optimal time/temperature path for reaction or crystallization may be obtained in this way more readily than in a continuous process. The batch process, given new impetus by computer control, is the major current example.

In the late 1960s and early 1970s, there was a burst of interest in controlled cycling of multistage fractionation devices, including distillation. This was represented as providing almost twice as many stages of separation in a given multistage contactor or more than doubling throughput at constant stages of separation (Schrodt et al. 1967) vis-à-vis SS operation. However, practical obstacles to commercialization do not seem to have been overcome, except in special cases like pressure-swing adsorption.

However, in the case of flow reactors, potential USSP advantages in productivity or yield are more promising. A number of techniques are available for pulsing. The chromatographic reactor, for example, processes feed pulses and offers GC-type displacement of reaction products and reactants from each other that may circumvent restrictive equilibria. Experimentally, conversion of butylene to butadiene at 85°–200°C well beyond equilibrium has been demonstrated with this technique (Roginskii 1964).

Feed-forced cycling (e.g., alternating reactant proportions or feed temperature variation) seems to achieve productivity advantages via a hysteresis-cycle storage effect of catalyst surface species. One of the publicized examples is the cycling of SO_2 vs. O_2 in SO_3 synthesis over vanadia (Silveston et al. 1973). The optimal pulse period is short compared to major time constants of the reactor. Some of the pulsing schemes advanced are difficult to realize in practice.

Gains in selectivity via USSP techniques may turn out to be more attractive than conversion increments. Since reactors are relatively cheap compared to fractionation equipment, an increase in reactor complexity and cost as part of USSP can be quite acceptable, provided that expensive elaboration of the associated fractionation equipment is not entailed as well. When and if a USSP scheme achieves commercialization, process dynamics and computer control will take on profoundly new aspects. The breakthrough probably awaits a brilliant experimentalist, not more paper/computational exercises.

Many so-called continuous processes are actually cyclic in that drift of performance occurs, necessitating cleaning, regeneration, or other periodic restorative procedures. Some of the same tools used for dynamic process identification can be used to detect catalyst deactivation trends, changes in fluid mechanics inside a reactor, and so on. The use of time series analysis to detect the onset of critical fouling in a high-pressure polyethylene reactor is a sophisticated example (Hwu and Foster 1982). Statistical filtering to infer the development of hot spots in multitubular reactors containing many parallel reactor tubes, from thermocouple bundles in just a few tubes, has been described by Smart (1981).

In the next chapter, we move out of the time dimension, to consider the compelling nonlinearities of phase equilibrium that sharpen the problems and opportunities of fractionation, before plunging into Chaps. 10 and 11, where nonlinearity and dynamics return together as a fearful duo, presiding over the world of reactors.

References

Aris, R., and Amundson, N.R. *Chem. Eng. Sci.* 7 121–47 (1958).

Beaverstock, M.C., and Harriott, P. *I&EC Proc. Des. Dev.* 12 401–7 (1973).

Beck, M.S., and Plaskowski, A. *Cross-correlation Flow Meters—Design and Application* Adam Hilger (Bristol, U.K.) (1987).

Bilous, O., and Amundson, N.R. *AIChEJ.* 1 513 (1955).

Bozenhardt, H.F. *Hydrocarbon Proc.* 67(6) 47–50 (1988).

Buckley, P.S. *Techniques of Process Control* Wiley (1964; reprinted 1979); *Design of Distillation Column Control Systems* ISA (1985).

Buckley, P.S., et al. *Chem. Eng. Prog.* 71 83–4 (1975).

Cholette, A., and Cloutier, L. *Can. J. Chem. Eng.* 37 105 (1959).

Fortuin, J.M.H., et al. *Chem. Eng. Sci.* 35 439–45 (1980).
Gould, L.A., et al. *Automatica* 6 695 (1970).
Harriott, P. *Process Control* McGraw-Hill (1964; reprinted 1983).
Hwu, M.C., and Foster, R.D. *Chem. Eng. Prog.* 78(7) 62–8 (1982).
Laspe, C.G. *ISA J.* 3 134 (1956).
McDonald, K.A., and McAvoy, T.J. Paper 69a, AIChE annual meeting (Nov. 1985).
McNeill, G.A., and Sacks, J.D. *Chem. Eng. Prog.* 65 33–9 (1969).
Mosler, H.A., et al. *I&EC Proc. Dev. Des.* 6 221–5 (1967).
Nisenfeld, A.E. (ed.). *Instrumentation in the Chemical and Petroleum Industries* 10 ISA (1974).
Nisenfeld, A.E., and Cho, C.H. *Instr. Technol.* 23(11) 41–6 (1976).
Rijnsdorp, J.E. *Chem. Eng. Prog.* 63(2) 97 (1963); *Automatica* 3 1 (Oct. 1965).
Rijnsdorp, J.E. U.S. Patent 3,344,040 (to Shell Oil Co.) (Sept. 26, 1967).
Rijnsdorp, J.E., and Maarleveld, A. *I. Chem. E. Symp. Series* 32 6–38 (1969).
Roffel, B. and Chin, P. *Hydrocarbon Proc.* 66(12) 40–2 (1987).
Roginskii, S.Z. *Kinetika i Kataliz* 5 490–5 (1964).
Schrodt, V.N., et al. *Chem. Eng. Sci.* 22 759–67 (1967).
Silveston, P.L., et al. *Can. J. Chem. Eng.* 51 623–9 (1973).
Smart, A.M., Jr. *Chem. Eng. Prog.* 77(11) 61–3 (1981).
Smith, O.J.M. *ISAJ.* 6(2) 28 (1959).
Wahl, E.F., and Harriott, P. *I&EC Proc. Des. Dev.* 9 396–407 (1970).
Westerterp, K.R., et al. *Chemical Reactor Design and Operation* John Wiley (1984). (See vector diagram on p. 343).

9

Classic Problem Elements: Nonlinear Phase Equilibria

Heresies are experiments in man's unsatisfied search for truth.

H.G. Wells

Chemistry is about strong interactions; so, in a way, the nonlinear combination of factors should be the expected norm. Yet, many technologists start with an assumption of linearity.

Nonideality vs. Nonlinearity

Nonlinearity can be a difficulty or an opportunity. Let us start by understanding some of the potential difficulties. Broadly stated, they stem from the fact that a component's behavior in one context is quantitatively different from that in another context. This deviation can occur even inside a single distillation.

Example: Stripping of an EtOH/water Azeotrope From Dilute Solution

EtOH is much more volatile than water when it is the minor component at the bottom of an EtOH/water stripper. But as EtOH climbs

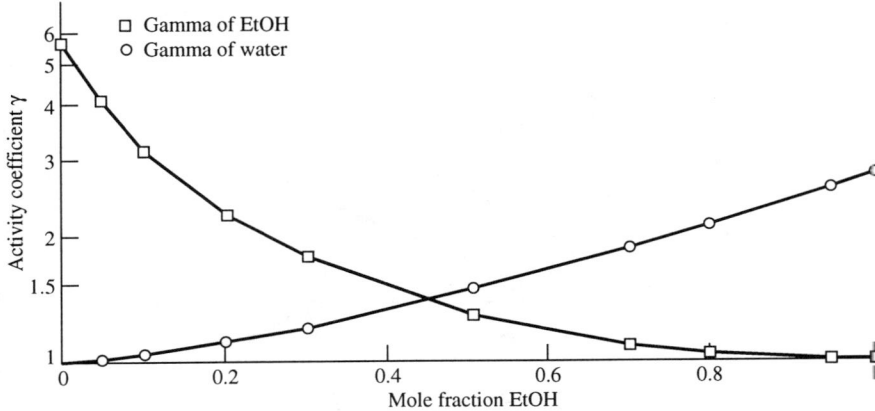

Fig. 9-1 Activity Coefficients in an Ethanol/Water Binary

strongly in concentration in passing out of the stripping section and becomes the major component in the rectifying section, there is a reversal of the situation at the bottom. Now EtOH shows ideal behavior, while water, as the minor component in a "sea" of ethanol, shows a mild "positive" deviation. The result is that further enrichment of EtOH is blocked by azeotrope formation. Fig. 9-1 gives the activity coefficient γ vs. mole fraction pattern of the binary. Fig. 9-2 shows the compositional and temperature profiles vs. tray number in ascending the column from the water bottoms discard to the azeotrope approached in the rectifying section (see Fig. 7-1). The broad inflections that characterize both temperature and composition profiles are basically caused by the non-linearities.

Most γ-value behavior to be described in this chapter involves positive deviations like those shown in Fig. 9-1. The transition from enhancement of component A's volatility in a sea of component B to enhancement of B's volatility in a sea of A ensures an azeotrope[56] if A and B boil close to each other. Positive deviations from ideality are more common because the molecules of many fluids self-associate. When surrounded by foreign molecules, there is a liberating effect partly because this self-association is no longer a factor. The maximum deviation of the γ value above unity for EtOH and water

---

[56] Novice technologists confuse azeotropes with compounds or complexes like hydrates. But azeo compositions shift with pressure because γ_a and γ_b vary in their individual ways with temperature.

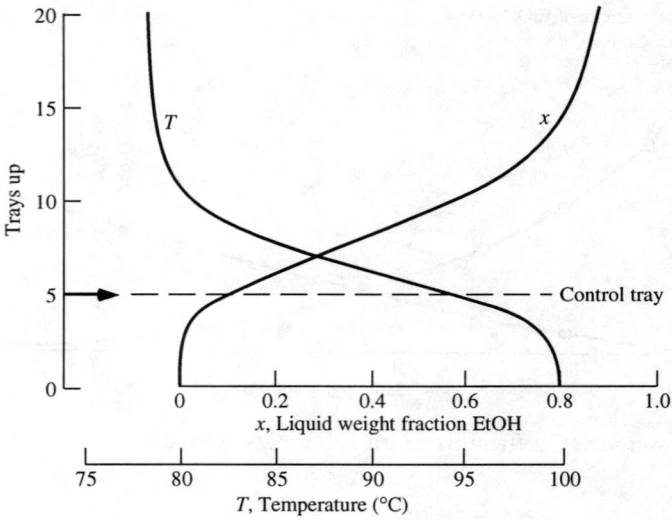

Fig. 9-2 Temperature and Composition Profiles in an EtOH Stripping Column

occurs when they are infinitely diluted (reaching a value γ^∞) by the other molecule. This is usually but not universally the case. The γ^∞ values for ketones, alcohols, and other organic oxygenate compounds in water rise sharply with the carbon number of the molecule because of the incompatibility of the hydrocarbon end of the molecule with water.

We must distinguish nonlinearity from nonideality. Henry's law, that $p = Hx$ (to define the solubility of a gas in a liquid as a function of its partial pressure) is an expression of nonideal mixing. Ideal mixing (Raoult's law), would give H the (constant) value of the vapor pressure p^0 of the pure component. However, H can be greater or less than p^0 and still be essentially a constant over the small range of gas solubility x (essentially infinite dilution) we may be dealing with; this constitutes a nonideal but linear system.

Utilization of Nonlinear or Nonideal Behavior

Nonlinear behavior can be usefully employed to render an azeotrope or near-azeotropic binary separable. Fighting "fire with fire," we add a third component that interacts differently with the two components. Extractive distillation (ED), in which a descending, heavier third component modifies the environment for the separation of the other two components, is one extensively used approach. Another is the use of a more volatile entrainer third component, which travels up the rectifying section from the feed, azeotropic distillation (AD).

The limiting condition of infinite dilution can be directly used to characterize the phase behavior of many trace components, whose behavior is nonideal, by the deviation of γ^∞ from unity, but still linear in the sense that the deviation from ideality is maintained constant in the "field" of the problem. A binary like H_2O/benzene, with limited mutual miscibility, is an example.

Example: A Nonideal Limited-miscibility Binary

The γ value of trace water in a benzene liquid phase will be given roughly by the reciprocal of its mole fraction solubility in benzene (since water must have the same activity in both the equilibrated water-rich and benzene-rich phases), approximately 60 at the atmospheric boiling point of benzene. Thus, in a drying stripper where we are taking water from, say, 200 ppm in the benzene feed down to 10 ppm in the dried bottoms, the water will display a constant relative volatility $\alpha = 30$ ($= \gamma_w^\infty p_w^0/p_b^0$) with respect to benzene over the entire column because it is at essentially infinite dilution throughout.[57] The constant H in Henry's law for trace water in benzene at 80° C will likewise be about 30 atm/mole fraction, γ times the value predicted for an ideal solution.

Another example of utilizing nonideal behavior is in effluent treatment. Many relatively heavy trace toxic organic impurities in effluent water can be stripped up to concentration levels that permit their economic purge by virtue of their limited solubility in water. A useful generalization is that, in a homologous series of organics in water, solubility in water decreases more rapidly with increase of carbon number than does the vapor pressure. Hence the effective stripping volatility out of water will *increase* as molecular weight increases, a counterintuitive result. Of course, there are practical limits to this, as when the organic compound is insoluble or heavy enough to plate out at metal surfaces.

---

[57] It should be noted, however, that water solubility in hydrocarbons increases strongly at higher temperatures, so that nonlinearity as well as nonideality will be exhibited by water in fractionation from a wide-boiling mixture of hydrocarbons.

Another example of a linearized, infinite-dilution nonideality is the behavior of monomers and other small organic molecules in high polymers. Aside from any nonideal chemical interaction aspect, the entropy consideration in the large difference in molecular size will itself produce a substantial correction to Raoult's law. (Because of the gross difference in mole weights, volume or weight fraction is more appropriate than mole fraction for defining the activity of the solvent or monomer in these systems.) While treatment with the Flory-Huggins equation or another comprehensive correlative equation (Grulke et al. 1985) may be required, I have found a rule of thumb that serves for initial approximations where, for example, we are trying to ensure removal of the last of a monomer from a polymer during final processing. This rule assigns an activity of four times its *volume* fraction in the polymer to the residual monomer.

Information Resources on Nonlinear Behavior

Reducing a problem with nonlinear phase equilibrium effects to one or more binary combinations and estimating the endpoint γ values of these binaries will give some structure to a problem. Our first try should be for system-specific data. Modern correlations of individual binaries (Gmehling et al. 1979; Hirata et al. 1975) are recommended. If a particular binary cannot be found, try correlations (ASOG, UNIFAC) based on functional group contributions to molecular interactions (Deal et al. 1959; Derr and Deal 1969; Fredenslund et al. 1977).

In desperate cases, γ vs. mole fraction pattern can even be inferred from two data: the azeotropic composition and boiling point!

Any two-parameter correlative equation (e.g., the binary Van Laar equation) can be used, inferring the parameters from γ_a and γ_b values for the two components at the azeotropic composition x_a, x_b ($= 1 - x_a$), obtained from:

$$K_a \equiv \frac{y_a}{x_a} = \frac{\gamma_a p_a^0}{P} = 1; \qquad K_b = \frac{\gamma_b p_b^0}{P} = 1$$

The above discussions have neglected effects of elevated pressure, such as gas-phase *fugacity* corrections, usually obtained by an equation-of-state method using *second virial coefficients* and summed up in a θ-denominator correction factor to the K value (Edmister 1969). θ corrections are usually smaller than the liquid-phase γ corrections to ideality and can be guessed for many trouble-shooting purposes, but they should not be ignored. θ ranges from 0.5 to 2.0; it will be slightly above unity for light hydrocarbon components whose critical temperature is below system temperature, and below unity for polar components (e.g., ED solvents), whose vapor pressure is well below system pressure.

Temperature Effects on Activity Coefficients

Calorimetry of heats of mixing gives directly the trends of γ values with temperature. An endothermic heat of mixing implies γ values decreasing with temperature. Heat of mixing may range from 0.2 to 2 kJ/gmol in nonideal systems, not quite negligible with respect to heats of vaporization. γ values of most binaries move toward unity as temperature rises and mixtures become more ideal. However, γ values of alcohols and other oxygenates in water increase with T in the low-temperature range, peak in the 50°–150° C region, and then decline. There are few comprehensive treatments of γ values vs. T; however, Lu et al. (1981) give temperature dependence of the group contribution parameters in the ASOG (Analytical-solution-of-groups) correlations γ prediction.

Intuitive Feeling for Nonideality

Even hydrocarbon binaries have deviations from ideality that merit attention. While the butene/butane binary has positive deviations of only a few percent at the ends of the composition range, paraffin/diene pairs have endpoint γ values of about 1.07, and a more disparate binary, benzene/cyclohexane, has endpoint γ values of 1.4 and 1.5, respectively. Mixtures of light and heavy hydrocarbons have much larger positive deviations from the entropy effect mentioned earlier. Propane in n-decane, for example, has a γ^∞ of about 7.

With practice, a feeling for hydrogen bonding and other molecular interactions provides a personal additional resource. I hope that novices and senior technologists will gain from this chapter an appreciation for γ values as far more than just fudge factors.

Entrapment and Bulges

The nonlinearity exemplified by EtOH/water produces a kind of compressive "warp" in a distillation column, EtOH being thrust upward at the bottom yet pushed down at the top by reversal of the volatilities. The sharp temperature and compositional breaks in the upper stripping section (Fig. 9-2) from this nonlinearity make such a column difficult to control (and even to stimulate). This sensitivity might seem of no consequence; EtOH is readily stripped from the aqueous bottoms, and the top product will always end up within a small percentage of the azeotrope composition.

However, as soon as a third component enters our considerations, the latent possibilities for problems begin to materialize. An able process design manager, Al Dierl, once asked me why plant alcohol/water separations never seemed to follow the temperature profile predicted by computer programs. The rationale for this shrewd observation took shape for me when I noted the operators' practice of continuously or periodically "milking" EtOH/water strippers via side-draw taps on various strategic trays.

Example: Water-immiscible Heavies Trapped in an EtOH/water Stripper

These streams were cloudy, indicating a second phase, even though the EtOH/water binary is totally miscible. It turns out that many hydrocarbon and other trace impurities,[58] with limited solubility in water but good miscibility in EtOH, tend to become trapped in the temperature break zone where tray composition is changing rapidly from aqueous to predominantly ethanolic. Even a trace of lube oil can be trapped this way! A differential in solubility and lower volatility than EtOH seem to be sufficient to drive a component up from the bottoms but thrust it down where its environment is ethanolic. This bulge phenomenon was introduced in Chap. 7. It is such buildups of trace components in a narrow zone that distort temperature profiles in real alcohol/water columns. In effect, an EtOH/water binary separation is a romantic fiction of the designer's art.

If nothing intervenes, the bulge accumulation may induce secondary complications. For example, Ross foaming[59] may disrupt the column as the trapped component approaches phaseout.

---

[58] Heavy (ca. C_5) alcohols, termed *fusel oil*, impurities in both fermentation and petrochemical ethanol, are trapped in EtOH/water fractionation, much as indicated above (Black 1982).

[59] Of course, the lower trays of an alcohol/water stripper will be foaming anyway from the Marangoni phenomenon discussed at length in Chap. 7. Entrapment can extend this to higher trays.

Magnitude and Position of Steady-state Bulge Concentrations

I have used the approximation $KV/L = 1$ to estimate the entrapment point of various potential impurities in such a nonlinear binary field. Where there is a rapid transition from $KV/L \gg 1$ to $KV/L \ll 1$ in going up the column, the entrapment "notch" will be sharp and buildup of a bulge correspondingly greater than when there is a mild transition through a $KV/L = 1$ region. Estimation of the bulge position of an obscure odor body is illustrated in the following example.

Example: t-Nonyl Mercaptan Entrapment in Alcohol/Water Separation

Fig. 9-3 illustrates the behavior of a member of the renowned "cat's urine" odor body family, t-nonyl mercaptan (Maarse and ten Noever de Brauw 1974), in an alcohol/water binary. From solubility behavior as described above, the mercaptan, although nominally the heaviest component, will rise in the aqueous stripping section of the column. However, as alcohol becomes the main component, the K value ($=\gamma p^0/P$) for the mercaptan will fall with its sharply falling γ value until KV/L becomes < 1 in the upper part of the column (just below the near-azeotrope top product). The point where $K = 1$ (since $V/L \approx 1$) defines the region of entrapment (at about 60% mole alcohol). I generally estimate the log K^∞ value of the trace component out of binary mixtures of the major components by anchoring its known γ value in each major component on a log-linear plot and then sketching in a slightly concave curve connecting these endpoints. Fig. 9-3 was obtained this way.

A side draw at the bulge can help cope with such an odor problem. Success in this effort requires control of column V/L traffic and the binary composition profile, such that the bulge is held steady near the draw point; such control is not easy to attain (see Chap. 8).

Purging a Water Bulge from a C_2 Splitter Column

A side-draw purge of a water bulge in ethylene/ethane splitter columns to avoid buildup and icing at that point is often required in modern steam-cracking olefin plants (Gleich 1975). This is necessary

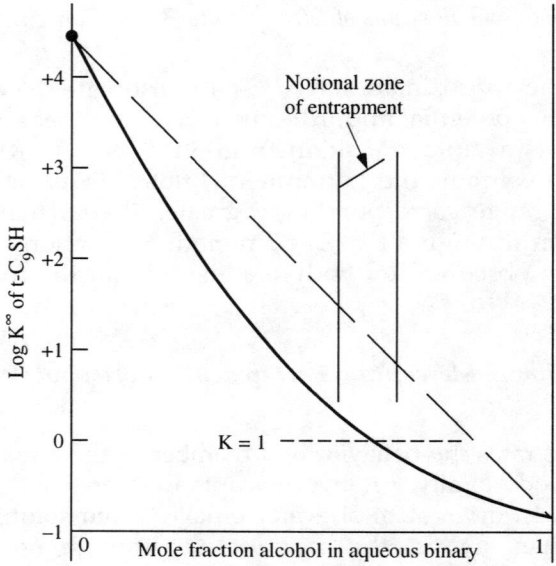

Fig. 9-3 Volatility of Mercaptan Odor Body in Alcohol-Water Binary

despite the presence of the upstream drier that a designer may have provided for minimizing water entry into the column. The following example shows how we can validate such entrapment with rough estimations, after the fact, or perhaps even in anticipation of a startup.

Example: Justifying Entrapment by Estimating the Nonideal Behavior of Water in Light Hydrocarbons

Although water is relatively insoluble in hydrocarbons, there is almost 10-fold greater affinity for olefins than for paraffins. Almost the same differential holds for solubilities of these hydrocarbons in water (see the following discussion of extractive distillation). The differential decreases somewhat with rising temperature and with rising carbon number. While water solubilities in ethane and ethylene are not readily available, we can obtain adequate estimates by extrapolation with respect to the carbon number of the alkane/alkene pair, using data for C_3 to C_7 hydrocarbons (*API Technical Data Book* 1963). The resulting solubilities of $5 \cdot 10^{-5}$ and $3 \cdot 10^{-4}$ mole fraction, in ethane and ethylene, respectively, at 32°F are essentially the inverse of the γ values we need for estimating K ($\equiv \gamma p^0/P$) for water as a trace impurity. At 32° F, a pressure of 315 psia and a water vapor pressure of 0.0886 psia, the estimated K values are 5.6 and 0.94 in ethane and ethylene, respectively.

We now switch to ice as reference state to extrapolate γ and p^0 to column temperature conditions. The ethylene-rich upper part of the splitter will be colder, about $-20°$ F, vs. about $+20°$ F in the ethane-rich bottoms. Vapor pressure of ice doubles every 15° F; the p^0 value shifts with temperature enhance K in ethane at the bottom sixfold relative to K in ethylene at the top.

Finally, the temperature sensitivity of liquid water solubility in alkanes amounts to doubling every 35° F vs. every 23° F for the alkenes (*API Data Book* 1963) and are assumed to be slightly greater for ice than for liquid water. This acts to narrow the spread in K values almost fourfold.

Out of all this estimation emerge K values for ice in the bottom and top of the splitter of 4 and 0.4, respectively. We have omitted θ vapor-phase nonideality corrections; however, these will not be large enough to contradict the obvious case for entrapment of water at some intermediate tray as long as V/L is not substantially greater than unity above the splitter feed.

The above discussion does not predict the tray location and magnitude of the bulge that may result from the entrapment, nor does it address hydrate equilibria in order to establish if buildup of water in the bulge will result in a hydrate rather than an ice phase. It serves, however, to establish a working hypothesis of entrapment if we are dealing with a known icing event in operation that constricts column throughput. Alternatively, it would cause concern to a design team prior to startup and spur a more exact calculation.

In my own experience with the problem, it was not anticipated for the prototype plant. The icing was fought with alcohol douches until the entrapment hypothesis was articulated and a side draw instituted. Once the problem became public knowledge (Gleich 1975), a wealth of publications appeared, giving fractionation calculations on the water bulge, equilibria, and kinetics of hydrate formation (Vysniauskas and Bishnoi 1985), and on how alcohols dissolve ice plugs. None were available to us when the problem first arose.

In hydrocarbon systems, water may be trapped simply by the temperature gradient (Kister and Hower 1986). This is related to solubility changes for water at elevated temperatures. Other recent articles (Grimma 1985) on entrapment fail to make clear how general a phenomenon it is in complex modern process trains with high-purity objectives.

Some publications attempt circuitous and unnecessary explanations of entrapment. For example, entrapment of crotonaldehyde (CA) in the EtOH-rich end of the EtOH/water binary was attributed (Ikari et al. 1986) to aliasing (see Chap. 12) of CA as its acetal or hemiacetal. I believe it could have been accounted for more simply by nonlinear phase equilibrium considerations. The shift in K^∞ value for trace CA as a function of pressure (i.e., temperature) and EtOH/water proportion shown by Ikari et al. requires only that γ^∞ for CA in alcohol decrease faster with temperature elevation than in water.

Advantages of Entrapment

Although in most cases distillation entrapment of minor components reduces operability by disrupting throughput and control, it can on occasion be used to advantage in concentrating an impurity for a side-draw purge. Carrying this out, however, requires sure control, as discussed above. A less general but potentially valuable use of bulges, akin to ED and AD, is to make a minor, intermediate-volatility component build up to major-component proportions in a restricted zone of the column in order to enhance the volatility of a problem trace component for better rejection.

Extractive Distillation (ED)

There is a tendency to think that an extractive solvent separates components by diminishing the volatility of one of them. Some solvents do work this way, as in the use of glycols to break an alcohol/water azeotrope (see the discussion of negative systems below). However, for the most part, the action stems from differential *enhancement* of volatilities of the components being separated. Alkane/alkene and alkene/diene separations by ED using a variety of solvents are now common in industry (see Chap. 7). In effect, the enhanced relative volatility (e.g., $\alpha = 1.6$ vs. 1.12 in conventional distillation) outweighs the capital and energy costs of handling the large solvent flow.

For insight into how polar extractive solvents separate paraffins and olefins of the same carbon number, consider the case of water. Solubility of alkenes in water is about five times that of the corresponding alkanes (Hopper et al. 1990). The vapor pressures will be close, with the olefin usually slightly more volatile. Therefore, paraffins in water may have an α value of 4 relative to the corresponding olefin vs. 0.8 in a hydrocarbon medium. Furthermore, with multiple isomers at a given carbon number, conventional distillation may be a total nonstarter, whereas ED can give generic separation between the olefins and paraffins as classes.

Of course, its low carrying capacity makes water impractical as an ED medium for hydrocarbons. It is, however, added to solvents such as acetonitrile to improve the α value, at some loss in solubility. Kyle and Leng (1965) have reviewed semitheoretical approaches using solubility parameters for choosing an ED agent. Choice of solvent for L/L extraction (e.g., removal of aromatics from paraffins) is discussed by Elkilani et al. (1989) in terms of balancing selectivity against carrying capacity and energy requirements for the separation.

Let us see how we can use nonideality creatively to make a difficult separation easier by finding an appropriate ED environment.

Example: Inventing a new ED Concept

Suppose that we have an alcohol/ketone azeotrope and are asked to improvise an ED separation. A quick generic sorting in such situations uses the Δ curves of Horsley (1973). These curves indicate that alcohol/ketone pairs must boil within 10°–15° C of each other to azeotrope, indicating a mild positive deviation system. Methanol/acetone is such a system. In Chap. 7 we showed the impracticality of purging trace MeOH from acetone as the azeotrope containing 15%w MeOH. Now let us seek a solvent that decisively enhances the volatility of MeOH relative to acetone.

Paraffin hydrocarbons provide a strong positive deviation for MeOH ($\gamma^\infty = 36$ in n-hexane) while enhancing acetone's volatility much less ($\gamma^\infty = 6$). Since MeOH, as a trace impurity, will be truly at infinite dilution, while the acetone will be at significant concentration, the enhanced relative volatility will be even larger than $\alpha = 36/6(1.8) = 3.3$ implies (1.8 being the vapor pressure ratio of acetone/MeOH). A few practical considerations must be kept in mind.

An alkane extractive solvent should be heavy enough that the subsequent recovery of acetone from the bottom product will be facile. From Horsley (1973), we must go up to at least n-octane to find a paraffin that does not azeotrope with acetone. We also need to consider the disposition of methanol rejected overhead. There will be some loss of octane in the overhead (MeOH azeotropes with n-octane), which is not too serious economically as long as MeOH concentration in the overhead is high enough for, for example, disposal by burning.

Heteroazeotropic Separations

The separation of the lower alcohols from water is usually done by azeotropic distillation (AD). The process of choosing an entraining agent to enhance the relative volatility of water is similar to that used above to select an ED solvent, proceeding from a knowledge of the γ^∞ value of each key in the candidate entrainer.

The much-studied separation of EtOH/water, which forms a homogeneous azeotrope, can be enabled by almost any hydrocarbon whose volatility will allow it to be stripped out of pure EtOH as a bottoms product. The choice among entrainers can be based upon overall energy requirements. Recently, we have seen substitutions, like cyclohexane displacing benzene or n-hexane, based on toxicity issues. The choice between AD and ED for a separation can be a narrow one. Chianese and Zinnamosca (1990) show that ED with a gasoline made up of C_7 and C_8 hydrocarbons is more energy efficient for drying EtOH for gasohol than the traditional AD with benzene as entrainer.

Doherty and coworkers (Doherty et al. 1989; Pham and Doherty 1990) have provided new insights into analyzing and optimizing AD as a unit operation. Their method uses *residue curve maps* (loci of batch distillation trajectories) on a ternary diagram (for the two keys and the entrainer) to sector it by separatrices (or *distillation boundaries*, in their terminology). This graphical approach may be more valuable for troubleshooting and process improvement than exact computational simulations (partly because of the difficulty of matching both V/L and L/L behavior in a ternary system with a single γ value correlation). Doherty et al.'s topological approach to synthesis and analysis of these systems stresses the phase separation locus of the ternary heteroazeotrope in the overhead condenser in relation to the distillation boundaries.

Multiple Steady States (MSS)

Much computing ingenuity has been devoted to the possible existence of two distinct solutions for distillations with the same boundary conditions. Mathematical difficulties can sometimes mirror real physical sensitivities in the system simulated. However, when MSS are reported for binaries whose key feature is a wide difference in mole weights (Jacobsen and Skogestad 1991), and when we are told that the MSS are found when flows are specified to the computer in kilograms but not when specified in moles, we can be fairly sure that the finding is an artifact of computer methodology.

Even in the reported MSS for the more nonlinear situation in ternary hetero-AD, one may suspect that seemingly trivial truncation errors in failing to close mass balance for the azeo agent in the computation are creating alternative solutions that are not physically meaningful. This feeling is reinforced by noting that the authors of such papers have reserved their talents for stepping tray equilibrium calculations through V/L/L situations and have not taken account of simple realities such as V/L tray efficiencies varying both with concentration and among components, and the poor contacting between the minor liquid phase and the other phases.

Correlative Equations for Activity Coefficients in Binaries

Early in my career, the Margules equation for correlating γ values in binary mixtures was being displaced by the Van Laar equation, which could achieve with two constants a data fit about equal to the one the Margules equation achieved with three. Since then there has been a progression to the Wilson equation (Wilson 1964), which has advantages in generating multicomponent system descriptions from the individual binary parameters without fitting additional parameters. However, this too is being superseded by the Nonrandom-two liquid (NRTL) equations (Renon and Prausnitz 1968) as the preferred correlative vehicle.

Efficiency of fitting, particularly better matching of midrange and infinite-dilution behavior, has been the driving force behind the evolution of correlative equations. Even so, it has proven difficult to fit adequately ternary phase equilibria involving two liquid phases and a vapor phase with any currently available correlative system. Such a fit is essential to ternary AD systems with a critical L/L phase separation in the overhead condenser. Also, enhanced fit with the newer equations has its price. For example, in the binary Van Laar, equation, the parameters A_{12} and A_{21} correspond directly to log γ_a^∞ and log γ_b^∞; in the Wilson and NRTL descriptions, the γ^∞ values are given by interaction of two or more parameters, and one loses an intuitive feel for the system.

For most troubleshooting estimation with quasibinary mixtures, the Van Laar equation will suffice. It can provide asymmetric endpoint γ values while maintaining thermodynamic consistency. (Verify the latter by plotting isothermal log γ vs. linear mole fraction; there should be equal areas under the respective log γ curves.) Although cataloging of endpoint γ values has greatly improved, we may sometimes be obliged to infer them from mutual solubilities or azeotropic compositions listed in compendia, as described above. Charts are sometimes used to implement the implicit algebra.

Endpoint γ values are directly useful in trace chemical investigations, but they can also provide a preliminary feel for the nature of a binary system interaction. For example, γ^∞ values >8 in a symmetrical binary imply a region of immiscibility, as exemplified by the hexane/MeOH binary at 45° C, with the γ^∞ values each being about 35. However, cyclohexane/isopropanol at 55° C is totally miscible, with endpoint γ values of 6 and 15, respectively.

A screening approach, using the little-known Δ and c tables in the appendix to Horsley's compendium (Horsley 1973), can frequently be useful when attacking a trace chemistry problem generically.

Example: Estimating Nonideal Behavior of Ethers in n-Propanol (NPA)

Horsley shows that the limits of azeotropy for ether/alcohol pairs are ± 33° C from the boiling point of NPA, which is 97.2° C. Isopropyl ether boils at 68.3° C and hence will form an azeotrope containing a trace of NPA. The vapor pressure of NPA at 68.3° C is about 0.276 atm. Hence the γ^∞ value of NPA in the light ether at 68° C will be about $1/0.276 = 3.6$. At the other extreme, isobutyl ether boils at 122° C and n-butyl ether boils at 142° C. Ethers as a class double in vapor pressure every 20° C in this temperature region. Therefore the asymmetric iso/normal butyl ether,[60] which has a vapor pressure of about 0.316 atm at 97° C, will form an azeotrope in NPA with a γ^∞ value $\approx 1/0.316 = 3.2$ for the trace ether. Thus, we can expect NPA/ether systems to be nearly symmetrical, with end-point γ values ≈ 3.

Pressure (Temperature) Shift Effects

Temperature shifts can affect separations through relative changes in either vapor pressures or the γ values of the components. The

---
[60] This ether is used simply to get a compound that boils near 130° C, at the upper bound for azeotroping with NPA.

vapor pressure of a polar, self-associating component like an alcohol increases more with temperature than that of a hydrocarbon; this is reflected in differing heats of vaporization. The decrease in self-association of alcohols with rising temperature, which underlies the above effect, has recently been documented (Hofman 1990).

Some industrially significant shifts in relative volatility stem from such divergent trends. For example, the acetone/water azeotrope disappears as pressure is reduced; commercial practice is to distill under slight vacuum to increase the α value further. This is due entirely to vapor pressure divergence, since the γ^∞ value of water in acetone actually rises as the temperature is reduced.

The coupling of columns for double use of condensing/reboil heats has become common in recent years as energy prices have grown. This usually involves increasing pressure in one column to create a temperature differential for heat transfer. Thorough attention to the effect of pressure (through temperature) on ease of separation is advisable before retrofitting one of these couplings.

Example: Extrapolation Error in Elevating Column Pressure

In one case in my experience, pressure rating limitations on one column made it seem obvious that we had to raise the pressure of the partner (heat donor) column. The result was a fiasco. The α value in the pressured-up column dropped so much that extra reflux needs canceled out all the energy savings. The error was not a subtle γ value error, but simply a faulty vapor pressure extrapolation. A probably feasible alternative, converting the low-pressure-rated column to vacuum had never even been considered!

Most vapor pressure curves are convex on a log p^0 vs. $1/T$ plot if a wide enough temperature range is involved. Thus a two-point fit for an Antoine equation representation of vapor pressure vs. temperature can cause trouble, especially if (as in the case cited) the points are the atmospheric boiling point and the critical point! Moral: A three-constant Antoine equation should be used wherever possible.

The Case for Adjusting Distillation Pressure

Often an improved α value from *lowering* pressure and temperature more than compensates for the increase in volumetric vapor load, *increasing* the capacity of a distillation column (contrary to intuition). Raising the column operating pressure helps condensing, and lowering it helps reboiling; the current optimum may differ from design pressure. "Floating" the pressure in a distillation column as a control concept for maximizing throughput vs. reboiler, condenser, and shell vapor-handling constraints (Rijnsdorp and Maarleveld 1967, 1969) is based upon such considerations.

We expect the solubility of a gas in a liquid to decrease as temperature increases. However, there is a reversal for dissolved gases above their critical temperature. Prausnitz (1976, 1986), continuing Hildebrand's pioneering work, has made estimation of Henry's law constant vs. temperature for gases in nonpolar solvents self-consistent and easy. He correlated the partial molal entropy of solution against nonadjustable parameters: the solubility parameter of the solvent and the energy of vaporization of the solute at its normal boiling point.

Negative Deviation (ND) Systems

There is a significant minority of cases in which negative deviation exists; this too can cause process complications. If there is a ND at one extreme of the binary composition range, there is almost sure to be a ND at the other end by virtue of the Gibbs-Duhem thermodynamic consistency rule. An example is binary mixtures of phenol and ketones, where their interassociation is stronger than their self-association. The γ^∞ values for acetone and phenol are both about 0.2 (Gmehling 1979; Goelles 1963). A γ vs. x plot based on the Van Laar equation and these γ^∞ values[61] is given in Fig. 9-4. For those intelligent enough to question why hydrogen bonding between alcohols and ketones does not parallel that of the phenol/ketone systems to create ND binaries, I give my commendation but have no answer!

The ND complicates the separation of acetone from phenol as coproducts of phenol synthesis. The two boil far apart, and a facile separation would be expected until nonideality is considered. The vapor pressure ratio is about 20/1; the γ/γ ratio reduces this by a factor of 4 to 5, not enough to create an actual azeotrope.

Example: Framing-in the Extent of ND

If we did not have system-specific data like the references cited above for acetone/phenol, we might proceed to frame-in the pos-

---

[61] At constant pressure, symmetry would be lost by the γ^∞ value of acetone moving upward at the higher-temperature, phenol-rich end.

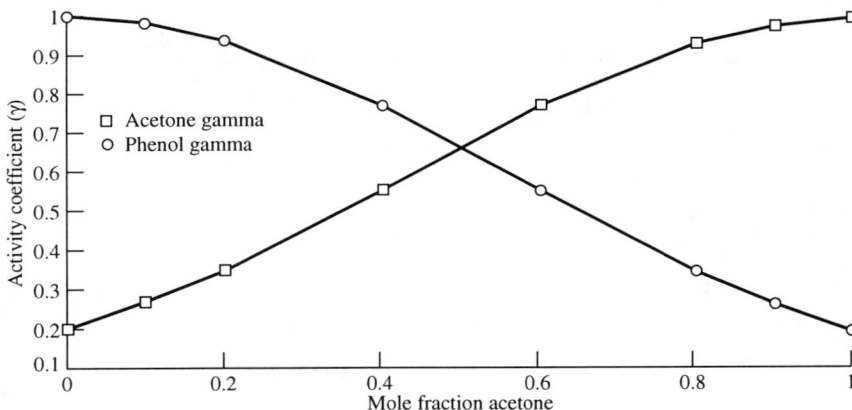

Fig. 9-4 ND Gamma Values in an Acetone/Phenol Binary

sibilities, akin to what was done above for ethers as a class in n-propanol. The Δ curve for phenol/ketone binaries (Horsley 1973) shows that only pairs boiling ≤20° C apart will azeotrope. A borderline azeotrope in such a set will have just a trace of the more volatile component (its volatility reduced by the ND) in nearly 100%m of the less volatile component at unit activity.

Thus, a trace of a ketone boiling 20° C below phenol will make up an azeotrope if its γ^∞ value ≈0.5 (since the vapor pressure of a ketone doubles over 20° C near the boiling point of phenol). Likewise, a trace of phenol in a 20° C higher-boiling ketone will make up an azeotrope if the γ^∞ value for phenol in the pure ketone is ≈0.5. These rough estimates using Δ curves do not necessarily contradict the more extreme ND cited above for acetone/phenol. The longer alkane tail on the higher ketones that boil near phenol and somewhat higher temperatures will move γ^∞ up toward unity.

-OH-containing Negative-deviation Systems

Water forms ND systems with glycols because of the strong interassociation, especially with polyglycols (Bestani and Shing 1989). As noted above, this property will suggest the use of a glycol to "drag down" water and break various water-containing azeotropes. Another application is the use of triethylene glycol for gas dehydration. The affinity facilitating absorption of a trace impurity into such a

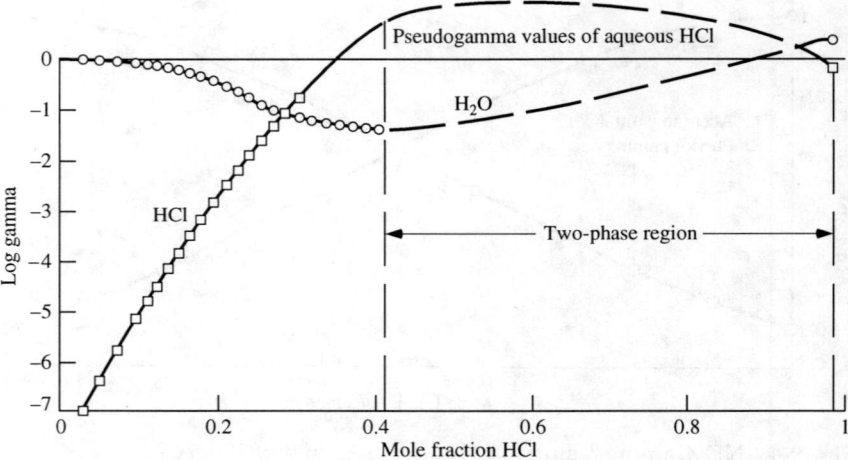

Fig. 9-5 Apparent Activity Coefficients in an HCl/Water Binary (□ Hcl; t, H₂O)

compound, of course, hinders subsequent regeneration of the sorbent.

The Aqueous HCl System

NDs are the cause of maximum-boiling azeotropes. The interaction of HCl with water is an important, if somewhat special, example of ND. HCl is near-ideal in mixtures with hydrocarbons, but in an aqueous phase, nearly complete ionization makes it act like a species with strong ND, greatly reducing its presence in a coexisting liquid organic or vapor phase. Pseudo-γ values like 10^{-4} may be used to describe its activity out of an aqueous layer at low mole fractions of HCl. Fig. 9-5 shows an approximate rendition of such pseudo-γ values for water and HCl out of the aqueous binary taken from a variety of literature sources. The behavior of water in the mixture starts out as a modest ND, intensified by the "salt" effect of ionized HCl, but is deflected toward a small positive deviation in the high-HCl range, where HCl starts acting like an organic chloride and an extensive two-liquid-phase region exists.

Several consequences important for troubleshooting flow from this fundamental behavior: One is the ready vapor-phase nucleation of

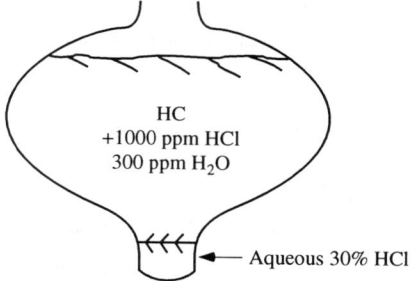

Fig. 9-6   Anomalous Phaseout of a Corrosive Aqueous HCl Phase

aqueous HCl mist droplets when a hot, dry gas containing traces of HCl meets atmospheric moisture. The vastly depressed activity of HCl in the prospective mist particles and, to a lesser extent, that of water provide a tremendous nucleation driving force for the new phase, which, of course, makes for extremely small liquid droplets. Such mist droplets (e.g., coming out of incinerators) have been known to pass virtually untouched through two scrubbing columns in series, hardly encountering the scrubbing liquor.

A similar effect can occur in distillation and reactor systems. For example, water may be well inside its solubility limit in a hydrocarbonlike liquid phase. Contact with dry HCl-containing vapor (perhaps generated by trace dehydrochlorination of an organic chloride in a reboiler) will pull some of this wetness out of the organic phase to create an aqueous HCl phase with significant corrosion potential. A little consideration of Fig. 9-5, for example, conjures up the scenario presented in the next example.

Example: Anomalous Phaseout of a Corrosive Phase (Fig. 9-6)

Water in a heptane (MW = 100) stream is at 100 ppmw, one-fourth of its solubility, hence at an activity of 0.25. Let us estimate the level of HCl that can nucleate a small aqueous HCl phase. It takes 20%m HCl in the aqueous phase to match a water activity of 0.25; its HCl activity of about 0.04 must then be matched by an HCl presence in the hydrocarbon phase. At low concentrations of water and HCl in the heptane, HCl will be near-ideal ($\gamma \approx 0.9$). Therefore it will take 4.5%m or 1.6%w of HCl in the hydrocarbon feed to give an activity of 0.04 and nucleate a trace aqueous phase.

But if water in the hydrocarbon feed were at a higher level, 80% of saturation, it would take far less HCl to cause an aqueous phaseout. An 11%m aqueous HCl phase gives a water activity of 0.8. But the activity of HCl out of such an aqueous HCl phase is very low, about 0.0003, because of ionization; hence it now takes only 110 ppmw of HCl on the hydrocarbon feed to nucleate a trace aqueous HCl phase, vs. 16,000 ppmw in the previous case.

Acid-base-type interactions can produce pseudo-ND systems such as the reduced volatility of alcohols and ethers in strong acids.

Example: ND of Ether in an Excess of Acid

In the diethyl ether (DEE)/H_2SO_4 binary, the ether, as an organic base, is being protonated by the acid, with a resulting reduction of volatility somewhat akin to that of HCl cited above. Apparent γ values for DEE vs. its mole fraction in a binary mixture with sulfuric acid, (calculated from the *International Critical Tables* 1930) are:

x	0.81	0.61	0.48
γ	1.06	0.48	0.09

Thus, the activity of DEE plummets as soon as there is stoichiometrically enough acid to tie it up.

Addition of a base like NaOH during distillation to reduce volatility of mercaptans by conversion to mercaptides can be taken as an example of profitably using a chemical ND.

Fixed-bed Sorption

Adsorption and other separation schemes like ion exchange, conducted in fixed beds on a transient basis, is an enormous field to which this book can give only a passing nod. The prediction of sorption equilibria and certain chromatographic wave effects of such beds are singled out for attention here.

Equilibrium Estimation on Activated Carbon

Activated carbon is one of the more versatile nonlinear sorbents available. A convex-type (e.g., Freundlich) adsorption equilibrium isotherm is useful in a cyclic regenerative process (the usual situation) based upon adsorption. There are remarkably few publications in the literature that help us make intelligent estimates of sorption equilibria among differing classes of compounds and within homologous series of gas or liquid sorbates.

On activated carbon, the sorbate/sorbent interaction is basically physicochemical. The field cries out for approaches similar to earlier homologous series γ^∞ correlations in liquid systems (Deal et al. 1959), leading on to functional group contribution methodology. The few published efforts that pattern broad trends (Abe et al. 1980, 1983; Giusti et al. 1974) tend to focus on effluent treating. A comprehensive predictive scheme would also include better chemical characterization of the functional groups that make carbon activated.[62]

On other sorbents, specific attention to chemical reactions may be needed. Fleming (1987) has provided a thoughtful paper on the energetics of adsorbing HCl on aluminas, going beyond the one- or two-parameter models used, for example, by the Langmuir model. He distinguishes two regimes of adsorption with minimal reversibility. Most important, he describes the action of Na-modified alumina, often used to increase the capacity of throwaway beds for HCl capture.

Dynamic Movements of "Fronts"

The mathematical modeling of unsteady-state adsorption is quite complex, although orthogonal collocation can reduce the dimensionality of the partial differential equations involved. It is usual to try to render numerical simulation results as charts displaying the effect of key parameters (Otten and Kast 1988). Vendors of sorbents like activated carbon tend to rely on scaled-down laboratory tests for guidance, since modeling here is fraught with more than the usual perils of leaving out an important physical factor.

Thermal effects are one such factor.[63] There are two distinct fronts traveling down the bed. One is a dynamic compositional adsorption

---

[62] As implied in the term *activated*, carbonyl and other functional groupings are present that may enhance chemical affinity of the adsorbate (Cookson et al. 1971).
[63] I am indebted to Chandler Barkelew for calling my attention to this method for sorting out wave phenomena in adsorption beds, which neglects dispersion and rate effects for a first approximation.

wave (CF)[64]; the other is a thermal wave (TF) from the heat of adsorption. The velocity of the latter is given by:

$$\frac{1}{V_T} = \frac{\epsilon}{F} + (1 - \epsilon)\frac{\rho_S C_S}{\rho_F C_F}$$

where ϵ = bed void fraction;
ρ = density;
C = heat capacity;
F = feed superficial velocity:

Subscripts F and S refer to feed fluid and bed solids, respectively. The CF travels with a velocity given by:

$$\frac{1}{V_x} = \frac{\epsilon}{F} + (1 - \epsilon)\frac{K(T)}{F}$$

where the adsorption constant K is a strong function of temperature and can be either >1 or <1.

Thus both the TF and CF velocities, V_T and V_x, will lag the feed velocity F/ϵ. Let us illustrate the utility of these approximations by evaluating the concern expressed that coincident TF and CF values can lead to very high local temperatures (Twigg et al. 1953).

Inspection of the two equations gives $K = \rho_S C_S/\rho_F C_F$ as the condition for $V_T = V_x$. Let us consider two possible avenues to coincidence. In the first, the TF is initially leading (K large). The CF will be self-broadening, since K declines with temperature, which is higher at the leading edge of the CF. The TF, by contrast, will probably be self-sharpening, since C_F, the strongest temperature-dependent term, will be lowest at the forward edge. Any ultimate overlapping of the TF by the CF will only mildly reinforce the thermal peak because of the broadening of the CF.

If, on the other hand, the CF leads initially, both it and the lagging TF become self-sharpening, and the feared superimposition of the two fronts is also not realized. The heat effect here is also diminished by the low K value implied in V_x leading.

Regeneration of Spent Activated Carbon

This is a major issue in effluent treating, where bed disposal is a problem. Steam regeneration fails to remove heavy sorbed impurities. Pyrolysis or partial burnoff is the usual recourse.

---

[64] The CF is a band whose leading edge is lean solid adsorbent and whose trailing edge is saturated adsorbent.

References

Abe, I., et al. *Bull. Chem. Soc. Japan 53* 1199 (1980); *56* 1002 (1983).

API Technical Date Book (1963). (See Fig. 9A1.1 for H_2O solubilities).

Bestani, B., and Shing, K.S. *Fluid Phase Equil. 50* 209–21 (1989).

Black, C. *Chem. Eng. Prog. 78*(12) 57 (Fig. 5) (1982).

Chianese, A., and Zinnamosca, F. *Chem. Eng. J. 43* 59–65 (1990).

Cookson, J.T., Jr. et al. *AIChE Symp. Series 124* 157 (1971).

Deal, C.H., et al. *Ind. Eng. Chem. 51* 95 (1959).

Derr, E.L., and Deal, C.H., *AIChE Symp. Series 32* 40 (1969).

Doherty, M.F., et al. *AIChE J. 35* 1585–91, 1592–1601 (1989).

Edmister, W.C. *Hydrocarbon Proc. 48*(6) 166–72 (1969).

Elkilani, A.S., et al. *Sepn. Sci. Tech. 24* 1095–1107 (1989).

Fleming, H.L. Paper 10d, AIChE Houston meeting (Mar. 9–Apr. 2, 1987).

Fredenslund, Aa., et al. *Vapor-Liquid Equilibria using UNIFAC* Elsevier Scientific (1977).

Giusti, D.M., et al. *J. Water Poll. Control Fed. 46* 947 (1974).

Gleich, W.A. U.S. Patent 3,921,411 (to Shell Oil Co.) (Nov. 25, 1975).

Gmehling, J., et al. eds., *Vapor/Liquid Equilibrium Data Collection* (continuing series) DECHEMA (Frankfurt) (1979).

Goelles, F. *Monatsh. Chemie 94* 1108–17 (1963).

Grimma, G.A., Jr. *Sepn. Sci. Tech. 20* 85–99 (1985).

Grulke, E.A., et al. *Ind. Eng. Chem. Proc. Dev. 24* 1036–42 (1985).

Hirata, M., et al. *Computer-aided Data Book of Vapor/Liquid Equilibria* Elsevier (1975).

Hofman, T. *Fluid Phase Equil. 55* 39–57 (1990).

Hopper, J.R., et al. *Chem. Eng. 97*(4) 177–82 (1990).

Horsley, L.H., *ACS Advances in Chemistry Series* No. 116, *Azeotropic Data III* American Chemical Society (1973).

Ikari, A., et al. *Ind. Eng. Chem. Proc. Des. Dev. 25* 859–62 (1986).

International Critical Tables 3 300 McGraw-Hill (1930).

Jacobsen, E.W., and Skogestad, S. *AIChE J. 37*(4) 499–511 (1991).

Kister, H.Z., and Hower, T.C., Jr. Paper 86c, *AIChE Miami Beach meeting* (Nov. 1986).

Kyle, B.G., and Leng, D.E. *Ind. Eng. Chem. 57* 43–8 (1965).

Lu, B.C.-Y., et al. *AIChE J. 27* 1022–4 (1981).

Maarse, H., and ten Noever de Brauw, M.C. *Chem & Ind. 24*(1) 36–7 (1974). For these authors, a dodecyl mercaptan, from propylene tetramer, was involved. Many explanations have been given to the "cat's urine" odor body by various observers; a *t*-alkyl mercaptan at the ppb level seems to be the common denominator of these episodes.

Otten, W., and Kast, W. *Chem. Eng. Technol. 11* 289–97 (1988).

Pham, H.N., and Doherty, M.F. *Chem. Eng. Sci. 45* 1823, 1837, 1845 (1990).

Prausnitz, J.M. *Ind. Eng. Chem. Funds.* 15 304–9 (1976); *Molecular Theory of Fluid Phase Equilibria*, 2nd ed. Prentice-Hall (1986).

Renon, H., and Prausnitz, J.M. *AIChE J.* 14 135 (1968).

Rijnsdorp, J.E., and Maarleveld, A. U.S. patent 3,344,040 (to Shell Oil Co.) (Sept. 26, 1967); *Instn. Chem. Engrs. Symp. Series* 32 6–38 (1969).

Twigg, G.H., et al. *J. Appl. Chem.* 3 198–206 (1953).

Vysniauskas, A., and Bishno, P.R. *Chem. Eng. Sci.* 40 299–303 (1985).

Wilson, G.M. *J. Amer. Chem. Soc.* 86 127 (1964).

10

Classic Problem Elements: Homogeneous Reaction Systems

Vision is the art of seeing things invisible.

Jonathan Swift

Anecdotal examples related to reactors are placed largely in other chapters. This and the next chapter provide a smörgasbord of insights and procedural advice. Emphasis is on flow systems, although much of what is said will apply to batch reactors as well. A major goal is to appreciate what opposing claims on his skills the designer faced before we as troubleshooters try reformulating a reaction system in the light of poststartup experience.

A Big Subject

It seemed neither feasible nor appropriate to provide here full coverage, with quantitative treatments, of effects and interactions that are readily available in standard texts (Carberry 1987; Himmelblau and Bischoff 1968; Levenspiel 1972; Westerterp et al. 1984). Rather, I have concentrated on the judgment aspects of chemistry and flow issues, which may give early insights and less ponderous ways of attacking reactor problems. In a few areas where my experience is extensive and where alternative sources are less available, such as in modeling free-radical reaction systems, more extended treatments are given.

This approach will not confer expertise; it is intended to embolden the technologist to take an independent position on reactor matters in PTS or process improvement activities complementary to those of experts who may be consulted. Qualitative insights take precedence over quantification. Order-of-magnitude estimates are invoked to gain perspective on key aspects of a reactor problem. The most valuable outcome of the sifting and sorting process advanced will often be to uncover an essential physical or chemical element, which was missing in our first conception or model of the reaction system.

It is not just with reactors that we deal here and in Chap. 11. Suppose that our problem is a finished product whose quality is being degraded by a trace reaction catalyzed by a steel or zinc surface. This problem deserves a full reactor treatment in terms of diffusional access to the wall, the nature of the catalytic sites, possibilities for metal substitution or "poisoning" of the existing wall surface to inhibit the unwanted reaction.

Division between Chaps. 10 and 11

The division is nominally between the homogeneous and heterogeneous aspects of reactors, but it was difficult to find a clean break. I have placed heterogeneous systems in Chap. 11 when the heterogeneous phase is a solid catalyst or where a diffusion/reaction "front" plays a major part in the transfer of a reactant from one fluid phase into an adjacent fluid phase. Material on two-phase systems where interactions are governed largely by fluid mechanics and reaction is confined to one quasifluid phase will be found in this chapter.

Entering Attitudes and Roles in Reaction System Support

The technologist should attempt to identify with the operative chemistry and physics in a very "physical" manner (e.g. visualize herself as literally perching on a catalyst dual site to obtain a feeling of how partial occupancy by a halogen gas-phase moderator can enhance reaction selectivity). The ability to sense phenomenological layers and move downward into the texture of a problem is one factor distinguishing a troubleshooter from a modeler.

I visualize an actual or implicit tripartite team in troubleshooting a reaction system. The technologist focuses on the more apparent aspects, such as fouling and attrition of catalyst or changes in recycle flows or composition. The reactor design expert addresses more elusive aspects, like flow instability or lateral heat transfer in catalyst tubes. He may be the original designer, present only in his works. The third member would be a laboratory chemist, skilled in analysis and bench-scale experiments, able to flesh out reaction mechanisms with help from the literature.

Working the interfaces between the three disciplines is a good way to begin the problem definition phase of PTS, certainly better than seizing on a false premise in order to let each member adjourn to partitioned problem-solving roles. Such a start can establish needed overlap between the technologist's intuitive appreciation of the problem and the mathematical insights that may be furnished by the reactor design expert. Let us begin with one example of activities appropriate to a longer-term reaction system support effort.

Example: Synthesis of Vinyl Chloride (VC) by Thermal Cracking

VC is usually made by thermal cracking of the precursor, ethylene dichloride (EDC), in parallel long runs of open tubular reactors through a furnace that supplies the necessary temperatures plus the endothermic reaction heat. This is a gas-phase free-radical reaction, seemingly quite simple. The problems that arise relate to buildup of fouling deposits, variable conversion, matching the conversion of parallel passes, side reactions that affect yields or product quality, and operability of the liquid quench at the reactor outlet. The division of action here might be as follows:

Reaction engineering expert: Formulate a free-radical kinetics model for the main reaction and a few minor reactions that highlights the effects of temperature path, radical initiator, and inhibitor compounds on productivity, yield, and trace impurity formation. Define the issues in the quench: effect of thermodynamics and rate of cooling on formation of tars and key product impurities.

Chemist: Provide analytical histories of trace components in the VC product and recycle EDC. Document chemical shifts in the quench and characterize the tars made there. Prepare initiators for plant injection tests to increase conversion at constant temperature.

Technologist: Log the effects of normal and test temperature profiles and tube metal temperatures to define regions of coking, shifts in furnace thermal efficiency, and strategies for equalizing pass conversions. Derive and track reactivity via an apparent rate constant, using a composite free-radical rate expression provided by the design expert. Relate reactivity shifts to the extent of coking and the levels of inhibitors and accelerators. Plan plant tests of supplemental initiators for cracking EDC.

In any particular reactor system for which we are trying to formulate a technical support strategy, one element may be more obviously troublesome than others, and there will be a temptation to highlight it. In reaction systems, however, interactions are the name of the game; efforts to isolate one factor can be nonproductive.

Chemistry vs. Fluid Mechanics

It seems to be common for chemical kinetics to be erratic when the fluid mechanics is well behaved (as in gas-phase heterogeneous catalysis), while the kinetics is often well behaved when the fluid mechanics is aberrant (as in liquid-phase catalysis involving an additional gas or liquid phase). Fluid mechanics/chemistry and homogeneous/heterogeneous divisions are used below to sector the field. Both of these distinctions are imperfect, so several hybrid situation must be discussed as well.

HOMOGENEOUS REACTORS WITH FLUID MECHANICS DOMINANT

Improved Realization of a CSTR for Reactor Improvement

Chaps. 6 and 8 indicated some likely shortfalls and tactical options in realizing an idealized mixing pattern such as the CSTR. The simplest rendition of an imperfect CSTR is a recirculated loop where the circulation is only a finite multiple (say, six times) of throughput. Models for imperfect CSTRs with short-circuiting and "deadwater" zones are available (Himmelblau and Bischoff 1968). In general, we focus attention on packets of the effluent that have a short

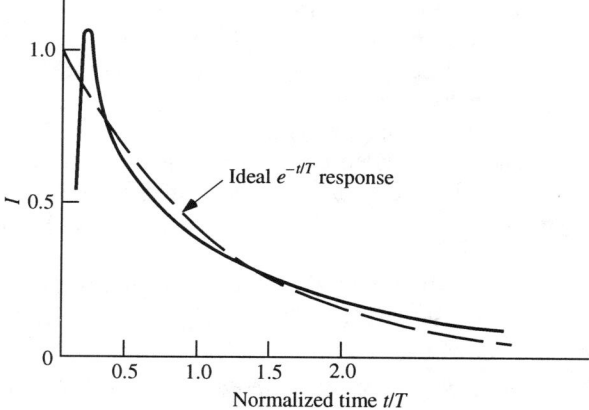

Fig. 10-1 Early and Late Deviations from Ideal CSTR Pulse Response

but finite or a much longer residence than the ideal CSTR residence time distribution (RTD). If we are troubleshooting an existing commercial reactor, tracer tests may well be warranted to characterize what kind of deviation from an ideal CSTR is operative.

Fig. 10-1 illustrates two deviations from the ideal CSTR response to a pulse input. These comprise an early spike of a small plug-flow bypass (see Fig. 6-2) and a long-residence hump above the expected exponential decay, corresponding to bleeding of material from a semistagnant zone. A log-scale impulse response plot will facilitate detection of the latter, which I believe is more important in CSTR considerations than a totally stagnant deadwater zone. Bypassing deviations from a pure CSTR RTD will affect selectivity little but may be significant in a control context. A longer-residence "tail" deviation may aggravate some obscure sequential reaction with quality implications; it merits additional study.

The complementary issue is: How closely do we need to approximate in a given situation the CSTR? While a CSTR configuration will be desirable for goals such as suppressing an undesirable high-order side reaction or for optimal management of heat of reaction, few of these require a perfect CSTR realization. In many cases a CSTR, followed by a short PFR to complete a reaction, makes a good combination.

Mechanical agitation in a vessel will not produce a CSTR if adequate power/volume is not provided. But CSTR shortfalls can occur even in small vessels, in non-Newtonian, high-viscosity systems.

The type of agitator may be critical to prevent a "coring" effect, where the outer regions are virtually unmixed. In some cases with mobile fluids, where no kind of internal agitator is safe or practical, jet stirring by feed momentum (e.g., via a Venturi eductor) can secure a recirculation ratio to feed of up to 4/1. In a liquid reactor where feed momentum alone would fall short, pump energy may be added via an external recirculation loop. In a G/L system, use of a draft tube can maximize lift/recirculation effects.

In the following example, insights into fluid mechanics laid the basis for a desired closer approach of a reactor to a CSTR.

Example: Gas-phase Reaction Requiring Increased Recirculation

In a high-temperature exothermic adiabatic gas-phase reaction, the premixed feeds entered at sonic velocity at one end of the open tubular reactor. Selectivity deteriorated each time production was increased (by enlarging the diameter of the entry pipe). It was inferred that entry mixing or the flow pattern in the reactor was deteriorating, but we lacked quantitative tools in reaction kinetics and fluid mechanics on which to base corrective action.

It was known that yield was lost to two side reactions. One was sequential to the desired main reaction; the other was parallel to the main reaction but was minimized by high temperature. Recirculation was therefore helpful for the second side reaction, since the large temperature rise could be pumped back to the inlet. Conversely, it was qualitatively clear that a plug flow reaction flow path would reduce the sequential by-product. Against these conflicting goals, we reasoned that something between PFR and CSTR extremes, but closer to the latter, would probably maximize overall yield.

Two developments gave shape to the remedial program. The first was a formal model of the main and side-reaction kinetics vs. temperature and concentrations. The second was our discovery of the Craya-Curtet treatment of fluid mechanics of jet-stirred vessels (Barchilon and Curtet 1964; Curtet 1958). This permitted the modeling of reactor fluid mechanics as an initial recirculation eddy, merging into a regular turbulent flow profile at the exit end of the reactor.

We came to understand that the ratio of feed nozzle diameter d to reactor diameter D determined the ratio of internal recirculation R relative to axial feed F; the smaller the d/D, the larger the R/F

and the better the reactor yield. Since the reactor volume was oversized and only nozzle diameter d had been increased as F went up, d/D had been increasing with the square root of throughput. Two options to improve yield by reducing d/D were devised: use of twin parallel reactors and/or shortening the reactors.

Practical Realization of a PFR

As with the CSTR, we begin with two extremes:

1. Situations where even an approximate realization of a plug flow reactor is difficult.
2. Chemistry situations where even minor deviations from plug flow may have a serious effect.

The first was covered in Chap. 6. Pitfalls in realizing a PFR in an open tubular reactor are forestalled by periodic remixing along the length. Long pipe reactors entail layout problems and costs; the constant challenge is to create an effective PFR via a series of 3 or more CSTR vessels.

Deviations from plug flow assume great significance in situations where close to 100% conversion of a limiting reactant is essential, the so-called zero-slip requirement. This topic was dealt with in an early publication (Edwards and Saletan 1967). Need for zero slip arises when recycle costs, downstream corrosion, toxicity, or like factors dominate. A velocity variation over the flow cross section of a plant-scale reactor that may seem minor can give much more slip or require substantially more residence time than laboratory batch results suggested. (Strong radial diffusion can even this out.) Suppose that we are seeking $k\theta = 6$ (θ here is dimensionless residence time; i.e., 6 reaction units or 99.8% conversion of the limiting slip component). If only 5% of the flow passes through in a channel with a velocity 1.5 times average, the slip will increase 40%, from 0.2% to 0.3%.

A second caution in configuring zero-slip reactors concerns extrapolating kinetics obtained from modest-conversion data to estimate the residence time required for zero slip. Even if non-first-order kinetics described conversion over the database, it is a reliable truism that the kinetics of a vanishing component will tend to move back toward first-order behavior. This may cause serious underestimates of required reactor length.

Example: Total Chlorine (Cl) Consumption in a Chlorination Reactor

Because of a facile and reversible dissociation of Cl_2 to form Cl atoms for the free-radical reaction, the kinetics of the overall reaction had been fitted as one-half order in Cl_2. However, as Cl vanishes, its dissociation becomes rate-limiting and the kinetics shifts toward first order in Cl_2. The tail-end section to go from, say, 97% consumption of Cl (covered by our developmental database) to the desired 99.95% consumption will be considerably larger than one-half-order kinetics had led us to expect.

Wall Effects in Tubular Flow Reactors (TFRs)

In our continuing pursuit of physical identification with the problem, the vessel wall deserves imaginative attention, as it affects flow patterns in a number of ways. RTD aside, we can think of a TFR as being either wall or particle drag dominated, while a CSTR is a way of escaping this dominance. Such an escape might be reflux cooling in a CSTR when removing reaction heat through a TFR wall poses special difficulties. Another way of escaping wall drag is slip at the wall, made possible when the fluid does not wet the wall. A further example of escape from a wall effect is use of axial injection of a disperse phase in a two-phase reaction system to delay its rapid coalescence as it impacts the wall.

Wall drag is not always a detriment. One opportunity, documented in the literature but little known, is enhanced wall-to-fluid heat transfer in the presence of a reaction,[65] effected by interaction of the reaction with velocity profile in the laminar flow boundary layer.

Entrance Effects

Entrance effects are flow transients accompanying the establishment of a steady velocity profile and boundary layer after entry into a pipe or vessel. In one striking case, corrosion from a minute saline/

---

[65] Strangely, the case has been made both for endothermic reactions (Stepanov 1969) like thermal cracking and for exothermic reactions (Pathangey et al. 1987) via product molecules leaving a catalyst surface with excess vibrational energy. Enhancements of 30–100% are cited.

HCl phase occurred selectively at the tube wall just after entry from a reactor plenum into a multitubular cooler. We attributed this to local thinning of the boundary layer. A positive use of the entrance effect is flow pulsing to disturb the boundary layer, both for reducing residence time variation and for heat transfer improvement (see Chap. 8). The enhancement of gas-to-wall heat transfer by fluidized solids can be viewed as a similar effect.

Axial Mixing

Axial mixing can be a major issue in certain fluid mechanics contexts. However, with respect to fixed-bed reactors, it has been somewhat overworked. Axial mixing (of mass or heat) requires *curvature* (i.e., an axial second derivative) to drive it. We should look for it as affecting performance at reactor entry (Wehner and Wilhelm 1956) or in flattening a temperature profile hot spot.

Nonaxial Flow Patterns

We have been considering (implicitly) only axial flow in pipes or flow entry perpendicular to the vessel wall. There are, however, other possibilities. Tangential entry generally discourages recirculation and makes possible other special heat transfer and mixing effects. Cylindrical and even spherical flow reactors, where the flow is directed radially outward, have found special niches where low pressure drop is highly valued.

"Loop" Reactors

This variant combines some of the virtues of CSTR and PFR configurations; it is particularly useful in staged additions in near-stoichiometric matching of two reactants. A high degree of turbulence, maintained by direct pumping or lift effects, facilitates mixing in of aliquots of reactants. Added utility occurs when high dilution enhances reaction selectivity. If the CSTR aspect hinders other reaction goals, several such loops can be arranged in series.

Fluid/fluid Mixing

If we wish to inject/mix a second reactant into a primary reactant flow, the same general principles apply as in inlet distribution. Subdivide the smaller stream into a large number of small packets, using inlet turbulence to assist these packets in breaking up into the main flow as quickly as possible. Transverse injection through multiple slits from a center pipe into an annulus is a fine way to do this, but the momentum of a packet must be kept low enough that it barely penetrates across the annulus. Directing incoming packets 180 degrees against the main flow is risky and promotes back mixing, which may not be desirable. The laboratory Meeker and Bunsen burners illustrate some of these mixing issues, the former as subdivision for coaxial mixing. The latter uses a single jet of premixed gases to complete turbulent mixing of fuel/air (see the discussion of the Craya-Curtet effect discussed above).

If the reaction is rapid, on the same time scale as G/G mixing (generally a matter of milliseconds, even with high energy input), and if this is harmful to selectivity, we may have to slow the reaction by using a lower temperature or less catalyst and accept whatever penalty results in terms of larger reactors. In many cases, the interaction of diffusion and reaction is a major consideration (see below).

If we stage reactant addition to a main stream in which little pressure drop is permissible, and if an external energy input device is precluded, design ingenuity must get maximum mixing effect from the entering jets.

Example: Accelerated Dissolution of an Injected Reactant

A limited-solubility reactant was injected into an aqueous reaction stream in stages along a pipeline reactor. High-pressure injection of the aliquots was used to break the reactant into fine droplets for rapid dissolution so that reaction (to a relatively soluble product) could take place before the next injection point, thereby maximizing selectivity. The original system gave 50-μm droplets. It took a surfactant, added to the reagent prior to injection, to obtain the <10-μm droplets required for fast dissolution.

HOMOGENEOUS REACTORS WITH CHEMISTRY DOMINANT

What do we mean by chemistry dominant? Not simply that fluid mechanics is well behaved, but also that special insights into kinetics are required. Examples include reactions with inverse dependence on temperature, autocatalytic reactions, and equilibrium-limited reactions. The issues here in first addressing the problem are these:

1. What is the right kinetics question to ask?
2. How do we get the information to formulate such kinetics?

From this large subject we present a few isolated examples, plus a more complete exposition of free-radical chain reaction systems.

Thermodynamics and Reversibility

Few technologists use the thermodynamics they have been taught to calculate reaction equilibria vs. composition and temperature. The timidity due to lack of practice is compounded by unsureness in correcting from reference states of free energy to the real state of nonideality in solutions and under pressure. In a system as simple as the (equilibrium-limited) condensation of acetone to form diacetone alcohol (DAA), I have found it useful to apply γ values of DAA in acetone, corrected for temperature and concentration effects, in calculating compositional equilibria (see the guidelines presented in Chap. 9).

There is a strong view in the literature that the use of γ value correctors in calculating the driving force for reaction kinetics does not work. I disagree and suggest that the preference for using only concentration for the mass-action-law driving force may stem from mishaps with the use of activities, such as when a minor component with high activity due to nonideal mixture behavior experiences diffusional retardation in delivering itself for reaction.

Kinetics Formulation

More important than nonideality corrections is framing the mass-action law in terms of the real participation in the activated intermediate, not formal stoichiometry. For example, we find that the kinetics for acid-catalyzed hydration of ethylene to ethanol correlates as follows:

$$k[C_2H_4] - \frac{k}{K_{eq}} \frac{[EA]}{[H_2O]} \qquad (10\text{-}1)$$

(Gel'bshtein 1960). Of course, in terms of equilibrium, all seems normal. But how did a reactant, water, end up in the denominator of the back reaction? Let us talk about activated intermediates and the quasi-steady-state assumption (QSSA).

Difficulties arise from our wish to force-fit kinetics as a single rate-limiting transition; in many cases, several steps in series are sharing control. This procedure made sense earlier, when we sought an expression that could be analytically integrated; with today's numerical integration capabilities, the only objection to multistep kinetics is that more parameters must be fitted. In addition, even when several steps are of comparable velocity, rate expression consolidation through the QSSA can still be achieved if the intermediate states are passed through rapidly enough to be at low concentration relative to the major species. Some consolidation *is* desirable to enhance our ability to reason about reaction trends.

The ethanol synthesis kinetics cited is based on protonation of ethylene to the $C_2H_5^+$ cation as the rate-limiting step. (The acid catalyst enters via the activity of H^+ but is combined into the rate coefficient k.) Water enters only in the subsequent hydration of the cation (formerly called *carbonium*, lately termed *carbenium ion*):

$$C_2H_5^+ + H_2O \rightarrow C_2H_5OH + H^+$$

Such mechanisms will not usually be known in advance. But once data indicate a formulation, it is rarely difficult to rationalize the facts with theoretical mechanisms. Helfferich and Chern (1990) and others have begun to make this a technically robust procedure.

The QSSA

The key aspect is rapid traffic through a small species (carbenium ion in the above case), whose time derivative approximates zero, because no storage is possible in the small population of activated intermediates in reactions. Equating the rate of formation to that of disappearance of these intermediates compresses two rate expressions into one. The QSSA, applied to the net formation of $C_2H_5^+$, with some further minor approximations, leads to eq. 10-1.

Methods to derive the number of truly independent reactions in a complex series are available. In free-radical chain reactions there may be many more than two steps, but there are also more low-concentration free-radical intermediates to which the QSSA can be applied to condense the rate expressions for the main and key side reactions. The QSSA may not hold exactly at the onset of a reaction or in a tail-end quench transition, but it is generally a useful tool. In heterogeneous catalysis, it can be applied to an adsorbed species on the catalyst surface (e.g., atomic hydrogen) to derive the complex rate expressions with denominator terms first popularized by Langmuir/Hinshelwood and Hougen/Watson.

Free-Radical (FR) Reaction Chains

I was just developing an understanding and use of FR reactions in my work when Ben Luberoff, editor of the fine cross-fertilizing journal *CHEMTECH*, belittled their usefulness, in reference to a series of articles he was publishing in the early 1970s on FR chemistry fundamentals. He implied that FR kinetics was not a quantitative tool for reaction engineering. However, I, along with others, have since used them to advantage[66] in at least a dozen industrially pertinent applications.

Why should one try to use FR kinetics for process support work? First, because of the amazing number of processes in which FR reactions are central components. Second, for the many counterintuitive insights into process improvement it provides. In the support of EDC cracking to vinyl chloride, the FR treatment indicates that the rate of VC formation (Barton and Howlett 1949) is proportional to

$$[EDC]^{1/2} [VC]^{1/2}$$

Instead of being simply proportional to the concentration of EDC, the reaction is (somewhat surprisingly) moderately autocatalytic in the product VC. This has obvious implications for the reaction monitoring and improvement tasks described above. This qualitatively distinctive representation arises from an induced homolysis mechanism of initiation (see below). Another example of counterintuitive insights obtained by the use of FR kinetic models is in minimizing fouling during the cooling of a pyrolysis effluent (see below).

---

[66] The excitement of FR work rests on our ability in most cases to establish a priori the preexponential and activation energy of the rate constants for the elementary FR reactions. It is true that estimation of preexponentials (Benson 1976) may have a factor of 3 uncertainty. However, temperature dependence is usually well predicted; the consolidated rate expression is correct in "shape"; and usually only adjustment of one or two parameters is needed, by fit to plant data, to make the whole usable as a fundamental descriptor.

But before discussing these examples, I would like to explore why homogeneous gas and liquid-phase FR chemistry, seemingly so inexact and nondiscriminating a vehicle compared to heterogeneous catalysis, has found wide application in the chemical process industries. Early applications involved liquid-phase polymerization, where the leverage of long chain length for each triggering initiation reaction was attractive. The need for better stereo (syndiotactic) control of polymer chain orientation via heterogeneous catalyst has displaced only partially this use of FR chemistry. Today, most thermal cracking reactions and many commercial oxidation and halogenations are conducted via FR mechanisms.

FR utility in so many monomeric reactions rests upon the ability to find energy "grooves" that direct the reaction, because our understanding of the elementary steps can be so much better a priori than in heterogeneous surface catalysis and because bulk turnover rates can be so much greater than in the latter. FR chemistry shares with heterogeneous catalytic reactions the ability to divide a large energy threshold into a series of small steps.

When we invoke FR reactions, however, we must choose our territory with discrimination. The FR oxidation of cumene to its hydroperoxide as part of phenol/acetone synthesis, for example, makes sense only because of the isolated tertiary hydrogen on cumene. Making phenol via oxidation of ethylbenzene or *sec*-butyl benzene, by contrast, would be too nonselective to be competitive without some heterogeneous catalytic "shaping." Similarly, VC is made by the FR cracking of dichloroethane, but dehydrochlorination of 1,2-dichloropropane to allyl chloride is much less productive and selective as a FR route.

Several generic weaknesses of FR reactions should be noted. They are very vulnerable to minor concentrations of inhibitors. In addition, heterogeneous surfaces (packing and even the tube wall) can facilitate chain termination via recombinations of radical species.

Anomalies in FR Initiation and Propagation

Although FR reaction chains are often initiated by readily dissociated additives (peroxides, azo compounds), dissociation of a major reactant also can provide initiation. However, there are some striking failures to explain the observed rate of FR reactions by initiation involving simple monomolecular[67] dissociation.

Homolytic Scission by Proximity of a Double Bond

Bukhman et al. (1968) invoked arbitrary wall-effect assumptions to justify much faster observed rates in FR chlorination of ethylene than could be explained by the (well-documented) simple dissociation of Cl_2. The anomaly is resolved by positing a bimolecular *homolysis* induced by the double bond of an adjacent olefin molecule.[68]

Example: Homolytic Scission in Initiation of Dichloroethane (EDC) Pyrolysis

A similar approach in VC synthesis disposes of monomolecular dissociation of EDC (either to Cl· and a chlorethyl radical or via C-C scission) as the initiation mechanism. The preferred alternative, bimolecular interaction of EDC with the double bond of VC (via either H or, more likely, Cl transfer) to yield a chlorethyl and a dichlorethyl radical as chain initiators, leads to the autocatalytic rate formulation already described.

Example: Anomalous Faster Chlorination of Ethane Than of Ethylene

The rate of chlorine attack on ethane is faster than on ethylene in an ethane/ethylene mixture, yet the chlorination of pure ethane is slower than that of pure ethylene (Chiltz et al. 1963); this reflects the role of the latter in the initiation step.

---

[67] At low pressure levels, the "monomolecular" gas-phase initiation is, of course, intrinsically bimolecular, requiring the collision with an inert molecule to become "hot" enough to dissociate.

[68] As explained by Pryor (1966, 1968) and Poutsma (1965), this energetically favored pathway probably involves a π electron from the double bond attacking the Cl-Cl molecule. The decisively lower energy threshold for the bimolecular scission more than makes up for the difference in preexponentials between monomolecular dissociation (about $6 \cdot 10^{13}$/sec) and 10^{10} L/mole sec for bimolecular initiation (this equates to about 10^9/sec in pseudo-first-order form at the normal gas-phase concentration of a major component). A difference of 63 kJ in activation energy at 500 K is a factor of $10^{6.6}$ in relative rate, for example.

There is a fairly direct ionic counterpart to such FR homolysis in liquid-phase reaction, as in Cl_2 dissociating into $Cl^+ \ldots Cl^-$ in the proximity of a double bond, with the former attaching to the double bond as a positive "onium" complex. Such a complex is a transient intermediate in the synthesis of ethylene chlorohydrin:

$$C_2H_4 + Cl_2 \rightarrow C_2H_4Cl^+ \xrightarrow{(+H_2O)} HO-CH_2-CH_2Cl + H^+$$
$$+ Cl^- \qquad\qquad + Cl^-$$

Acceleration by Retarded Termination Reactions

In FR polymerization reactions, a curious autoacceleration of the rate can occur when heightened viscosity from conversion to the polymer begins to retard the chain-terminating recombination of radicals, while initiation is still little affected (Hamielec 1983). However, physical complications (heat transfer, flow pattern) accompanying such high viscosities are probably more important than the chemistry anomaly for practical reactor engineering considerations.

Chain Branching

Occasionally there can be a two-for-one yield of radicals, as in

$$H_2 + O\cdot \rightarrow H\cdot + OH\cdot$$

This effect, important in combustion, is the counterpart of a nuclear chain reactor going critical. I have found it of qualitative use in my safety and environmental work for understanding some anomalies in explosion envelopes and atmospheric chemistry issues.

Radical Traps

An important anomaly in FR propagation reactions is blocking of a chain reaction by creating a radical species that cannot continue the chain. Most antioxidants function as radical traps; the new radical created by hydrogen abstraction from the antioxidant is trapped in a resonance energy "well." The added activation energy for continuing propagation will typically be 34–63 kJ. Such trapped radicals usually terminate by coupling with each other.

Example: Propylene as Inhibitor in FR Cracking of EDC to VC

The propagation steps in this chain reaction are a sequence involving chlorine atoms Cl· and dichlorethyl radicals:

$$Cl\cdot + CH_2ClCH_2Cl \to HCl + \cdot CHClCH_2Cl$$
$$\cdot CHClCH_2Cl \to VC + Cl\cdot$$

Now suppose that, in one of the many iterations of this chain, consuming EDC for the production of VC and HCl, the Cl· atom abstracts hydrogen from trace propylene instead of from EDC. The resulting $C_3H_5\cdot$ (allyl) radical is, in effect, a chain stopper because of the extra 52 kJ activation energy needed for it to continue the chain with H abstraction from EDC (almost a 10^{-4} factor on the rate constant relative to H abstraction by a simple Cl atom).

Side-reaction Possibilities for Trapped Radical Species

These resonance-trapped radicals can assume new life and initiate unwanted side-reaction chains in certain special contexts, as the following example shows.

Example: Allylic-initiated Condensation Reactions in a Quench

This relates to yield loss and tar formation in postreactor cooling, such as in the transfer-line heat exchangers that are part of steam-cracking processes for olefins production. Although the reaction chain in such cracking of hydrocarbons is carried by H·, methyl, and ethyl radicals, a population of idle allylic radicals ($C_3H_5\cdot$, $C_4H_7\cdot$, etc.), trapped by the above logic, builds up to concentrations exceeding those of the working alkyl radicals. The allylic radicals are positioned to do damage when the falling temperature in the postreactor quench shifts equilibria to facilitate C-C coupling to build polymers/tars. These differ chemically from the polynuclear aromatics or graphitic coke that are the only condensation products thermodynamically possible at high temperature.

The resonance energy hurdle of the allylic radicals is overcome by postulating cyclization as the first step in forming tars:

$$CH_2 = CH - CH_2 \cdot + C_2H_4 \leftrightarrow c\ C_5H_9\cdot\ (cyclopentyl)[69]$$

---

[69] The choice of C_3 for the allylic participant and C_2 for the olefin participant is illustrative of a range of possibilities. Dienes are particularly good candidates for such condensations (Nohara and Sakai 1988). I am indebted to a former colleague, André Lagendijk, for suggesting this reaction route.

The cyclopentyl radical formed can continue the chain by adding to an olefin; beginning the building of more complex molecules. Space prevents the full development of this scheme here, but after inserting estimated rate constants, two counterintuitive conclusions are reached:

1. For the first fraction of a second of quench, the allylic radical population is a decaying function of time only.
2. Provided that the quench temperature is near the thermodynamic ceiling temperature (about 650°F) for the condensation to cyclo-C_5 structures, lower temperature actually *increases* the condensation of olefins in this early phase. This may apply to fouling at a cooling tube wall even if the bulk temperature is safely high.

Of course, with further elapsed time and cooling to lower temperatures, the allylic population decays by self-condensation:

$$2\ CH_2 = CH - CH_2 \cdot \leftrightarrow C_6H_{10} \qquad \text{(biallyl or 1,5-hexadiene)}$$

The overall quench strategy suggested by this kinetic examination is one of dropping from pyrolysis reactor temperatures to a temperature of 750°–800° F over about 0.5 sec so that the allylic radical population can decay while cyclization condensation is still thermodynamically forestalled. Then a second stage of the quench can quickly drop the temperature close to a terminal 600° F.

The above examples may seem to require extensive arcane knowledge beyond the reach of a process technologist. But such answers can emerge if we start with a meaningful set of questions: What do I mean by a quench? What is harmful about lingering at reactor exit conditions? How low must I cool to stop bad things from happening? Why must I cool rapidly? Am I concerned primarily with yield loss or with fouling from quench postreactions? What advantages beyond speed might cooling with injected liquid give over slower indirect cooling? If I use indirect cooling, how important is metal surface temperature relative to effluent bulk temperature?

Example: Dilution Effects in Steam Pyrolysis

How do we know when a diluent affects reaction pathways simply by concentration reduction and when it has other roles? The use of steam in pyrolysis for olefins manufacture is a case in point. By observation, steam dilution diminishes the formation of tars and coke, and an economically optimized proportion of steam is always used. We can reason that, broadly speaking, the desired cracking reactions are 3/2 order in the vapor-phase concentration of the hydrocarbon feedstock (from the combination of chain initiation, termination, and propagation elements). On the other hand, the fouling side reactions are second-order in product olefin concentration level, possibly 5/2 order overall (see the example above). Thus we can see that tar formation may decrease significantly from dilution alone.

However, proactive PTS consideration will not stop here but will elicit other possible contributions of steam. These include:

1. Thermal moderation of endothermic cooling;
2. Free-radical chain intervention to form ·OH radicals;
3. Moderating the oxidation state of wall-surface metal atoms;
4. Decoking by the water gas reaction.

The first factor is self-explanatory in easing the need for heat delivery through the pipe wall. The second is useful primarily as a convenient way to organize treatment of the formation of lower aldehydes and other trace oxygenates that may pose quality issues. The third and fourth are discussed in Chap. 11.

Gas vs. Liquid Free-radical Kinetics

The a priori prediction of elementary FR reaction rates is surest in the gas phase. Occasionally we need to transpose an established system of vapor-phase kinetics into the liquid phase. If the second-order rate constant were unchanged for the liquid phase, higher molar concentrations in the latter would cause a heightened rate there. However, Lebedev and Tsepalov (1979) found that the bimolecular preexponential of homolytic reactions was about 40-fold lower in the liquid phase and gave a rationalization of this shift. An activation energy shift reduces the rate almost 10-fold more (Pukhovsky et al. 1985). Let us use a 300-fold attenuation in the FR rate constant on entering the liquid phase as a first try crude estimate.

Example: Runaway Chlorination in a Small Liquid Entrainment Phase

Thermal effects and rapid coking were noted in a gas-phase chlorination reaction whenever minor entrainment of liquid appeared in either the chlorine or hydrocarbon vapor feeds. We attempted to rationalize the effects on the basis of rate acceleration in the liquid phase, despite the relatively minor mass of that phase relative to the vapor. With molar concentrations of about 3 m/L vs. about 0.02 m/L in the vapor phase the prospective rate enhancement, even with our posited 300-fold attenuation of the preexponential, led us to conclude that even 1%w of liquid phase could be quite significant for a second-order FR side reaction. However, consideration of the mass flux to transport reactants into a runaway liquid-phase chlorination yielded a second condition: that entrainment be of near-micron size to create enough surface for transport.

Self-consistent Kinetic Models

Let us now examine in some depth another (non-FR) reaction system where attention to fundamentals strengthens kinetic modeling and provides an understanding of the trade-offs among process options.

Example: Minimizing an Undesirable Sequential Reaction

The alkylation of benzene (B) with propylene (P) to form cumene (C) is a classic acid-catalyzed reaction in which we seek to minimize net formation of the unwanted sequential dialkylation product di-isopropylbenzene (DPB) or simply D:

$$B \xrightarrow{+P} C \xrightarrow{+P} D$$

I use it to illustrate two points:

1. The usefulness of applying up front the almost-superseded[70] disciplines of physical organic chemistry and physical chemistry to buttress key mechanistic issues.
2. The usefulness of visualizing selectivity strategies via phase plane trajectories rather than plotting components vs. time.

In cumene synthesis, propylene is protonated by an acid catalyst ($C_3H_6 + H^+ \rightarrow C_3H_7^+$); the resulting isopropyl cation attacks benzene, forming an intermediate complex, which breaks down to give cumene and regenerate H^+. The cation can also attack cumene to form a sequential DPB by-product. These intermediates stem from the basicity of aromatic hydrocarbons, whose π electrons form a σ

---

[70] These disciplines have not vanished, but their computerization has deprived even practitioners of a qualitative feel for effects and has made the whole area obscure for the ordinary technologist.

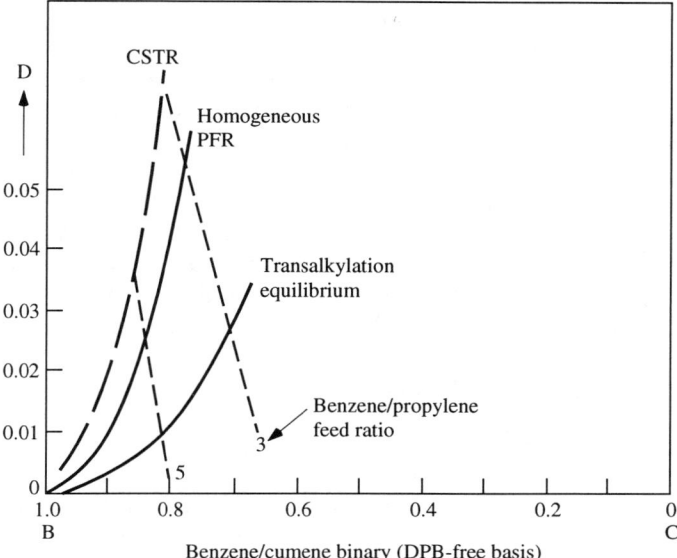

Fig. 10-2 Dialkylation By-product in Cumene Synthesis for Various Trajectories

complex with the attacking carbocation. Alkyl substitution of the benzene ring increases such electron donation, (March 1985; Reid 1954). Hence cumene is more basic than benzene and has a greater affinity for the isopropyl cation. Understanding these factors helps to validate the observed greater rate of the second alkylation (Condon 1948), which leads to an undesirably large yield of DPB by-product unless we use an exorbitant excess of benzene.

Dividing the rate expression for DPB formation by the rate expression for benzene consumption to make cumene, we get a differential expression for selectivity, with the time coordinate eliminated. This is the basis for a phase-plane plot[71] (Fig. 10-2):

$$\frac{dD}{-dB} = \beta \frac{C}{B}$$

where β, the ratio of the second alkylation rate constant to the first, ≈ 1.7 (Condon 1948).

---

[71] A phase-plane plot is a compression device, borrowed from control theory, for displaying trends of one variable vs. another where a time coordinate would add nothing to understanding.

Eq. 10-2 shows that staged addition of propylene reactant will not per se improve selectivity; only the B/P feed ratio (by keeping us from regions of high C/B) will help selectivity. However, a CSTR *will* have worse selectivity than an integral (batch or plug-flow) reactor, since it operates at the highest (terminal) C/B ratio. The trajectories of the effluents from homogeneous PFR and CSTR alkylations are shown vs. the B/P feed ratio in Fig. 10-2.

A second finding from this sorting of mechanisms and kinetics is that the alkylation is effectively irreversible (i.e., the intermediate $[C_3H_7 - C_6H_6]^+$ complex will decay to form cumene and re-form H^+, not propylene or the isopropyl cation). Similarly, DPB cannot be reverted to the desired cumene by simple dealkylation; the bimolecular transalkylation with benzene must be invoked:

$$\text{DPB} + \text{benzene} \xrightleftharpoons[]{H^+} 2 \text{ cumene}$$

This reaction is much slower than alkylation; this follows from the difficulty of protonating even the dialkylated DPB relative to facile protonation of the olefin in alkylation. If alkylation is carried out in the presence of a homogeneous catalyst acidic enough to effect transalkylation, the difference between the rates is so great that the alkylation/dialkylation trajectory will be essentially complete (with consumption of all propylene) before any significant transalkylation occurs. A cusp maximum of D formation is reached before the slow but steep transalkylation begins (See Fig. 10-2).

An approximate integration of eq. 10-2 shows that D formation along a plug flow trajectory will be roughly quadratic in C:

$$D_{PFR} \approx \beta C^2 / 2B \tag{10-3}$$

(with B in excess and quasiconstant), while the equilibrium formation of D after transalkylation has been effected will also have quadratic proportionality:

$$D_{xeq} \approx \frac{K_{eq} C^2}{B} \tag{10-4}$$

With $K_{eq} \approx 0.23$ (Popov et al. 1974).

Eqs. 10-3 and 10-4 are the basis for the PFR and equilibrium paths shown in Fig. 10-2. With Condon's value of β, DPB will increasingly diverge above its equilibrium level during alkylation. Only in a final slow transalkylation (with an appropriate catalyst) is this deviation corrected (along the B/P parameters of Fig. 10-2).

Cumene via Heterogeneous Supported Liquid-phase Catalysis (SLPC)

A fixed acid catalyst has handling advantages in a continuous reaction system (see the discussion of SLPC below). The *heterogeneous* alkylation of a mixed-phase flow of benzene and C_3H_6 over a SLPC catalyst[72] has been used by UOP/Allied (Pujado et al. 1976) and others. An unexpected bonus of this arrangement appears to be that the heterogeneous acid catalyst improves the selectivity (reduces β) of the forward alkylation reaction, presumably because the solubility of cumene in the inherently polar catalyst phase is lower than that of benzene. (One can deduce from the limited solubility data a 2- to 10-fold reduction from the homogeneous β of 1.7.)

However, an offsetting disadvantage of heterogeneous alkylation is that even the concentrated phosphoric acid present in a SPA system is too weak for practical catalysis of the transalkylation of benzene by DPB. The more selective forward alkylation with this catalyst, however, gives a PFR trajectory coincidentally close to the transalkylation equilibrium locus of Fig. 10-2. One can be misled into believing that the catalyst is accomplishing transalkylation.

Cumene Catalysts Capable of Transalkylation

With an acid catalyst strong enough to effect transalkylation, the entire selectivity loss to DPB can be recouped via fractionation of DPB from the alkylation product to recycle it to a transalkylation reactor. Indeed, the reversibility of DPB allows a design optimization wherein the benzene excess to alkylation is much reduced (Canfield and Unruh 1983). If the stronger catalyst is also used for alkylation, the reactor volume (or temperature) there can also be reduced.

However, some new difficulties arise in creating a homogeneous[73] reaction system around such a catalyst:

1. The heterogeneous advantage in slowing the sequential dialkylation may be lost.

---

[72] A SLPC of phosphoric acid (SPA) on kieselguhr or other silicaceous material is one version of such a catalyst. Despite all the activity during World War II in the United States in synthesizing cumene for aviation fuel, the only decent published treatment of phosphoric acid catalysis has come from Russian scientists (Kotsarenko et al. 1988, 1989; Rustamov et al. 1955). Because of chemical interactions with SiO_2, however, SPA is only an approximate example of a SLPC.

[73] In the present state of catalysis, the superacidity needed for transalkylation implies a homogeneous once-through catalyst.

2. If the transalkylation catalyst is also used for alkylation, a mismatch is created there between very rapid reaction and entry of propylene into the reaction mixture. The resulting high mixing intensity needed in propylene entry may make the alkylator a CSTR, with a resulting further increase in once-through DPB formation.

A strong acid catalyst capable of transalkylation, however, can ultimately overcome both disadvantages. The complete trajectory of a material-balanced system, where DPB is recovered and recycled to extinction, is shown in Fig. 10-3. Despite a circuitous path, zero DPB make is achieved at a B/P feed ratio substantially lower than that used in a heterogeneous alkylation-only system.

Subtleties like those cited above for cumene and VC synthesis are not usually available to the designer at the time of plant design. We can obtain deeper technical insights over an extended period of process support than was provided in the completed development at the inception of the process, but only if we ask good questions.

Some Remarks on Staging the Addition of One Reactant

An excess of one reactant can reduce side reactions and/or moderate the thermal path. However, it is costly in energy and capital to separate and recycle unreacted excess to the reaction. By staging the limiting reactant in injections along the reaction flow path, recycling needs may be reduced. An example of thermal control is multiple beds in hydrotreating where each addition of H_2 is followed by a cold-shot injection of cooled recycle reaction liquid.

In *cumene synthesis*, we have noted that staging of propylene would not be expected to reduce the DPB/cumene ratio in the final product. However, if there were an unwanted side-reaction of higher order in [C_3H_6], staging would help. Even in a DPB context, a single injection of propylene into a rapid reaction could lead to large local variations in the P/B ratio, which increase the overall DPB yield because of a concave dependence on the P/B ratio. Halogenation reactions show similar effects. Staging here simply provides a better "aspect ratio" for mixing in of the critical reactant.

In the *catalytic hydrogenation of ethylene* product from steam cracking, we seek complete reaction of acetylene impurity without hydrogenating ethylene to ethane (see the discussion of the mechanism in Chap. 11). A first appraisal might classify this case with that

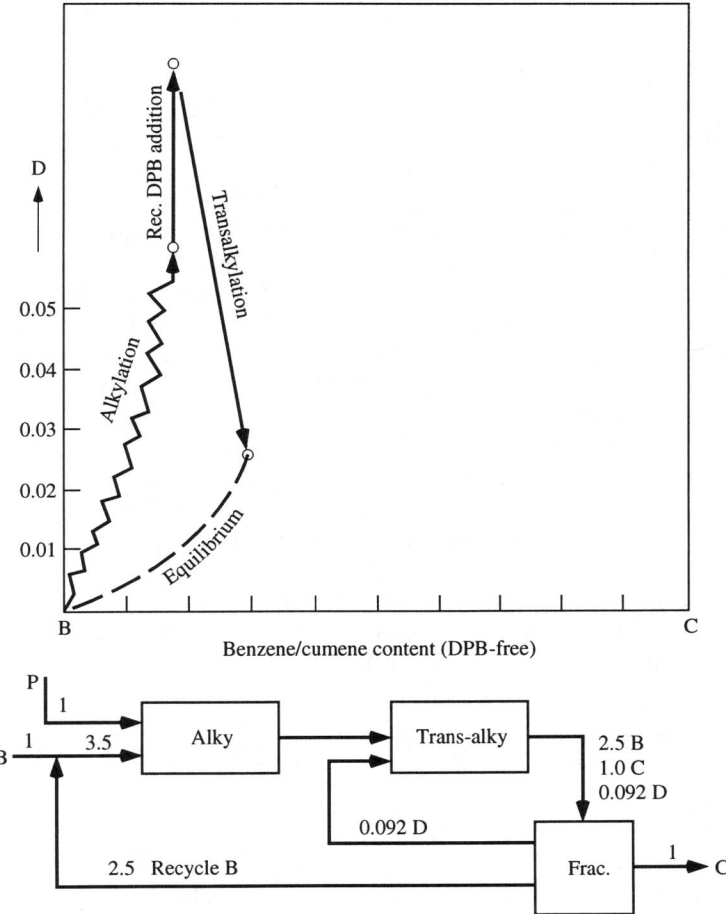

Fig. 10-3 Reaction Trajectory in a Cumene Synthesis System Capable of Transalkylation, Showing Recycle of DPB to Extinction

of cumene synthesis, concluding that H_2 staging cannot improve selectivity unless the sequential reactions have different reaction orders in H_2. However, there is a different reaction path here, a catalyst pore dimension as well as the axial one. Because of hydrogen's high diffusivity relative to C_2 compounds, it will be overrepresented in the depth of the pores, leading to nonselective hydrogenation there. Staging of H_2 or use of a gas-phase moderator can improve selectivity under such circumstances.

One-phase Striational Mixing/Reaction

Before getting into the interaction of mixing with reaction in heterogeneous systems, it is well to mention the unresolved state of the art with respect to mixing of one reactant (A) into another (B) in a one-phase system. This problem is important when the reaction rate is fast compared to the speed of blending of the several components or (as in atmospheric chemistry) when the dimensional scale of the initial striations or packets is quite large. Even with turbulent eddies as the mixing mechanism, we will need about 1 msec on a laboratory scale[74] or 10 msec on a plant scale for turbulent dissipation of such packets after injection. There is no closed computational solution for this kind of turbulence/second-order reaction interaction, and it taxes our grasp of micromixing as well. Ou and Ranz (1983) have framed in the problem via a modulus involving the striation thickness s, which discriminates the extremes of mixing control (as in a fast acid-base reaction) or slow reaction (the usual design situation). For mobile fluids and 100-μm-thick striations, their analysis puts us in an intermediate gray area for reaction times between 1 min and 1 msec. Here we need to know at least the diminution of s with time in order to try using a rough stretching-cell computation proposed by Sternling and Wendt (1974).

HOMOGENEOUS LIQUID-PHASE CATALYSIS

Homogeneous Ionic Catalysis: Some Insights

It should not be thought that only free radicals can produce chain reactions. Cationic and anionic chains are used in polymerization. A nonpolymerization system that I recommend for conceptualizing ionic chain reactions is aliphatic alkylation of isobutane by n-butenes. A problem here is that olefin is required to start the reaction by its protonation but olefin/olefin reaction is bad for gasoline quality. Therefore a key early step required is hydride (H^-) transfer to involve isobutane as a central reactant:

$$CH_3-C^+H-CH_2-CH_3 + H_3C-\underset{CH_3}{\overset{CH_3}{C}}-H \rightarrow n\text{-}C_4H_{10} + t\text{-}C_4H_9^+$$

---

[74] Regenfuss et al. (1985) were able to achieve liquid mixing in only 10 μsec in the laboratory, in a turbulent wake behind a spherical obstacle.

The chain reaction now propagates by the *t*-butyl cation adding to *n*-butene to create trimethylpentyl cation; hydride (H^-) transfer to the latter from isobutane forms iso-octane product and launches another *t*-butyl cation. With hydride transfer so central, much process improvement focuses on additives to promote it (Kramer 1986).

Acids, Bases, and Nucleophiles

We have touched on some key differences in ease of protonation in acid catalysis in our discussion of cumene. There is on the whole a reassuring generality: Rates of a given reaction tend to be the same in aqueous solutions of HCl, H_2SO_4, or H_3PO_4, with the same value of Hammett[75] acidity function H_0 but differing concentrations. We can expect the log of rate constant k for a reaction to be linear in $-H_0$, with a slope of 1.0 ± 0.2. In nonaqueous media, shifts in acidity of protic acids can occur. For example, small levels of H_2SO_4 in dry phenol/acetone mixtures have much stronger Hammett acidity than in water, but the difference drops rapidly as moisture is added (Alexandrov et al. 1984). Such media shifts can also alter the acidity differential between organic and inorganic acids.

Other Variants of Acid Catalysis

In acid catalysis, one distinction worth making is in reactions where a network or grouping of acid centers,[76] rather than a single protic acid site, is involved. This occurs at vanishingly small water concentrations in strong inorganic acids and, reputedly, in acid catalysis by sulfonic acid resins (Thornton and Gates 1974). Nonprotic or Lewis acidity (e.g., BF_3) can also be important (Jensen 1980). A scale for measuring acidity without proton transfer is elusive; more reliance is placed on performance in probe reactions. Note also that Lewis acidity is a broad category: Simple cations such as Mg^{2+} and Li^+ can be Lewis acid catalysts in certain contexts.

Catalysis by Bases

Catalysis by bases, though not nearly as common as acid catalysis, is finding increasing application in isomerization, oligomerization,

---

[75] Other acidity scales have been advanced, but $-H_0$ is adequate for most needs. Gel'bshtein et al. (1956) give temperature corrections to H_0.
[76] Our discussion here unavoidably touches on solid catalysts.

hydrogenation, cyclization, and other reactions (Pines and Stalick 1977). A special advantage is the absence of many of the side reactions that occur in acid catalysis. We must distinguish base catalysis from general electron-donating "nucleophilic displacement," with widely differing rates attached to particular anions (Johnson 1967). The latter is important in reactions involving carbon/oxygen, as in hydrolysis of carbonyl and cyclic ether compounds. I was drawn into this area in dealing with ring-opening side reaction of epichlorohydrin, where minor molar concentrations of nucleophilic inorganic anions can greatly augment the hydrolytic action of water (Shvets and Aleksanyan 1974).

Nucleophilicity and "hard/soft" cation and anion indices are available to impose some order on these effects for the ordinary technologist. Another feature of base and nucleophile catalysis worth noting is their nonlinear kinetics, often containing quadratic terms in nucleophile concentration [N] or [OH$^-$] (Buttersack et al. 1986).

Superacids and Superbases

These extensions of acid/base catalysis may seem remote from problem solving, but they are already installed in plants (Tanabe 1989) and will soon be among the challenges to support people.

Measures of Superacidity

Reactions in 100% H_2SO_4 (edge of the superacid region) correspond to a Hammett acidity $H_0 = -12$. Aliphatic (e.g., isobutane) alkylation uses sulfuric acid or hydrofluoric acid (HF), with acidity about 3 units short of this. Acidities up to -20 (eight orders of magnitude stronger than 100% H_2SO_4) have been demonstrated in laboratory liquid-phase systems, and up to -16 for heterogeneous catalysts, but systems stronger than $H_0 \approx -12$ do not yet seem to be in practical use.

Superacid catalysts are often two-component, a Lewis acid plus a Brönsted (protic) acid. A caution in dealing with superacids is that organic compounds we may not think of as bases, such as ketones, ethers, and olefins, can quasineutralize or drag down the acidity of the catalyst considerably. Fărcaşiu et al. (1989) have shown the utility of hexamethylbenzene as a base against which to measure the acidity of candidate superacids. A broad scale of hydrocarbon basicity has been given by Olah (1985). Since a reactant can lower the acidity or basicity of a catalyst in use, rates of probe reactions such as hydride transfer may be a surer measure of catalyst strength.

Zeolites as Superacids

The notion of zeolites as superacid catalysts is widespread but warrants some skepticism. There are other less dramatic ways of explaining some cited cases of zeolite performance. For example, many acid-catalyzed hydrocarbon reactions over zeolites are likely to have an activation energy E of about 84 kJ. When a zeolite gives seemingly superacid behavior at 250° C, temperature alone may account for 3.5 orders of magnitude of rate relative to reaction at 100° C. It is the higher-temperature, gas-phase operation that is characteristic of zeolite use. Such conditions exclude many applications that require moderate temperatures and liquid-phase operation.

Solid Superbase Catalysts

There is a broad H^- basicity scale, with superbasicity beginning above 26 (arrived at by symmetry with superacidity, which begins 19 units below $H_0 = 7$, the neutral point in water). The base strength required in hydrocarbon reactions can be gauged by the thermodynamics of creating a hydrocarbon anion. Streitwieser et al. (1980) have provided a broad-range scale of hydrocarbon acidities. CaO (prepared by calcining $CaCO_3$) is at the edge of the superbase region. To get well into superbase behavior, one uses something like metallic sodium, vaporized onto alumina or magnesia (Yashima 1978). As this field ripens, solid catalysts will probably come to the fore more decisively than in the case of superacids.

Media Effects on Reaction Rate

Often we have information on an ionic reaction in an aqueous medium and seek some basis for extending that information to other medium-polarity, nonaqueous media. Usually the reaction will be greatly diminished in magnitude but still significant enough to warrant the generation of rough kinetics. I have found that a semilog plot of rate constant vs. dielectric constant ϵ of the medium is an adequate predictive vehicle from a three-point database in ϵ.

Example: Loss of Epichlorohydrin (ECH) during Wet Shipment

An interplant transfer of wet ECH involved a trace water layer. Hydrolysis kinetics in the water layer is well known (Pritchard and Siddiqui 1973; Shvets and Aleksanyan 1974). However, because of its dominant volume, we also needed to predict the hydrolytic loss in the ECH layer, which is saturated with a small percentage of water. Our means of defining the hydrolysis rate in nonaqueous media was laboratory experiments in two organic media (other than ECH to avoid analytical problems). For example, dioxane, acetone, and water, with ϵ = 2.2, 20.7, and 78.5, respectively, bracket that of ECH nicely.

Example: Reactant-induced Polarity Change in an Organic Layer

Another case involved a largely aqueous stream where the main reaction took place. An unwanted side reaction took place in a small organic layer, between the (mildly polar) main product (B), distributed between the phases, and another component (C), whose low dielectric constant ϵ placed it largely in the organic phase. When we roughed out the kinetics of the organic-phase reaction in the laboratory, we were amazed to find that the second-order rate constant increased strongly with the concentration of B. In effect, B was doing this by raising the ϵ of the organic phase appreciably when it was present there in significant concentration! This made the side reaction escalate[77] rapidly as major product B accumulated.

Reactions with Zero or Negative Temperature Dependency

It should not come as a surprise that a number of important reactions have these unusual temperature dependencies. For example, recombinations of free radicals have zero or slightly negative activation energies. There is no way that a reaction can have a significant negative energy of activation. Therefore, when we encounter a negative temperature dependence, as in some cationic polymerizations, we should seek an explanation in terms of an activated

---

[77] Despite its ionic nature, this reaction was confined to the organic phase by the solubility characteristic of reactant C. It is a variant of Case 1c in the fluid mechanics categories presented below.

complex whose formation equilibrium has a strong negative temperature dependence. That is, we are looking for a denominator term of a rate expression (such as one consolidated by the QSSA methodology) that increases strongly with temperature.

HETEROGENEOUS REACTIONS WITH FLUID MECHANICS DOMINANT

The term *heterogeneous*, as used in this chapter, connotes systems where reaction chemistry occurs in one quasi-fluid phase, while a key reactant or product resides primarily in another fluid phase, and fluid mechanics issues separate the two.

Liquid-Liquid Systems

Following are some of the classic goals, tools, and stumbling blocks in achieving maximum productivity or selectivity in a fluid/fluid heterogeneous environment, where fluid mechanics is negotiable:

Case 1: The desired reaction is confined to one liquid phase (I), but a key reactant R resides preferentially in a second phase (II). The desired product P may or may not reside preferentially in (I). There are three subcases.

1a. We wish to maximize productivity by preferential retention of (I) in the reactor, with enough presence of (II) that the reaction is not limited by mass transfer of R. (II may also be a gas phase here.) This situation applies particularly when the reaction in (I) is relatively slow. Phase transfer catalysis is a chemistry answer where the insolubility of R in (I) is extreme.

Preferential Phase Dropout

Some Case 1a L/L reactors may tread a delicate balance in agitation between mass transfer needs and achieving selective dropout of (I), further complicated when drifting surface tension changes the droplet size of (II). The problem in an upflow reactor of this kind has been described in Chap. 6. A more tractable case, because particle size is stable, is when (I) is a heavy solid, as in Raney Ni hydrogenations or fluid-bed reactors where we seek to avoid settling out or entrainment, respectively.

1b. An undesirable side reaction of R with P can occur in (I), with P residing primarily in (II). Mass transfer must be restrained relative to the main reaction rate. Avoiding higher oligomer formation in acid-catalyzed dimerization of isobutylene is an example. A useful expedient here is the use of a coiled reactor tube to minimize turbulence in the two-phase system.

1c. An undesirable side reaction of R with P occurs in (II). We seek to expedite dissolution of R into (I) without facilitating back diffusion of P into (II). This could take the form of high interphase mass transfer early in a pipe reactor, with reduced transfer late in the reactor.

1d. In a reaction of R_1 and R_2 to make P, a side reaction of R_1 with itself in (I) is to be avoided; yet R_1 partitions better from (II) into (I) than does R_2. Alkylation of isobutane by butylenes is an example. Unlike Case 1b, this calls for maximum agitation to ensure that (I) is not "starved" of R_2 (Sprow 1969).

Case 2: R resides in (I), but there is a sequential reaction of P in that phase to a waste by-product B, which may be minimized by removing P to an inert liquid or vapor phase (II), provided expressly for its removal, as rapidly as it is formed. These approaches are termed *extractive reaction* and *reactive distillation*, respectively. Which one we use will depend upon the volatility of P relative to R in (I).

Case 3: The reaction phase is a fixed, heterogeneous, liquid-like catalyst (I). Two subcases with important issues are as follows:

3a. (I) belongs to the growing class of acidic or basic ion-exchange resin bead catalysts. This is a halfway house between homogeneous and heterogeneous catalysis. Gel-type resins resemble a viscous, homogeneous liquid catalyst phase. Pellet swelling is strongly affected by the reaction medium, and with it external fluid mechanics factors like pressure drop and bed expansion and internal effects like effective diffusivity. If the resin is "macroporous," it is close to a solid heterocatalyst (see Chap. 11).

3b. The ultimate in the logic of preferential retention of a liquid reaction phase is supported liquid-phase catalysis (SLPC). It combines the advantages of stable fluid mechanics and describable (quasihomogeneous), relatively stable reaction kinetics. SLPC, however, does not entirely escape from the limitations of fluid/fluid systems. Entrainment of catalyst off the support, loss of active catalyst ingredients by volatilization, and capillarity effects in holding the liquid catalyst in place while permitting ready access of reactants are some critical issues for performance. SLPC is more widely used than is usually appreciated: in oxidizing SO_2 to SO_3, in hydration of olefins to alcohols, in oxychlorinating hydrocarbons and the Deacon oxidation of HCl, alkylation of benzene, and olefin oligomerization. Some catalysts that we think of as solids are molten salts at reaction temperature.

Useful technical reviews of SLPC have been provided by Villadsen and Livbjerg (1978) and Villadsen (1981). A central consideration is the optimum loading of active material in (I); too much decreases reactivity by blocking diffusional access of the mobile phase (II) (gas or liquid). Maximum activity at 40–60%v pore volume fillage is typical (Hesse 1982). The diffusional resistance of liquid pockets in the pore structure depends on the pressure level and the nature of the mobile phase.

Maintaining SLP catalyst activity in the reactor by minimizing leaching or volatilization of ingredients into the product effluent may entail building up in the reactor vessel either P or by-products that foul or degrade the catalyst system. This imposes some limitations on catalyst formulation and on the choice of pressure and temperature. Some form of continuous or periodic online makeup of depleted catalyst is sometimes improvised to sustain a long catalyst run. Devising SLPC systems to heterogenize the widely used homogeneous Wacker oxidation and (metal/ligand) hydroformylation systems is a goal of some university researchers (Stegmüller and Hesse 1988; Scholten et al. 1985), but designers and academicians generally have given scant attention to SLPC as a standard catalyst type.

Micellar Reactions

In a Case 1 reactor productivity situation as defined above, with mass transfer limiting rates, interfacial area rather than preferential retention of the reaction phase is required. Micellar reactions provide that area while still giving us some freedom of maneuver in deciding which is the external phase. In micellar catalysis, a reactive counterion that is also a surfactant (RCS) draws the substrate reactant into intimate contact in a pseudophase of micellar "domains" of ionic aggregates. The catalysis becomes strongest as the RCS exceeds its critical micelle concentration (CMC) in the parent solution (Romsted 1984).

Example: An Unwanted Micellar Reaction

Micellar chemistry was made real for me only after it manifested itself in a system involving olefin reactions in a strong-acid liquid catalyst (I). A nuisance side reaction formed higher olefin oligomers by the reaction of protonated simple olefins residing just inside (I), with olefin oligomers in the adjacent hydrocarbon phase (II) (see Case 1c). The latter had quite low solubility in (I), so that the side reaction took place in the boundary layer of (I), with a rate proportional to the interfacial area between the phases. Our strategy for reducing the oligomerization side reaction was to minimize the emulsion-type interfacial area between (I) and (II).

This situation, which held for processing lower olefins, altered dramatically when we fed somewhat higher carbon number olefins. Here even the simple olefin had minimal solubility in (I), and the unwanted oligomerization "adopted" (if we can thus personify the adversary) a different strategy: (I) became the core of "inverse micelle" globules, surrounded by esters or alcohols of the olefin oligomers, whose long hydrocarbon tails gave the packing factor needed for inversion to hydrophobic micelles. These penetrated (II) to install widely available (I) *inside* the bulk of (II). Our strategy to reduce oligomerization by minimizing the interfacial area of bulk phases was no longer appropriate to the higher olefins.

Reactive Absorption of Impurities—Treaters

Many reactors do nothing more than remove an undesirable impurity from a gas or liquid by contacting with a reagent phase to absorb the impurity and convert it to a disposable species. Acid-base neutralization, oxidation, and chelation are common embodiments. Mass transfer efficiency aspects have been covered in Chap. 7. Providing intimate contact without incurring massive subsequent phase separation problems is the main fluid mechanics issue. There is a secondary issue relating to the aspect ratio of the two phases, which requires recirculation or tray design or baffling variants. If we are dealing with a neutralization reaction where excess reagent eliminates any equilibrium considerations, cocurrent flow options are opened up that simplify aspect ratio and other considerations. Some important chemistry issues are discussed in the next chapter.

High-performance Scrubbing and Treating

In cases where scrubbing is used to remove a component from a gas stream, a very high degree of removal (say, 99.99%) is sometimes required. A little thought will indicate that fluid mechanics alone may not provide the needed performance. For example, a typical distillation tray will have only slightly more than one gas-side transfer unit; therefore, only about 70% removal per actual tray would be expected, even if the chemistry of the scrubbing liquid ensured a very favorable absorption equilibrium and made liquid-side resistance to transport negligible. Eight such stages would be required for 99.99% removal. We conclude that in a balanced approach, fluid mechanics ingenuity (tray layout, etc.) should focus on gas-side resistance and bypassing effects, while chemistry should aim at substantially reducing transport resistance on the liquid side (see the discussions of reactive enhancement of diffusion in Chap. 11).

Interphase Resistance

There has been a continuing debate as to whether material at a L/L interface could constitute a significant additional barrier to mass transfer. It is certainly true that maximum transfer occurs during the formation of a liquid droplet (as in flow through an orifice) and drops off substantially as the interface ages. Today's more insightful (but not yet quantitative) treatment focuses on the dynamic viscosity and surface tension/elasticity of the interface. It has been shown, for example, that surfactant molecules at the G/L interface reduce absorption of SO_2 into aqueous caustic, not in terms of physical blockage but rather by stiffening the interface against turbulent deformations (Hikita et al. 1977).

On the other hand, reaction-assisted high absorption fluxes can produce turbulence cells that enhance absorption even for nonreacting components (Linek 1972). Janakiraman and Sharma (1982, 1985) have shown that micron-sized carbon particles can substantially enhance absorption of a gas into a scrubbing liquid. The mechanism is apparently adsorption of hydrophobic compounds on the particles, which, because of their small size relative to the diffusion film (50–100 μm thick), can function as either an extended interface or a Trojan horse, depending on one's taste in images. The effect is similar to the micellar acceleration of L/L reaction systems cited above. On balance, material that goes preferentially to an interface seems likely to promote transport unless it acts to stiffen it.

In Chap. 11 we deal with multiphase systems involving solids, completing the fluid mechanics story, and then pass on to chemistry-dominated systems, beginning with reaction-modified transport.

References

Alexandrov, A.S., et al. *Zh. Fiz. Khim.* 58 1265–7 (1984).
Barchilon, M., and Curtet, R. *Tr. ASME* 86 777–87 (1964).
Barton, D.H.R., and Howlett, K.E. *J. Chem. Soc. 1949* 155 (1949).
Benson, S.W. *Thermochemical Kinetics* Wiley (1976).
Bukhman, F.A., et al. *Kinetika i Kataliz* 9 970–8 (1968).
Buttersack, C., et al. *J. Mol. Catalysis* 38 365–83 (1986).
Canfield, R.C., and Unruh, T.L. *Chem. Engineering* 32–33 (1983).
Carberry, J. *Chemical Reaction and Reactor Engineering* Marcel Dekker (1987).
Chiltz, G., et al. *Chem. Revs.* 63 355–72 (1963).
Condon, F.E. *JACS* 70 2265–7 (1948).
Curtet, R. *Combustion & Flame* (London) 2 4 (1958).
Edwards, W.M., and Saletan, D.I. *Ind. Eng. Chem.* 59(3) 37–42 (1967).
Fărcașiu, D., et al. *JACS* 111 7210–13 (1989).
Gel'bshtein, A.I. *Dokl. Akad. Nauk* 132 384 (1960).
Gel'bshtein, A.I., et al. *Dokl. Akad. Nauk* 107 108–11 (1956).
Hamielec, A.E. *Chem. Eng. Commun.* 24 1–19 (1983).
Helfferich, F.G., and Chern, J.-M. Paper 54a AIChE Chicago Meeting (Nov. 1990).
Hesse, D. *Ber. Bunsenges. Phys. Chem.* 86 746–53 (1982).
Hikita, H., et al. *AIChE J.* 23 538–44 (1977).
Himmelblau, D.M., and Bischoff, K.B. *Process Analysis and Simulation* Wiley (1968).
Janakiraman, B., and Sharma, M.M. *Chem. Eng. Sci.* 40 235–47 (1985); 37 1497 (1982).
Jensen, W.B. *The Lewis Acid-Base Concepts* Wiley (1980).

Johnson, S.L. *Adv. in Phy. Org. Chem.* 5 237–330 Academic Press (1967).

Kotsarenko, N.S., et al. *Kin. i Kataliz* 29 190 (1988); 30 1117–22 (1989).

Kramer, G.M. U.S. Patent 4,560,825 (to Exxon); also *Tetrahedron* 42 1071 (1986).

Lebedev, Ya S., and Tsepalov, V.F. *Dokl. Akad. Nauk* 252 1424–7 (1979).

Levenspiel, O. *Chemical Reaction Engineering* Wiley (1972).

Linek, V. *Chem. Eng. Sci.* 27 627–37 (1972).

March, J. *Advanced Organic Chemistry*, 3rd ed. Wiley (1985) pp. 220, 452.

Nohara, D., and Sakai, T. *Ind. Eng. Chem. Res.* 27 1925–9 (1988).

Olah, G. *Superacids* Wiley (1985).

Ou, J.-J. and Ranz, W.E. *Chem. Eng. Sci.* 38 1005–13 (1983).

Pathangey, B., et al. Paper 80g AIChE New York meeting (Nov. 15–20, 1987).

Pines, H., and Stalick, W.M. *Base-Catalyzed Reactions of Hydrocarbons and Related Compounds* Academic Press (1977).

Popov, V.E., et al. *Neftekhimiya* 14 364–7 (1974).

Poutsma, M.L. *JACS* 87 2161, 2172 (1965).

Pritchard, J.G., and Siddiqui, I. *J. Chem. Soc. Perkins Trans.* 2 452–7 (1973).

Pryor, W.A. *Free Radicals* McGraw-Hill (1966), pp. 184–6; *C&EN* 46(3) 70–89 (1968).

Pujado, P.R., et al. *Hydrocarbon Proc.* 55(3) 91–6 (1976).

Pukhovsky, A.V., et al. *Intl. J. Chem. Kinetics* 17 1321–32 (1985).

Regenfuss, P., et al. *Rev. Sci. Instrum.* 56 283–90 (1985).

Reid, C. *JACS* 75 3264–8 (1954).

Romsted, L.S. *Surfactants in Solution*, Vol. 2, p. 1015, ed. K.L. Mittal and B. Lindman, Plenum (1984).

Rustamov, Kh.R., et al. *Zh. Fiz. Khim.* 29 1945, 2113–19 (1955).

Scholten, J.J.F., et al. *J. Mol. Catal.* 31 107–118, 371–83; 32 77–90; 33 119–28 (1985).

Shvets, V.F., and Aleksanyan, L.V. *Zh. Org. Khim.* 10 2027–30 (1974).

Sprow, F.B. *I&EC Proc. Des. Dev.* 8 254–7 (1969).

Stegmüller, R., and Hesse, D. *Chem.-Ing.-Tech.* 60 1074–5 (1988).

Stepanov, A.V. *Teor. Osn. Khim. Tekhn.* 3 934–7 (1969).

Sternling, C.V., and Wendt, J.O.L. *AIChE J.* 20 81–7 (1974).

Streitwieser, A., Jr., et al. *Comprehensive Carbanion Chemistry*, chap. 7, ed. E. Buncel and T. Durst Elsevier (1980).

Tanabe, K. *New Solid Acids and Bases and Their Catalytic Properties* Elsevier (1989).

Thornton, R., and Gates, B.C. *J. Catalysis* 34 275–87 (1974).

Villadsen, J. *NATO Adv. Study Inst. Ser. Review, Ser. E* 51 541–76 (1981).

Villadsen, J., and Livbjerg, H. *Cat. Rev. Sci. Eng.* 17 203 (1978).

Wehner, J.F., and Wilhelm, R.H. *Chem. Eng. Sci.* 6 89–93 (1956).

Westerterp, K.R., et al. *Chemical Reactor Design and Operation* Wiley (1984).

Yashima, T. *Petrotech* (Tokyo) 1(1) 27–33 (1978).

11

Classic Problem Elements: Heterogeneous Reaction Systems

Double, double toil and trouble Fire burn and cauldron bubble
                                               Shakespeare's Macbeth

We continue the discussion of fluid mechanics–dominated systems begun in the last chapter, with extension to systems containing a solid phase.

Three-phase Reactor Systems

Slurry Reactors (SR) and Trickle-flow Reactors (TFR)

A *slurry* reactor is one configurational answer to the need to contact a solid catalyst with L/G reactants, as in batch and continuous hydrogenations. Calderbank et al. (1959, 1961a, 1961b) offer correlations of both the relatively tractable solids suspension and the more demanding G/L dispersion/mass transfer issues. A SR has advantages in heat transfer to moderate exothermic effects and in the ability to employ primitive powder forms of catalysts instead of formulated pellets. It differs from a TFR in its CSTR behavior and high liquid/catalyst ratio. An important difference is whether the liquid is simply a vehicle for suspending the catalyst, with gaseous reactant feeds, or is the reaction substrate itself, as in hydrogenation of unsaturates. There are a number of insightful reviews (Hofmann 1983a; Satterfield 1970).

Excessive impeller speed for the sake of G/L contact can attrite catalyst. If the tasks can be split by holding catalyst in a cage in adequate contact with the liquid, addressing proper dispersion for G/L contact is simplified. Proprietary devices like the Buss reactor achieve intimate contact (a pump-around catalyst slurry absorbs gas in a Venturi-like entry) with, it is claimed, little attrition.

The TFR abandons mechanical agitation and seeks to manage with less intense contacting, with PFR behavior and relief from agitational complications as rewards. In addition to liquid/catalyst contacting, flow distribution is a key issue in scale-up and scale-down and in maintaining irrigation patterns over run time. Catalyst size and high liquid flows preclude countercurrency in a TFR. The gas phase is quasistagnant in flow. Ng and Chu (1987) review a number of critical fluid mechanics issues, such as wetting/pore penetration of the catalyst by liquid; other more process-oriented reviews include Satterfield (1970), Weekman (1976), and Hofmann (1983b).

Reactive Distillation (RD)

This is a new and interesting alternative to TFR layouts to address three-phase problems. In exchange for achieving countercurrent V/L movement to provide fractionation, complications are accepted in configuring the catalyst. The cage concept mentioned earlier takes the form of a fiberglass cloth wrapping in the CD-TECH proprietary technology for manufacturing methyl t-butyl ether (MTBE) over ion-exchange resin catalyst (Smith et al. 1982; 1984). Enhancing reaction rates to match distillation-type mass-transfer rates is essential to broaden the application of RD (see Chap. 15).

We now turn back to the multifluid-phase systems, begun in the previous chapter, for the case where chemistry predominates over fluid mechanics or interacts strongly with it.

HETEROGENEOUS REACTIONS WITH CHEMISTRY DOMINANT

Diffusion Enhanced by Reaction

This applies to systems removing a component A from a source phase (II) by reaction with a reagent B in a reaction phase (I). Four regimes are usually distinguished: For a very slow reaction (1), the absorption rate of A is governed entirely by chemical kinetics. As the reaction rate grows (2), the diffusion of A through the boundary layer (BL) in (I) is essentially in series with its reaction with B through the bulk of (I). As the reaction becomes a bit faster (3), the "resistances" are still in series, but transport of A through the BL is enhanced and much of the overall reaction occurs there. For an "infinitely" fast reaction (4), both A and B react to "extinction" in a plane within the BL. Here enhanced transport through the BL entirely defines the absorption rate of A.

Here we are concerned with a situation where the reaction has become fast enough to interact with diffusion through the BL in an accelerative way (3). This can be visualized as sharpening the concentration gradient in a way that is equivalent to thinning the BL[78] and diminishing mass-transfer resistance within (I).

This situation was opened up by Hatta in a little-noted publication in 1932, followed by Danckwerts and Pigford in the early 1950s, but came into general chemical engineering consciousness only with advances in numerical computing capabilities in the late 1950s. These computations provide plots showing the enhancement of mass transfer by reaction for varying reaction orders as a function of a modulus M, which embodies the ratio of chemical kinetics to diffusion (Astarita 1967; Brian 1964; Hikata and Asai 1964; van Krevelen 1958). The parallel to catalyst pore diffusion will be discussed below.

The enhancement[79] of the apparent mass transfer coefficient k_L for the BL inside (I) is shown in Fig. 11-1 for a second-order (irreversible) reaction between B, residing in (I), and A, migrating into (I) from (II). The abscissa M is defined as $\sqrt{kC_B D_A / k_L^0}$ where the second-order rate constant k has been converted to a pseudo-first-order one by multiplying by C_B. The enhancement grows with M but levels out at a value of roughly $1 + Z$, where the parameter $Z \equiv D_B C_B / D_A C_{Ai}$ measures the availability of B for reaction in the BL.[80] For a reversible reaction $A + B \rightleftarrows P$, the asymptote becomes $1 + KZ/(K + 1/C_{Ai})$, where K is the reaction equilibrium constant in reciprocal

---

[78] We use the quasi-steady-state film model here. It is simpler and gives virtually identical results (Chang and Rochelle 1982) to the more fundamental surface renewal representations of diffusion. There is a mild preference for the latter in a few selectivity cases.

[79] Over the "physical" coefficient k_L^0 ($= D_A/\delta$), predicted by hydrodynamic correlations. D_A is diffusivity in the BL of thickness δ.

[80] C_{Ai} denotes the equilibrium concentration of A at the interface with respect to (II). Note that the acceleration of mass transfer is confined to (I); transport of A out of phase (II) may become limiting.

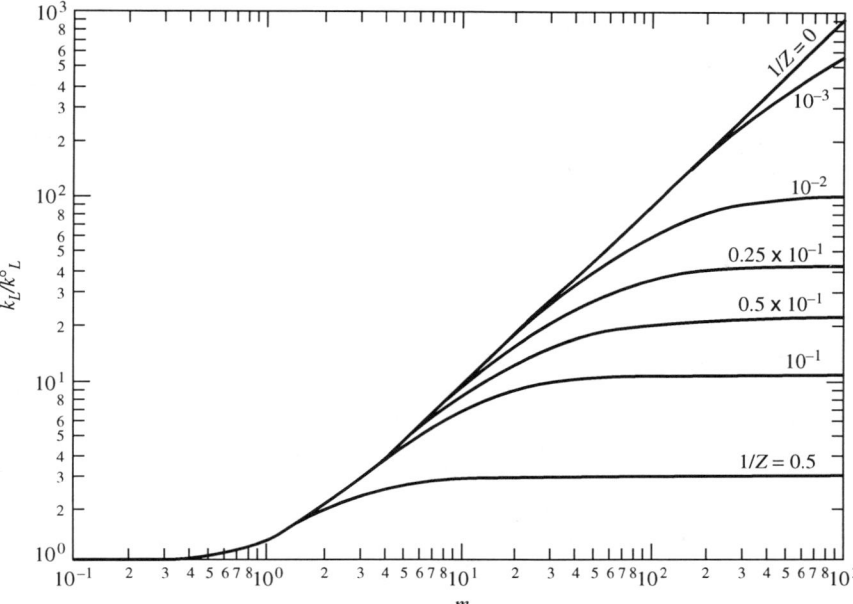

Fig. 11-1   Enhancement of Interphase Mass Transfer by Chemical Reaction

concentration units. I have used this in my own work to show strong differences in reactive absorption rates among n-butene isomers because of differing chemical equilibria.

A situation as simple as chlorine absorption in water involves reaction-accelerated diffusion (Brian et al. 1962, 1966; Vivian and Whitney 1947). The reaction in question is the hydrolysis of chlorine: $Cl_2 + H_2O \rightarrow HOCl + HCl$, which (via ionization of these acids) allows the water to store more chlorine equivalents than a simple Henry's law saturation would permit. Reactive acceleration rationalizes the strong effect of temperature on Cl_2 absorption via square-rooting of a chemical rate that doubles every 10° C. Of course, there are other reactive absorbents that can speed chlorine absorption even more aggressively, such as acidic $FeCl_2$ solution (Hikita et al. 1975).[81]

Academic workers sometimes use a reaction as a marker to infer the interfacial area in G/L or L/L contacting (Hikita et al. 1975).

---

[81] This oxidation reaction may account for much of the role of iron chloride in assisting the reaction of chlorine with ethylene, iron being oxidized by chlorine and reduced by ethylene in a redox cycle.

Reversal of this technique has been used to measure the kinetics of very fast liquid-phase reactions. By converting an apparatus with a well-defined interfacial area to absorption/chemical reaction, kinetics can be more easily mapped via the square-rooting effect on rate constant in M. Huq and Wood (1974) absorbed C_2H_4 into a laminar jet of aqueous HOBr to establish hypobromination kinetics in this way.

Absorption of Acidic Impurities By Basic Reagents

The base is used not just to neutralize the acid but also to accelerate its absorption, as indicated in the section on high-performance capture of SO_2. A "swamping" effect (see Chap. 4) acts to interfere with the capture of weak acids like CO_2 and SO_2 when more strongly acidic compounds are being simultaneously absorbed. The latter tend to react faster with the absorbing base and deplete it in the boundary layer where all the action is occurring, allowing some of the weaker acid to avoid capture. Some process schemes make use of such effects for example, to allow CO_2 to pass through while capturing nearly all of the stronger acid, H_2S (Gioia 1967).

Scrubbing of CO_2 has been notably improved by the chemical insight to use sterically hindered amines (Sartori and Savage 1983). Bulky alkyl groups next to the amine group destabilize a carbamate[82] ion intermediate, leaving more free base to enhance mass transfer. Mixtures of hindered and nonhindered amines can give a balanced system (see the discussion in Chap. 9 on the dangers of too good an absorption equilibrium) in terms of the need to regenerate the scrubbing liquor by stripping.

Two-step Chemistry to Improve Absorption of Acidic Impurities

On the other hand, poor capture of mercaptans heavier than methyl mercaptan MeSH by aqueous caustic stems not from acidity differences (which do play a role in MeSH vs. H_2S absorption) but from low solubilities of the higher mercaptans (Yabroff 1940). One former chemistry answer to this problem, use of "solutizer" solutions (where

---

[82] It is amazing that understanding of CO_2 absorption by amines is still incomplete! Sensitivity to the importance of the transition from carbamate to bicarbonate (see the COS story below), which underlay the hindered amine development, may be only a beginning.

added oxygenate organics like isobutyrate anion improved the solubility of mercaptans in caustic), has been displaced due to its high cost. One current alternative treating process (MEROX) removes mercaptans by absorption in caustic, with concurrent catalyzed oxidative coupling to convert them to disulfides; these spring back into the hydrocarbon phase, freeing caustic for further acid-base reaction (Leitào and Rodrigues 1990).

The difficulty in the capture of COS relative to other light sulfur compounds has been attributed by some to solubility differences. However, the best judgment seems to be that one step in the surprisingly complex chemistry, hydrolysis to the thiocarbonate anion, is favored by the use of dilute caustic (Phillipp and Dautzenberg 1965; Sharma 1965). There are also specific adjunct (nucleophile) catalysts, such as aluminate anion (Sample and Miller 1948), which accelerate this step.

Phase Transfer Catalysis

Suppose that, in a Case 1a situation (see the discussion at the end of Chap. 10) where limited solubility of R in (I) governs, we have agitated the system to ensure that R from (II) saturates phase (I) that is, fully played our fluid mechanics card. Chemistry, in the form of a phase transfer catalyst (PTC), can sometimes further enhance the rates (Starks 1980). Consider the hydrolysis of an organic chloride, R'Cl, to a carbinol in aqueous base (I) in which it is extremely insoluble. The PTC concept is a lateral thinking reversal. Instead of conveying R'Cl into (I) to react with the OH^- there, we bring the OH^- into the organic phase (II), where R'Cl resides, via a quaternary ammonium hydroxide carrier $R_4N^+OH^-$, whose bulky R groups confer significant solubility there. The PTC shuttles back and forth at the interface between the phases, exchanging an OH^- for a Cl^- in (II), then reversing the process with aqueous base in (I):

$$R'Cl + R_4NOH \rightarrow R'OH + R_4NCl$$
$$R_4NCl + NaOH \rightarrow NaCl + R_4NOH$$

PTC can be applied in a variety of physical systems (S/L, G/L, and L/L) in a variety of chemical reactions (general nucleophilic displacement reactions, oxidations, alkylation, etherification). Even when it enhances the rate substantially, there may be a number of drawbacks. The first is cost; even though the PTC agent is reused many times, like a true catalyst, its turnover rates are such that a significant (about 1%v) concentration usually is required, which is lost in the effluent. A second problem is that the very quality that makes the PTC agent an interphase shuttle also makes it a good detergent; postreaction phase separation may not be a trivial task. A third consideration is the toxicity of many of the best PTC agents, such as crown ethers. Diffusion/reaction interaction, with all of its complexities, is still present with PTC (Evans and Palmer 1980).

CATALYSIS ON A HETEROGENEOUS SURFACE

Choosing a Catalyst for a Reaction

Most publications on catalyst performance focus on the catalyst, with applications appended. There are fewer reviews that proceed from the desired reaction (including selectivity issues) to the appropriate catalyst. From a troubleshooting standpoint, our entry question will often be one like: Why am I using supported Ni for this hydrogenation, rather than a Cu chromite or supported Pd catalyst? The criterion may be a minor side reaction (e.g., decarbonylation in hydrogenation of a ketone over Rh).

Matching Substrate to Catalyst Functionality

Suppose that we need to shift the double bond of a branched olefin. Both acidic and basic catalysts have capabilities. The former give polymerization side reactions and would normally be avoided. However, an acidic catalyst may be preferred if we seek to create a t-olefin by shifting the double bond to the branch point because of the thermodynamic preference for a tertiary carbocation in protonating the olefin. (To maintain selectivity, we will want to provide a dual-catalyst system, in which the second functionality moves the t-olefin to the desired end product as rapidly as it is formed.)

If, on the other hand, we seek to move a double bond to the end (l-position), a strong base is preferred because of the greater stability of the terminal anion. Today we can find an appropriate strong-base catalyst for even the low acidities of hydrocarbons.

In PTS activities, a catalyst substitution may preempt our thinking. Yet these are subtle matters; unexpected side effects of such a retrofit are a prime source of new crises. I recommend checking the physical aspects of the catalyst, operating mode, and possible chemistry "patch" jobs, such as catalyst or vapor phase moderators, before making outright replacement the main line of attack (see the account of the unhappy experience with catalyst substitution in Chap. 3).

Correlations in Heterogeneous Catalysis Applications

Suppose that we seek to use an established catalyst system on a homologous reaction for which we have no data. Type II of Boudart's (1961) classification of catalyst correlations applies. Common patterns of Type II correlations with respect to the preexponential A and activation energy E of the Arrhenius rate constant are as follows:

1. A values are similar but there is a pattern of shifting E;
2. E values are similar, but there is a pattern of shifting A; or
3. Compensation applies: higher E values go with higher A values.

I have used such correlations to estimate interpolatively kinetics for hydrogenating a particular ketone from kinetics data for lighter and heavier ketones available in the literature. Even with no need for such a projection, such correlations can help us establish the physical validity of a pragmatic kinetic model.

A more complex "volcano" pattern is sometimes found in Boudart Type I correlations, where a single reaction is studied on a series of catalysts. This is a peaking rather than a monotonic progression of a kinetic parameter as the catalyst is varied. Such a reversal obliges us to reach for a physical conception of the activated complex of catalyst and substrate. Formic acid decomposition to CO over various metal catalysts is the classic example. The heat of formation of the metal formate was made the index for correlation.

One of the earliest and perhaps most famous Type I correlations plots activation energies for dehydrogenation and dehydration of alcohols over metal catalysts against an abscissa that tries to convey the energetics of the activated complex of the substrate with the surface sites (Balandin et al. 1958, 1960a,b). In its original form, the abscissa was simply lattice spacing of metal atoms. Today we find geometric parameters being used to correlate bifunctional (acid and base!) activity of mixed metal oxides as a "template" for the reactant molecule (Morrison 1977; Tanabe 1976, 1978).

Langmuir/Hinshelwood, Hougen/Watson, and their Successors

Forty years ago, modeling heterogeneous gas-phase catalytic kinetics by the methods of Hougen and Watson was at its zenith. By considering individual adsorption, surface reaction, and desorption steps and premising one of these as rate-limiting, alternative reaction models with three or four parameters were generated. Regression fitting, with constraints on parameters for physical reality, was used to single out one model as best. Later, a certain amount of pessimism about the "knowability" of catalytic kinetics set in.

People began drawing ± 2σ "confidence envelopes" around the defined H-W parameters and found that they were very broad. Also, awareness grew of how *heterogeneous* a real catalyst surface might be; there might be a range of sorption and activation energies, not single values. Finally, formulations consolidated around a single step might fail to hold over the temperature/pressure range required (see the mechanism shift in Takezawa and Toyoshima's (1966) study of NH_3 decomposition).

Nevertheless, I propose that PTS be carried out with kinetics models (when mathematical models are warranted at all) that are equivalent to a three- or four-parameter H-W model.[83]

Example: Alcohol Dehydrogenation in Heated Converter

Consider dehydrogenation of secondary alcohols to ketones, abundantly studied in the literature. Results over a number of catalysts, such as ZnO (Sheely 1964), brass (Hayashi 1957), and supported Cu (Lokras et al. 1970), indicate that a H-W model, with a numerator first-order in alcohol and denominator $(1 + \Sigma K_i p_i)$ with two "inhibition" terms for alcohol reactant and ketone product, is a minimal formulation.

We validate the denominator terms as follows: The alcohol term makes the reaction zero or negative order in alcohol at high partial pressure, matching data on initial dehydrogenation rate vs. pressure. The ketone term shows preferential adsorption of ketone, as confirmed by the falloff in initial rate when ketone is added to feed. But the *order* of the denominator returns us to the basic question: Is this a one-site or dual-site mechanism? While a few investigators

---

[83] The consolidation of multiple steps in one rate expression gives {1 + Kx} type denominator expressions both in LH-HW heterogeneous kinetics and in homogeneous free-radical sequences, although the denominator terms, of course, have different physical meanings.

have fitted their data with the former, our physical intuition almost demands multiple sites,[84] with adsorbed ROH passing an H atom to an adjacent vacant site to launch the surface reaction. Thaller and Thodos (1960), choosing the squared denominator term over brass catalyst, also found a shift in controlling mechanism. Below 650° F, the surface reaction controlled the rate:

$$r = \frac{kp_A}{(1 + K_A p_A + K_K p_K + K_H p_H)^2}$$

Above this, the desorption of product hydrogen controlled the rate:

$$r = \frac{kp_A}{p_K (1 + K_A p_A + K_K p_K + KK_H p_A/p_K)}$$

This makes intuitive sense in that increasing temperature usually makes the chemical rate outpace diffusion. Both formulations convey that the reaction is less than first-order in alcohol, except at very low pressure, and that the rate is inhibited by product ketone.

Of what use is such a model? One possible use is in optimizing the heat input pattern to a flow reactor; this endothermic reaction needs elevated temperatures to push an unfavorable chemical equilibrium. It is desirable to add heat to the reacting stream in a pattern matched to reaction rate to avoid local overheating. The kinetics here need be only good enough to link to heat transfer calculations as an endothermicity (heat sink) predictor.

Thus, for PTS, the kinetic formulation need not be the last word, but it must predict trends for the variable in question. In process support, I have seen a succession of limited scope "parochial" reaction models used on one process over a period of years to meet the changing scope of problems; this often makes more sense than trying to develop a complex, comprehensive model. Systematizations to make more consistent choices of reaction order, number of sites, and so on in pragmatic kinetic models have been offered theoretically by Bjornbom (1975) and experimentally, via radiotracers, by Happel and Walter (1986).

Reactor Scale-down

It is quite difficult to maintain similitude with the commercial scale even for a purely fluid mechanics issue in small-scale laboratory

---

[84] Kolboe (1967, 1969, 1972), in a careful study over ZnO, concluded that at least three active sites were involved.

studies. With complex intermeshing of chemistry with fluid mechanics, similitude is virtually impossible. We must choose a limited number of factors to honor. For example, we usually resort to crushing catalyst for use in small-diameter laboratory reactors. This escapes the heightened axial velocity near the wall that arises when there are only a few pellets per tube diameter; however, it may miss the internal diffusional effect present in full-size pellets. Radial heat transfer will also be altered from what it would be on a commercial scale. In addition (for a multitubular reactor), if we maintain catalyst pellet size and commercial tube diameter in laboratory work but reduce tube length in order to feed (and dispose of) less material, the lower velocity falsifies fluid/pellet convective contact effects. Weekman (1974) reviews the problems and options.

With abundant resources, we can study each of the fundamental elements separately, define reaction kinetics isothermally on a few catalyst pellets in a Berty (1983) or other laboratory differential reactor where contacting similitude is maintained, and then add thermal and other effects, via a mathematical model, to simulate the commercial bed. More often we would prefer to use a model inferentially, to derive kinetics from commercial data, inserting fluid mechanics and heat transfer as knowns. Existing simulations tools, however, leave much to be desired in this area.

Catalyst Structure

Supported Metal Catalysts

Economics makes it pointless to have a solid metal catalyst. In the case of a nonsupported metal catalyst like Raney Ni, we obtain a high degree of porosity by the method of formation. More often, we will be dealing with metal dispersed on an oxide support. In the case of an expensive metal like Pt, Pd, or Rh, high dispersion as small crystallites is desired for maximum utilization.[85] Note, however, that high dispersion of a low weight loading of a noble metal is purchased at the price of a large area of bare support, which may

---

[85] Rarely are isolated atoms desired. A crystallite of about 100 atoms, perhaps reflecting the value of certain geometric surface "kinks," is near-optimal. Yates and Sinfelt (1967) showed that Rh crystallites 1 nm in size (about 25 atoms) were obtained for ≤0.3%w loadings on SiO_2 but that maximum activity per gram of Rh required loadings of 0.5–1%w, giving 1.5- to 2.5-nm crystallites (about 100- to 200-atom clusters).

encourage unwanted side reactions. For an inexpensive metal like Ni, extensive coverage can be obtained to suppress such side effects. Where surface coverage is incomplete, experimental quantitation of active metal sites via chemisorption of hydrogen, CO, or oxygen is desirable and is superior to electron micrography. X-ray diffraction is suited only to larger crystallites.

After the state of organization of the metal on the support is understood, we may need some idea of the conformation of the reactant as it is chemisorbed on the metal site(s). In acetylene hydrogenation on Pd, is it laid down in the σ-bond position (i.e., flat) or in the ethylidene position (tail up) (Yates 1985)? This turns out to be critical to the role of H_2 in the oligomerization side reaction.

Why would we use a noble metal catalyst like Pd in preference to a Group 8 metal like Ni? Many chemical advantages could be cited: ability to hydrogenate the benzene ring, interaction with carbonyl group substrates, and so on. An idea of the interplays that may be involved is given in the following example.

Example: Hydrogenations of Doubly Unsaturated Compounds

Generally, noble metals will give a much higher ratio of the first hydrogenation rate constant to that of the second (to full saturation) in the sequence

$$D \rightarrow M \rightarrow S$$

than will catalysts like nickel or cobalt. This gives noble metals dominance in, for example, the selective hydrogenation of trace acetylenes in purification of individual olefins from steam cracking plants. Here complete conversion of low-level impurity D ($\equiv C_2H_2$, for example) demands very high selectivity for the first step to avoid significant conversion of the desired olefinic product M ($\equiv C_2H_4$) to the saturate S ($\equiv C_2H_6$). However if D were a major component that we wished to convert completely, mostly to M, but with good H_2 utilization, allowing some conversion to S, a noble metal catalyst would be the wrong choice. It might seem that holding the H_2/D ratio in the feed at slightly above unity would enforce the desired selectivity split. However, the very fast first-stage reaction threatens thermal runaway (discussed at the end of this chapter).

Example: Pore Diffusion and Slip in Hydrogenation of Acetylene

Acetylenes removal is a special case of a zero-slip requirement in reaction completion, since it rests on selectivity. Vendor literature for the Pd catalysts involved indicates a narrow window of operating temperature for maintaining essentially 100% removal of acetylenes. Lower temperatures permit slip by an incomplete reaction. Higher temperatures cause slip by making pore diffusion gradients steeper, diverting H_2 to olefin hydrogenation, as discussed under staging in Chap. 10. Later in this chapter, we will show how surface chemistry also plays a role in selectivity.

Bimetallics

Bimetallics are an interesting frontier in metal catalysts. These may be true alloys or sometimes just uniform vicinal deposition. Alloying of Pd can produce (at some loss in activity) a supported catalyst with improved selectivity in acetylene hydrogenation vis-a-vis sequential and polymerization side reactions (Léger et al. 1985).

Catalyst Supports

This is a broad subject; Satterfield's text (1990) is recommended for a full exposition. Alumina and, to a lesser extent, silica are the basic supports, but there are many others. Alumina is very versatile (Wefers and Bell 1972), but a bare γ-Al_2O_3 support that has a surface area on the order of 200 m^2/g and facilitates metal dispersion usually has acidic and basic functionalities (Malek et al. 1983) that may foster side reactions. Titania shares this characteristic. Many metal oxides are organized as spinel structures. Rather than view an attached -OH group (conferring base functionality) as an add-on, it is helpful to view something like γ-Al_2O_3 as a "defect" oxyhydroxide (Soled 1983), with water capable of moving into or out of the spinel structure, as in $H_2Al_2O_4 \leftrightarrow Al_2O_3$.

Thus, low-temperature calcination retains some water, yielding an alumina rich in basic sites, while high-temperature calcination provides acidic functionalities. Doping alumina with a base (Na, Ca), used to overcome support acidity, fuses these cations into the spinel structure. (Even Ni, supported on Al_2O_3, can become largely fused into the support as a spinel $NiAl_2O_4$.) An alternative to the γ form is α-Al_2O_3, which has a much lower surface area.

SiO_2 supports are more inert but less easily impregnated. They have low initial mechanical strength but tend to form a coarser pore structure under hot, humid conditions. Silica spheres, introduced by Shell (de Jongste 1984), are an interesting variant in terms of strength, shape, and medium surface area. Activated carbon is quite inert, but is erratic in impregnation and difficult or impossible to regenerate due to temperature and oxidation constraints.

Methods for fine dispersion of a metal catalyst have been reviewed by Andrew (1979), with the support viewed primarily as an obstacle to reagglomeration. He indicates the different effects of three preparation methods: wet precipitation, reduction, and impregnation. An alert troubleshooter will seek information on preparation basics from the catalyst vendor. Andrew's generalized time/temperature way of plotting metal agglomeration may be useful in troubleshooting a particular metal catalyst's deactivation cycle.

Arithmetic of Catalyst Structure

Pore structure is determined experimentally by penetration methods, using fluids such as Hg, oils, or N_2. Facility in cross-calculating catalyst parameters from such primary measurements is valuable for the troubleshooter. For example, if we are dealing with an activated carbon with a high specific surface of 1000 m^2/g, common values of internal void fraction ϵ and apparent pellet density ρ indicate an average pore diameter of about 1.2 nm. Such fine pores have implications for diffusional restrictions, capillary filling with liquid, and so on. This is why one feels more comfortable with support surfaces of 120–180 m^2/g. The arithmetic flows from an image of a pore as a long cylinder of diameter d whose surface, relative to the internal voids, is therefore $4/d$; the catalyst specific surface is $4\epsilon/\rho d$.

The arithmetic of catalyst loading on the support should also be familiar: If (for cost reasons) we have only 0.3%w Pd on an alumina support of 150 m^2/g, the support will be 99% bare, even assuming 0.1 nm^2 of surface coverage per Pd atom (monatomic dispersion):

$$\frac{\text{Covered}}{\text{Total}} = \frac{6 \cdot 10^{23} \text{ atoms/gmol} \cdot 0.1 \text{ nm}^2/\text{atom} \cdot 0.003 \text{ g Pd/g cat}}{150 \text{ m}^2/\text{g} \cdot 10^{18} \text{ nm}^2/\text{m} \cdot 106.4 \text{ g Pd/gmol}} = 0.01$$

On the other hand, a loading of 20%w, common with Ni catalysts, can give a high percentage of coverage.

Catalyst Dopants and Promoters

Much of this technology for activity or selectivity enhancement is proprietary. We distinguish *dopants* (low-volatility inorganic additives) from *vapor phase modifiers* (*VPM*), which can be added continuously or intermittently during a catalyst run. Alkali doping to reduce surface acidity has already been cited. Some other purposes of alkali doping (Mross 1983) include reducing the volatility of catalyst components; achieving a SLPC catalyst by lowering the melting point; helping coke removal by the water/gas reaction (KOH) (Brown et al. 1988); and improving the selectivity of metal oxide catalysts, as in the use of Cs on AgO catalyst for ethylene oxide synthesis.[86]

Even the strong acidity in zeolites exhibited by isolated alumina groups in a sea of silica can be considered an example of doping. The image recalls our visualization of infinite-dilution activity coefficients. Such metaphors can be valuable for understanding qualitatively key features of the technological terrain.

It is not uncommon to find low levels of classic catalyst poisons such as S, Cl, or CO used as moderators for performance improvement.[87] Thus S is used to enhance stability and selectivity of the Pt catalyst in reforming; volatile halogen compounds figure in ethylene oxide and acrolein synthesis; and trace CO is used in acetylene hydrogenation over Pd. There are, of course, cases where modifiers are used to enhance activity for the main reaction, at a loss in selectivity. Chlorination and hydrochlorination of alumina surfaces to enhance acidity is one example (Malinowski et al. 1983). Where a VPM gains over a dopant is in flexibility—being able to supply it, renew it, or withdraw it during a catalyst run. It is a frontier whose possibilities are far from completely realized.

Function of a CO VPM in Improving Selectivity

My exposure to use of CO as a VPM in C_2H_2 hydrogenation over Pd led me to inquire more deeply into the surface chemistry. The following treats fundamental insights from the literature:

---

[86] The utility of K_2O in the Fe catalyst for NH_3 synthesis has been attributed to easier electron transfers and a favorable effect on the strength of chemisorption (Satterfield 1990). Metal oxides used in catalysis are often semiconductors; it is reasonable that trace metal dopants might affect performance via electron availability.

[87] The most common case is a sacrifice in rate for a selectivity gain. This relationship results from the VPM occupying catalyst sites in a way that blocks a multisite side reaction.

A special case of VPM is the use of reversible CO_2 treatment to stabilize supported Ni, ammonia synthesis, and water/gas shift catalysts which are pyrophoric, for safe shipping (Krylova et al. 1988).

Example: "True" Surface Chemistry in Acetylene Hydrogenation

The quotation marks convey the unreachability of an ultimate understanding of surface chemistry in each heterogeneous catalyst system we confront. We have given examples of partial insights that may suffice for addressing side reactions (ethane formation and the fouling hydro-oligomerization) that are major process problems in hydrogenation/removal of trace C_2H_2 from ethylene. However, the work of Weiss (1977, 1981, 1984a,b) over many years has provided an in-depth understanding of the mechanisms in this important industrial reaction. Let us summarize his findings here and see how they increase our ability to solve real problems in this system:

Weiss has shown that there are three distinct modes of chemisorption of C_2H_2 that must be taken into account:

1. Dissociative adsorption to form surface $\cdot C \equiv CH$, which makes possible polymerization to C_4 + polymers;
2. Reactive ethylidene-type adsorption onto two vacant sites plus an adsorbed H atom to form

$$H_3C-C\overset{\displaystyle *}{\underset{\displaystyle *}{-*}}$$

which is responsible for the ethane by-product directly from acetylene;
3. Associative adsorption onto dual sites, from which the C_2H_4 desired is formed by successive reaction with adsorbed H.[88]

We start, then, with the goal of minimizing the first two modes. The first route is hindered by Cu or CO to decrease the required availability of adjacent Pd sites. The second mode is blocked by a C_2H_2 partial pressure ≥ 130 Pa, sufficient to blanket the Pd sites. However, Weiss showed that both bare alumina sites and deposited polymer can lead to the formation of ethane from ethylene by adsorbing ethylene, which can react with "spillover" H from Pd sites. (The polymer also harms selectivity by pore blockage/diffusion effects.) This discovery resolves a controversy between academic workers, who dwelled on "direct" formation of ethane from C_2H_2, and the sure observation of those in industry that more

---

[88] His experiments used radio-tagged atoms and Pd catalyst variants, work beyond what a single process support laboratory is likely to undertake. But one can take advantage of the open literature.

ethane was often formed than could be accounted for from C_2H_2 disappearance.

When C_2H_2 pressure falls below 100 Pa, on its way to extinction (or in the pore depth interior), CO can be helpful by competing with ethylene and hydrogen for Pd sites to restrain ethane formation (Karanth and Luss 1975; McGown et al. 1978). Weiss's work also established why reactions tend to be slightly above first order in H_2 but zero order in acetylene (as long as it blankets the sites). Selectivity thereby increases somewhat as H_2 pressure decreases (see the earlier remarks on staging hydrogen).[89] The contribution of such fundamental work on this system centers on the role of H_2 and of aging by fouling in affecting selectivity, and it explains why a degree of fouling that reduces selectivity markedly does not greatly affect the overall rate of acetylene disappearance.

Forestalling Catalyzed Coking by Sulfur as VPM

To shed more light on the control of side reactions by VPMs, we return to the chemistry of how S affects coke buildup in pyrolysis, begun in our discussion in Chap. 10 of steam dilution.

Example: Roles of S, Steam, and H_2 in Limiting Coking

The net formation of wall deposits in pyrolysis is the summation of tars (polynuclear aromatics) generated in the boundary layer of the main gas flow and adhering to the wall, plus coke generated catalytically by wall metals, less the removal of C by the water-gas reaction. Here will be concerned only with the last two factors. Fe, Ni, and other metals can catalyze both coking and decoking. The coke-forming process (see the discussion of deactivation below) and water-gasification are closely linked because the nature of the former creates metal inclusions in wall-generated coke "whiskers."

While a certain amount of decoking can occur noncatalytically by the water-gas reaction of graphite, catalytic decoking by metal inclusions can be desicive.[90] The relative levels of H, steam, and H_2S (as representative of S components) critically affect this catalytic

---

[89] Much earlier, Bond and Sheridan (1952) had shown that H_2 promoted both side reactions, fouling via a free-radical-like hydropolymerization, and ethane formation when Pd crystallites were large enough to store H_2 in a bulk β-hydride phase dissolved in the Pd.

[90] Baker et al. (1972, 1982) compare the effects of hydrogen, steam, and oxygen environments on Ni-catalyzed gasification of graphite.

augmentation of decoking. In simple dehydrogenations over supported metals, sustained catalyst life usually requires some critical partial pressure of H_2. This may work by forestalling occupancy by polyunsaturated entities that are coke precursors. In cracking, where the steam/oil ratio is typically 1/1 w/w (set by economics), the gross rate of coke formation exceeds the concurrent removal by gasification, but not by a huge margin. We may have gross C formation of 5–20 mg/hr/cm^2, with gasification of one-half to three-quarters of that amount. The resulting *net* rate of coke formation is therefore quite sensitive to process parameters.

If we take steam and H_2 as defining, respectively, the oxidation and reduction tendencies of the metals, lower net coke laydown at high steam levels may reflect both enhanced gasification and lower coke generation, the latter from surface metal passing into higher oxidation states. To assist steam in pyrolysis and in other systems where steam is not process-compatible, we turn to S, since Fe and Ni have a much greater propensity for sulfidation than for oxidation. The sulfiding of surface metal atoms for suppressing catalytic coke formation is described via equilibria such as the following:

$$FeO + H_2S \leftrightarrow FeS + H_2O$$
$$Fe + H_2S \leftrightarrow FeS + H_2$$

Chemisorption equilibria indicate that, at 400–900 K, a Ni surface[91] can be half-sulfided at as little as 1 ppb H_2S/H_2 (McCarthy and Wise 1980).

Many naphtha and all gas-oil feeds to steam pyrolysis contain enough S to sulfide surface metal heavily. The issue of optimal S level arises only for a sweet feed such as ethane. Why (aside from corrosion and removal costs) should we avoid using too much S?

Several factors are conducive to an optimum dosage. A modest sulfidation blocks the metal graphitization mode of generating coke in the tube wall without killing catalytic decoking by metal clusters in the coke whiskers (Trimm et al. 1981). Gasification should be concurrent with steam pyrolysis, lest encapsulation reduce accessibility of the metal inclusions for post-run decoking. With a perfectly sweet feed, however, an enhanced water-gas reaction may offset most of the increase in coking but form amounts of CO that pose down-

---

[91] Ni sulfides more readily than Fe (Rostrup-Nielsen 1968). Both are present in the alloy tubing likely used in steam pyrolysis furnaces. Note also that formation of a nickel sulfide surface layer is much more favored than that of bulk Ni_2S_3 (Bartholomew 1987; Katzer et al. 1982). Hence the use of bulk thermodynamic properties to estimate poisoning effects is not necessarily conservative.

stream processing problems, as in the use of by-product H_2. Exxon patents (Struth et al. 1972) using low S levels (3 ppm) appear to have a stronger technical basis than those advocating >50 ppm S.

ACCESSIBILITY/PORE EFFICIENCY IN HETEROGENEOUS CATALYSIS

The effect of diffusional retardation within a porous catalytic pellet upon reaction rates was developed independently by Damköhler, Thiele (1939) and Zeldovich at almost the same time. The Thiele modulus (U.S. terminology) ϕ from which the effect can be quantified includes path length L, pure reaction rate constant k, and effective diffusivity D of the reacting fluid within the catalyst. Catalyst effectiveness η (overall reaction rate relative to what would occur if surface concentrations prevailed throughout the catalyst) is given for a first-order reaction by [ϕ^{-1} tanh ϕ]. We can apply the pore efficiency concept to both gaseous and liquid reacting systems, provided that we use an appropriate diffusion coefficient (see below).

An enormous literature has provided a profusion of subcases of geometry or reaction sets. Perhaps the most valuable was the numerical solution for kinetics with Langmuir-Hinshelwood denominator (adsorption) terms (Roberts and Satterfield 1966). Other useful cases correct for counterdiffusion and for chemical equilibrium (Satterfield 1970). Let us recapitulate here a few simplifications in applying the pore efficiency concept to PTS.

1. Differences in pellet geometry can be bridged by using pellet volume divided by external surface as the characteristic dimension. Thus a one-sided slab of thickness L will translate into spherical pellet geometry by substituting $R/3$ for L.
2. The Weisz (1957) demonstration that observed rate dN/dt can be used as a quick test of the nonexistence of significant pore efficiency correction by

$$\frac{R^2 \, (dN/dt)}{DC_0} \equiv \Phi < 1$$

where C_0 is the surface concentration of reactant is much less difficult to use than the Thiele modulus criterion:

$$\phi \equiv (R/3) \cdot \sqrt{(k/D)} < 0.5$$

which requires inferring true reaction rate constants before use.
3. If we are well into the diffusion-retarded regime ($\phi \geq 2$), the efficiency can be approximated as $\eta \approx 1/\phi$. In this region, since the temperature effect on diffusivity D is usually slight, reaction kinetics *as falsified by diffusional retardation* will have an activation energy E about one-half of the true value for the reaction.
4. For gas flow through fixed catalyst beds, thermal resistance will usually be most significant between pellet surface and gas, while mass transport resistance will tend to be inside the pellet. Such external thermal resistance can result in enhanced reaction, for reactions with high exothermicity or activational energy, with the pellet perhaps 2° –20° C hotter than the gas.

The above defers exact calculation of pore efficiency in favor of rough sorting of the status of the diffusional retardation regime. There are cases where a fairly exact treatment of pore diffusion *is* needed as a subroutine of a numerical integration running in one or more major reactor dimensions (e.g., length and radius). Calculus inside calculus, which itself is inside some boundary value iteration, can be taxing even for today's digital computers. In such a case, good results can be achieved with the collocation approximation for converting the pore efficiency problem into an algebraic subroutine (Villadsen 1984).

Ameliorating Pore Diffusional Retardation

The economic thrust for a more active catalyst in a developing technology will tend to create a pore efficiency problem. The natural countermove, using smaller pellets, may encounter excessive pressure drop (for downflow) or fluidization constraints (for upflow). Shape innovations that maintain strength and pellet size while extending surface/volume reconcile these conflicting factors. First, rings were introduced as an alternative to spheres and cylinders. More recent innovation has brought the ceramic honeycomb (Danowski et al. 1988; Kainer and Reh 1989) used in automobile exhausts, trilobes (de Bruijn et al. 1981; Pereira et al. 1988), and other ingenious variants. Bunching active material near the pellet's exterior by preferential impregnation is another practical expedient (Smith and Carberry 1975). (This lowers the Thiele modulus even though k increases, since the un-square-rooted L term shrinks.)

Bifurcation and All That

There are certain systems (such as autocatalytic reactions and non-isothermal, highly exothermic systems) in which the reaction rate can be greatly enhanced by transport resistance. Their potential for instability and multiple steady states warrants attention (e.g., the automobile catalytic converter). But commercial reliability of a strategy that deliberately invokes diffusional resistance to increase the rate of a major reaction is likely to be low. The profusion of mathematically oriented literature in this area need not claim much of the troubleshooter's attention.

Diffusional Retardation of a Catalytic Reaction vs Reactive Enhancement of Mass Transfer

Although we speak of a retardation of reaction rate in catalyst pore efficiency vs an enhancement of mass transfer in a reactive liquid phase absorbing reactants from a second phase, their basic similarity is obvious. In both cases, absorption of the reactant increases only as the square root of the reaction rate.

Kulkarni and Doraiswamy (1975) appear to have been the first to try unifying the two in a common framework by considering a gas/film/bulk liquid system. Their parameters for graphical mapping are not easy to grasp physically. Hulburt (1976) has given a more compact algebraic formulation, with a map (see Fig. 11-2) that helps relate the many regimes to physically visualizable parameters. Only one new parameter[92] is added to combine the separate treatments in a smooth continuum from a gas/solid catalyst case to the gas/liquid diffusion/reaction system considered earlier in this chapter.

The following regimes are mapped on Fig. 11-2: For large β (top), we recover the Thiele formulation for porous catalyst efficiency E. For values of $m\delta > 0.5$ (RH edge), we are in the fast reaction regime

---

[92] To follow the unification effected in Fig. 11-2, one must understand the somewhat elusive β parameter. It is the ratio of the volume in the film boundary layer to the nonfilm volume of the bulk liquid. β ranges from very low values for wetted packing to infinity for a porous catalyst, which is inherently all film volume. Hulburt's $m\delta$ is in effect a Thiele modulus in which film thickness δ is the length parameter. The enhancement index E is the absorption flux relative to purely physical mass transfer, with no A yet in the bulk liquid. Hulburt gives guidance for iteratively estimating/defining the operative regime in a given situation. For example, k_L^0 ranges only from 0.01 to 0.1 cm/sec for a wide variety of contactors; hence δ ($=D/k_L^0$) will be in the range of 1 to 10 μm.

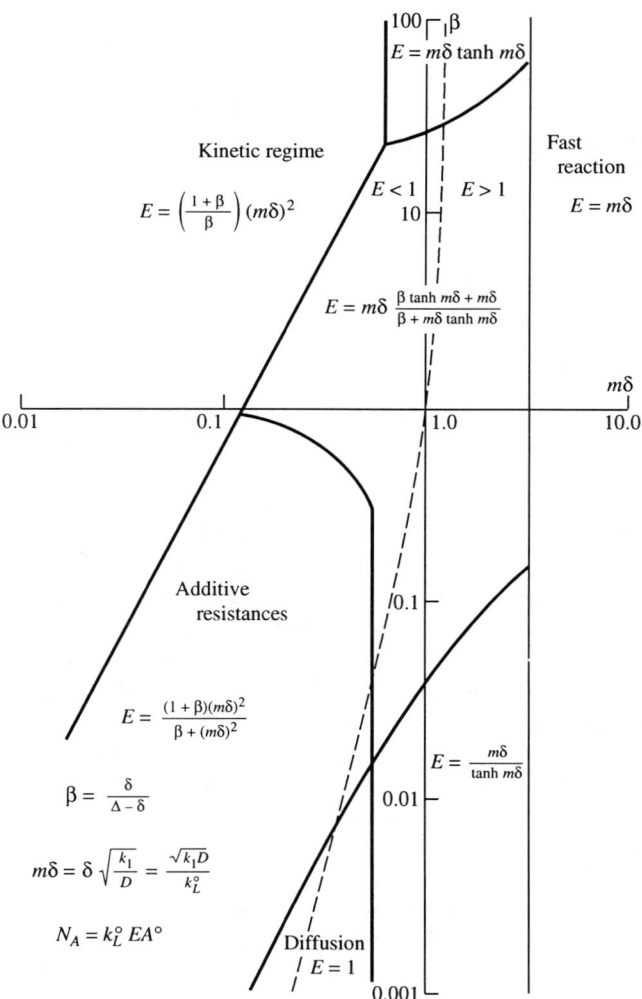

Fig. 11-2 Unified Mapping of G/L/S Diffusional-Reaction Coupling (From Hulburt 1976)

(regions 3 and 4 of the liquid diffusion/reaction interactions outlined near the beginning of the chapter), with transport enhancement $E \approx m\delta$. For very small β and $m\delta < 2.5$ we have pure diffusion control (regime 2), in the LR segment of Fig. 11-2, with enhancement E in the range 1–2. For very small $m\delta$, we have pure kinetic control (regime 1), in the UL segment of Fig. 11-2, with $E \ll 1$, although effectiveness in the Thiele sense is obviously near unity. Finally, in the lower center of Fig. 11-2, we have a zone of resistances in series (region 2 + 3) and $0.1 < E < 1$.

Selectivity Strategies in Diffusion/reaction Situations

Selectivity can present more difficult and commercially more important issues than gross rates limited by diffusion, where we have shown several response options. Wheeler (1951) gave the first treatment of how diffusional effects in fixed catalysts affect selectivity. His mathematical results for three categories of two-reaction systems make intuitive sense when we read them. In parallel reactions, *increasing* the spread between rates is possible only when the undesired reaction is higher order and precursor concentration is reduced by the diffusional gradient. In sequential reactions where the intermediate product is desired, selectivity is reduced.[93]

In dealing with sequential reactions in absorption/reaction or porous catalyst systems, where the initial product is the preferred one, we are simply trying to limit the damage.[94] In the latter case, surface-biased impregnation can help. Reducing catalyst activity uniformly in the pellet can be used in two ways: reducing the Thiele modulus throughout the reactor or doing so zonally, at hot spots where reactivity most outstrips diffusion.

The parallel development for diffusion in G/L absorption/reaction was begun by Chermin and van Krevelen (1961) and continued by van de Vusse (1966). The lowering of yield of the desired intermediate is similar to the G/S case. Sequential hydrogenations, oxidations, and chlorinations of organic substrates are common examples. In the case of hydrogenations in porous catalysts, the damage is amplified by faster relative diffusion of hydrogen into the pores.

---

[93] Surprisingly, the whole selectivity reduction occurs over $1 < \Phi < 3$.
[94] It is conceivable to prefer the final product in a sequential reaction. But we would use high conversions and/or a CSTR configuration to achieve this, not diffusional adjustment.

In the liquid systems, our options for retaining a selectivity close to that without diffusional limitation are broader than in porous catalysts, since there are four distinct regimes as reaction rate increases relative to mass transfer. We try for better mixing or take advantage of second-order kinetics by increasing the concentration of the substrate relative to the treating agent (Cl_2, O_2, etc.). This might include staged addition. I have been involved in a project where we deliberately slowed the reaction rate as a last resort.

Approximate analytical solutions that give us a basis for a selectivity strategy in liquid systems have been obtained by Onda and coworkers (1970) for a variety of competing reactions, parallel and reversible as well as sequential.

Strategies for Parallel Reactions in Porous Catalysts

For porous catalysts, the diffusional effect on selectivity has been extended by Roberts (1972) for a variety of parallel reactions. While there are only a few "openings" where diffusional effects can increase selectivity, we must be ready to capitalize on such opportunities. Thermal effects shift selectivity, just as in a homogeneous system, since any distortion of the concentration profile of the precursor is common to the two parallel reactions.

Estimating Diffusivity

We have already talked about defining the pore structure of catalysts. Let us show how to estimate effective diffusivity D_e in such structures and interphase transport parameters in absorption/reaction systems.

In experimental laboratory definition of D_e in a catalyst specimen, the steady-state Wicke-Kohlenbach method is the best procedure for gases. Yang et al. (1977) reviewed this method, suggesting a transient step integral procedure for D_e in the range 10^{-2}–10^{-8} cm^2/sec. If a catalyst can be studied only in a plant setting, estimation methods may have to suffice. Pulse techniques for measuring D_e, while temptingly convenient, are not reliable in either setting.

In porous catalysts, the usual "dusty gas" model treats the pellet as a continuum, with the gas (or liquid) penetrating it with a D_e lower than D for unconfined fluid by virtue of the geometry shift from pores to the pretended continuity of the pellet. D_e varies as about the 1.2 power of the internal void fraction ϵ and inversely with pore tortuosity τ; this reflects meandering that makes the pore path longer or shorter than the pellet radius used in a Thiele modulus (Satterfield 1990; Whitaker et al. 1980). If we lack a specific basis for estimating τ, the rule of thumb that $\tau \approx 10\epsilon$ can be invoked. Pore diameter enters the considerations in several ways:

1. Diffusion of gas molecules is hindered in the Knudsen regime, where the mean free path is long and molecules make frequent impacts on the pore wall. Low pressures make a Knudsen regime more likely; diffusivity here is proportional to pore radius, and mass flux becomes proportional to pressure. The threshold below which Knudsen diffusion takes over is about $P \cdot d = 100$, where P is in atm and d is pore diameter in nm. The transition is thus at a 20-nm pore size if pressure is 5 atm, but at 2 nm if pressure is 50 atm. Critical pore size grows with the square root of gas mole weight. Liquid diffusion has a crude parallel to the Knudsen regime based upon a different restrictive mechanism (Ternan 1987).

2. Capillary effects, whereby a concave meniscus of a liquid substantially lowers vapor pressure, may lead to condensation at temperatures and pressures where only vapor might have been expected (e.g., a 4-nm-diameter pore can lower vapor pressure by a factor of 3). In some systems, such condensation within the pores can cause an abrupt, sharp drop in catalyst effectiveness.

3. Decreasing pore size will not necessarily decrease porosity, but tortuosity will tend to increase, thereby somewhat reducing D_e, even if molecular size is not constraining diffusion.

Bulk diffusivity of a gas (for medium-sized molecules at ambient temperature) ~ 0.1 cm^2/sec at one atm and shifts inversely with pressure. Bulk diffusivity of a mobile liquid is four orders of magnitude lower, about 10^{-5} cm^2/sec. Absent better means, use of a D_e value within the catalyst pellet about one order of magnitude less than bulk diffusivity is suggested. However, we must always ask if the catalyst in question conforms to the image of a maze of undulating and interweaving hard-walled pores of circular cross section.

Diffusional Retardation in Resin Catalysts

For example a D_e value of 10^{-6} cm^2/sec may apply, as expected, in a macroporous ion-exchange resin acid catalyst for liquid reactants;

note, however, that this diffusivity applies only from the pellet's external surface to the microspheres that contain the sulfonic acid groups within them. A D_e value $<10^{-13}$ cm^2/sec may apply here from the highly cross-linked structure (Ihm and Oh 1984). Thus, even though the microspheres have a small diameter (3 nm), only those sulfonic acid groups at or near their surface may be available for catalysis. Microsphere pore efficiency, subject somewhat to swelling, is essentially in series with transport down the macropore. I recall cases where removal of an acidic impurity from a hydrocarbon medium by a macroporous anion-exchange resin required a space velocity 100-fold lower than for capture of the same impurity from a polar solvent that swelled the resin appreciably.

For a gel-type resin, the dusty gas continuum image is more valid than that of interweaving pores. D_e in the gel phase can be substantially lower than the 10^{-6} cm^2/sec typical of liquids in a porous solid. However, D_e, and hence catalyst efficiency, will vary widely with molecular size and solvent swelling action in these lightly crosslinked resins and much more strongly with temperature than in an inorganic catalyst. Duda and Vrentas (1979, 1980, 1981) give a useful orientation on free-volume theories of diffusion in polymers.

With inorganic solid catalysts, dimensions are relatively stable; the only parallel with resins is when we are close to a Knudsen-type regime, when D_e varies strongly with the size of reacting molecules. For example, H_2 may have substantially easier catalyst access than other molecules in a gas-phase hydrogenation, while heavy by-products in a liquid system may accumulate in the catalyst pores, with blockage or even equilibrium-limiting consequences. In zeolites, where even the mechanism of diffusion is in question, D_e is impaired three orders of magnitude from bulk gas-phase behavior, as a rough generalization, dropping further as molecular dimensions increase toward the critical size of the channels.

Heterogeneous Acidic and Basic Catalysts

Ion-exchange Resin (IER) Catalysts

Acid and hydrogen-transfer catalysts are the workhorses of catalytic conversion. The former, involving mostly homogeneous catalysts, has badly needed heterocatalysis because of corrosion and related problems. Early attempts at heterogenization used SLPC catalysts such as SPA, discussed in Chap. 10. Only with macroporous cation-exchange resin (CER) in H^+ form did a heterogeneous catalyst emerge that was suitable for broad liquid-phase use under moderate conditions.[95]

Anion-exchange resins (AERs) as basic catalysts, while occasionally useful, suffer from their limited stability against elevated temperature and oxidants. CER catalysis, particularly that provided by sulfonic-acid resins based on crosslinked polystyrene, is much broader. However, even these are constrained by temperature, losing their acid functionality[96] in months at temperatures $\geq 120°$ –$160°$ C (depending upon the reaction medium). Deactivation by stray cations in the feed is an obvious consideration (see the discussion of regeneration below). Depending upon the presence of polar components that detract from acid strength, they can give an acidity of 2–5 in negative Hammett acidity ($-H_0$) units, equivalent to 30–60%w aqueous H_2SO_4.

Perhaps their most successful application has been in synthesizing (MTBE), the methyl t-butyl ether additive to motor gasoline, from isobutylene and MeOH. The fit here is particularly good because the easily protonated $i-C_4H_8$ needs only modest acidity for commercially useful rates and thermodynamics allows high conversions to the Me ether at modest temperatures. Methanol, while an acidity depressant, serves to swell the resin usefully.

The beads of IER catalysts are smaller than typical inorganic solid catalysts. This does not prevent pore diffusion from being a limitation in many applications. Buildup of heavy by-product molecules can also reduce accessibility in some cases. However, there are applications where diffusional resistance enhances selectivity by trapping heavy by-products so that they build up within the resin to block further formation by equilibrium limitation. The small bead size is not optimal for pressure drop or for reactors where good radial heat transfer for cooling is required; upflow arrangements of resin catalyst beds are often used, even though some fluidization may occur in this mode. Carefully assembled gridwork supports like Johnson screens are a must for beds of resin beads.

---

[95] This is not to minimize the role of silica-alumina and zeolites in catalysis; however, a high temperature gas-phase environment adds complications that are unacceptable in many situations.

[96] Much better thermal stability than that of ring-sulfonated polystyrene has been shown by alternatives like sulfonated polyorganosiloxanes (Ono et al. 1987). Deleted acid groups can do additional damage by concentrating in a downstream location and causing side reactions.

Gel-type IER in Raschig-ring form, with better pressure drop/accessibility characteristics, seems to be no longer available. Another unfilled need is a medium-acidity resin, between the available sulfonic and carboxylic types. These gaps illustrate the limited vendor help we can expect in small-volume uses of products whose major application is elsewhere (here, water treatment). This is a general caution in catalyst work with commodity chemicals.

Metal Oxide Catalysts

With supported metals, the emphasis tends to be on H transfer reactions. With metal oxides, significant attention should go to acid-base behavior. For example, supported Ag catalyst used in oxidation reactions is covered with a monolayer of O. It has recently been recognized that the basicity of the Ag-O surface is important in determining the tenacity with which various reactants and products are held. For example, the relative acidities of two olefins may cause a strong difference in epoxidation selectivity.

Bifunctional Acid-base Catalysts

Tanabe and Nishizaki (1977) have pointed out the high productivity and selectivity of solid bifunctional catalysts, combining relatively weak acid and base sites in a particular "ensemble" geometry. Metal oxides are a common embodiment. Tanabe cites a GE patent for selective ortho-alkylation of phenol by MeOH: With MgO as a dual-site catalyst, phenol is anionically attached in a "cocked" fashion to strongly basic sites of the oxide, a position sterically favorable to ortho attack by an adjacent site on the MgO surface. On an alternative catalyst like silica-alumina, the excessive acidity of the alumina sites causes the benzene ring of phenol to π-bond and sit flat on the surface, a position not conducive to selectivity.

$BiMoO_3$ is a productive mixed oxide catalyst. Schuit and Gates (1983) use it as a platform to address generally multifunctional catalysis.

Deactivation of Catalysts

There is a variety of mechanisms for impairing catalyst activity. We can have external surface coverage, pore mouth blockage, pore diffusion restriction for certain larger molecules, site coverage, or chemical alteration of functionality. The first three stem from fouling deposits, which can originate in a side reaction within the reactor or from material entrained in with the feed. Chemical alteration of sites is a broad category; it includes feed poisons that chemically combine, in situ chemistry to defunctionalize the active site, or thermophysical effects like sintering of metal crystallites. The last is fortunately reversible in some cases (see the discussion of regeneration). Defunctionalization, as in desulfonation of a CER catalyst or volatilization of a SLPC key ingredient, is not.

Various empirical equations purport to describe the rate of deactivation of, for example, a supported metal or acidic catalyst. When these are simply formulas for the time trajectory of decay, unrelated to the environment factors, they are of no use to a troubleshooter, even deflecting her from the careful focus on problem definition that is particularly required in a complex phenomenon like catalyst deterioration.

S poisoning is more damaging than Cl, C, and O poisons because of the latter's ability to leave via gasification. The number of Ni sites reported to be deactivated per S atom varies widely, but two is a modal average. A fundamental approach is to study S chemisorption on metal crystallite faces (Peralta et al. 1981). True poisoning requires electronic-type bonding. It generally takes more than the presence of a single organic molecule on a site to produce foulant-type blockage. Bartholomew (1987), in a review of Ni poisoning (primarily by S but including Se, P, As, Zn, NH_3, halogens, and C_2H_2), attributes activity loss to a combination of blockage and longer-range electronic effects, the former dominating only at high surface coverages.

I give more attention here to the fouling mode than to loss of functionality and physical deterioration; the latter are not pleasant, but they are easier to diagnose and their remedies are straightforward, if costly. In searching for the definition of a fouling problem, we should try to move away from morphology, linking it backward in time to a formation chemistry, and forward to the exact way in which activity is lost. For example, the chemical reducing action of a carbonaceous deposit, rather than a blockage effect, might be deactivating a metal catalyst by preventing it from moving between valence levels.

In describing carbon deposits, we should try for a more precise label than a *coke*. There are three main forms—amorphous, filamentous, and graphitic—each with a distinctive mode of formation, which may not be distinguished in casual inspection (Baker and Chludzinski 1980). In general, both the nature of the precursor compounds and the catalytic environment in the reactor strongly influence the rate of fouling and the structure of the carbonaceous deposits.

For example, the formation of amorphous carbon may be significantly catalyzed by acidity of the catalyst or support. What began as a polymeric species created by such acidity, with about 2H per C, might go through several stages of H stripping/cyclization, perhaps with the aid of the very metal sites it is fouling, before it is observed as graphitic "coke," with perhaps 0.5 H left per C. Frusteri et al. (1987) have argued the converse: that coking on Pt reforming catalyst begins on metal sites by forming an unsaturated intermediate, which is then catalytically condensed by support sites.

Trimm and co-workers (1973, 1975) and Baker et al. (1972, 1982) began delineating the fascinating story of filamentous C generation in hot tube walls or catalysts containing Ni or Fe about 20 years ago. Olefinic hydrocarbons (particularly acetylenes) are active precursors, but even compounds like acetone can create C. The driving force for C filament growth is fairly well understood, although there is some dispute as to whether an unstable metal carbide phase is the coke precursor (Alstrup 1988; Geus et al. 1985; Yang and Chen 1989). While the phenomenon relates primarily to high-temperature reactors, the insights about Fe oxidation state and other chemistry factors in C formation (see the example in Chap. 2) have wide application.

Fouling of Cooled Reactor Effluents

The thermodynamic vulnerability cited in Chap. 10, whereby polymerization side reactions "take off" in cooling the effluent of a gas-phase reaction, need not involve free radicals.[97] This is one reason why designers have preferred direct contact quenches to indirect cooling. The thermal and mass dilution of the injected liquid, plus its dissolving effect on polymers, provides a cleaner design. Unfortunately, it is less energy efficient than indirect cooling, which can, among other possibilities, generate low-pressure steam.

---

[97] For example, HCl plus metallic corrosion products of the containing piping can constitute acidic polymerization promoters.

Minimizing Buildup of C Deposits

Let us review the options for slowing or forestalling the buildup of C deposits. These include using additives or enhancing latent tendencies for coke removal. In the latter category, metals with hydrogenolysis capabilities, like Rh and Ru, have some ability to gasify fouling C deposits in the presence of H. Even C dissolved in Ni metal lattice can be so gasified (Gaidal and Kiperman 1974). Among the additives, we have already discussed the role of VPMs such as S and steam in controlling the buildup of deposits. A new technology is developing to place Si and other elements into or on ferroalloy tube metal walls to interfere with the sequence that forms filamentous coke (Baker and Chludzinski 1980; McGill and Albright 1987). The role of such additives seems to consist in either reducing the solubility of C in the metal or acting as a barrier to decomposition of the intermediate carbide into coke.

In Situ Regeneration

In many cases, catalyst value relative to run length will not warrant the capital to provide the often elaborate facilities needed for in situ regeneration. Such spent catalyst is either discarded (entailing environmental problems) or sent to a reclaimer for regeneration or recovery of metal values in it. Assuming in situ regeneration, let us describe four categories of reactivation.

1. *Volatilization removal of coke and tars*
   a. *Burnoff of fouling coke by steam/air.* This operation involves complex effects not subject to direct observation; there are many ways to do it wrong. The troubleshooter needs to grasp three important factors: (1) the wave phenomenon whereby a temperature rise in a narrow zone can exceed manyfold the adiabatic temperature rise for consumption of all O_2 in the regenerant gas (Schulman 1963; Westerterp et al. 1988); (2) the transition from a reaction-rate-limiting to a diffusion-limiting regime; and (3) the difficulty of determining when C burnoff has been completed.

   The rate of C burnoff (in fraction per hour per atm of O_2 partial pressure) varies, according to the nature of the catalyst and support, 1000-fold (Satterfield 1990). The lower end of this range (4–40 at about 800 K), where the catalyst does not contain transition metals or otherwise assist burnoff, merges into the òxidation rate for graphite. Obstacles to O_2 access, like pore plugging or gas flow maldistribution, can slow burnoff. As temperatures exceed 900° F, the kinetics-controlled regime (activation energy of 150–170 kJ, rate proportional to the C mass and O_2 pressure) changes to a diffusion-controlled one.

The first stage of a managed burnoff will likely be O_2 starved at moderate temperatures, passing to a higher-temperature diffusion control, with $[O_2]$ still at low levels for control. Only in the final phase is high temperature combined with excess O_2. Inclusion of steam in the regenerant gas assists combustion (probably through the water-gas generation of H_2 and CO) and is not merely an alternative to N_2 for dilution of O_2. In the first phase, the tube wall support temperature may be about 800° F. With O_2 diluted all the way back to 0.5%v, rising ultimately to perhaps 2%v as the exotherm slacks off, any local temperature rises above this support temperature can be limited to <300° F, which is safe with respect to the tube wall and the catalyst.

Monitoring the CO_2 in effluent gas, sometimes suggested for determining the completion of C removal in the burnoff, is of doubtful value in this respect, but it can guide increases in temperature and $[O_2]$ to push combustion in the later stages of regeneration.

b. *Hydrocracking removal of tars.* A method known as *hot hydrogen stripping* is used in some hydroprocessing catalyst applications. It combines volatilization of tars with a significant degree of hydrocracking to convert foulant molecules to more volatile species (Somorjai et al. 1982). This approach requires early regeneration before sequential coking can occur.

2. *Chemical removal of functional poisons.*

   a. *Regeneration of IER catalysts.* Neutralization of basic or acidic sites on IER catalysts, by anions or cations, respectively, is reversible poisoning, but some constraining factor must be kept in mind: (1) Regeneration by passing an excess of caustic soda or acid through a resin bed is effective but not efficient; regeneration will be incomplete at economic usages of regenerant. (2) The equilibrium for regeneration of CERs to the H form from pickup of polyvalent metals can be rather unfavorable. (3) Loss of activity will be more than proportional to neutralization of functionality; from (1) and (3), practical run length may be a cycle between 75% and 50% of full functionality, with regeneration three times more often than a naîve design might provide. (4) Water and other impurities intrinsic to the regeneration reagents may complicate the in situ process (e.g., elaborate re-drying of the resin may be required to restore high activity).

Some forms of chloride poisoning of supported metal catalysts can be ionically regenerated by an aqueous caustic wash or with methanolic caustic when water contact is bad for the support.

b. *Removal of S from supported metal catalysts.* Because of the low partial pressure of S required for surface coverage of many metals, regenerating sulfided Ni in situ with H_2 is impractical. Oxidation can form the sulfate, a useful conversion in some cases but one causing unwanted acid catalysis of side reactions in others (Mathieu and Primet 1984). In one situation, I had hoped to obtain a pseudoregeneration of S-poisoned Ni by heating to allow migration of S from the surface into the interior bulk metal, until I understood how strongly thermodynamics favors surface sulfidation (Katzer et al. 1982). For other metals, such as Pt, greater but still incomplete regeneration from sulfidation is possible with H_2.

3. *Solvent removal of fouling tars.* This is not done as often as its simplicity warrants. A careful survey of solvent and temperature options for maximum solubility will lay the groundwork. More frequent regenerations will be involved but may considerably increase ultimate catalyst life.

4. *Chemical redispersion of agglomerated metal catalyst* a. *Oxidative breakup.* With aging, dispersed metals tend to agglomerate and lose activity, irrespective of fouling and other problems. A rule of thumb is that a temperature (K) equal to 0.4 of the bulk metal's melting point marks a threshold above which sintering by purely thermal effects is likely to dominate catalyst life (see Andrew's time/temperature plots cited above). This rests on a general observation that a microcrystallite may have an effective melting point half that of the bulk metal, with bulk properties inapplicable until we deal with clusters of at least 100 atoms. Of course, the presence of O_2 or Cl in the gas phase may also affect sintering.

With some expensive metal catalysts that have lost activity, the only recourse had been return to a reclaimer for chemical recovery from the support. However, in recent years, commercial in situ techniques for redispersion involving chlorination or oxidation have been developed (Birke and Engels 1977; Straguzzi et al. 1980). Oxidative redispersion seems to break up metal clusters by the expansion involved in conversion to the oxide (Dadyburjor 1980); a follow-up reduction with H_2 then restores active metal crystallites. A loss of effectiveness for Pt redispersion seems to occur at temperatures above 550° C (Fiederow and Wanke 1976), which supports the sintering rule of thumb cited above.

b. *Redispersion by mobility of metal chlorides.* Chlorination to secure fluidity and some volatility[98] of the agglomerated metal (via HCl, CCl_4, etc.), followed by reduction, has a secure place in re-forming and other Pt catalyst technology (Levinter et al. 1974). Many features of these regeneration schemes are kept as proprietary secrets and are not reported in the literature.

Reactive Removal of Impurities

Guard Beds

A sacrificial or regenerable bed to remove a feedstock impurity before it can poison a catalyst warrants more use than it gets. For example, the complexity of managing scrubbing of traces of H_2S or mercaptans is often allowed to stand in the way of protecting a supported metal catalyst from deactivation. The advent of activated ZnO that gives fair performance even at modest temperatures offers a new option for an unattended guard bed for H_2S. Removal of inorganic chlorides as trace HCl can be achieved in several ways; however, in my experience, HCl has a perverse tendency to "organicize" itself as RCl (e.g., by addition to a double bond) and to pass through beds or scrubbers designed to capture HCl. An innovative method for RCl interception is suggested in Chap. 12.

---

[98] Surface diffusion may be the metal redistribution mechanism (D'Aniello et al. 1988). Cl may have a second important role in decoking Pt/Al_2O_3 reforming catalysts: generating H atom spillover from the support (to initiate hydrocracking at Pt sites) when enough Cl (about 0.5%w) is present to displace about half of the $-OH$ sites on the alumina with Cl. At this level, sustained regenerability has been shown in methylcyclohexane dehydrogenation (Frusteri et al. 1987).

Trace metals and other poisons such as As, Hg, Pb, P, and V may be naturally present in feedstocks to supported metal catalysts. Upstream elimination may not be cost-effective; the emphasis must be on guard beds, often working on the same principle by which they poison the catalyst (e.g., amalgam formation). The guard bed differs from the reactor it protects in being small, run at ambient temperature conditions, and configured for ready removal and replacement.

Another impurity often requiring removal down to the sub-ppm level is O. The old laboratory standby for deoxygenating, hot Cu filings, is not adequate for current quality needs. Hydrogenation to water over Pd or Pt catalysts is workable, given an adequately controlled hot catalyst regime. A variety of supported Cu catalysts have recently become available, reputedly capable of bringing O to the ppb level.

A special type of guard bed that I believe has merit, but has never been implemented, is one based on a "sponge" concept. Suppose that a process system that must be kept extremely dry has a small but definite capability to reject entering water. If a slug of water enters during startup, this capability is overwhelmed and corrosion, catalyst failure, and other problems may result. However, if we had a bed with medium adsorption capability for water (e.g., a sulfonic IER in H or K form), it could take up the slug and then release it slowly into the system as poststartup dryness is established, at a rate consistent with the overall rejection capability of the system envelope. Prospective clients have been uncomfortable with the idea.

While we tend to classify a guard bed as a removal operation, akin to treaters or scrubbers, it can also be a reactor, converting the impurity to an innocuous form that remains in the process stream.

Reactive Purification (RP)

This unit operation merits broader use. It converts the impurity to a form that may not need removal or is easier to separate from the crude product. Placing it near the end of the process helps catch impurities formed in process; front-end placement can intercept feed impurities that impair process operability. Conversion of mercaptans to disulfides is a RP. See Chap. 12 for more examples.

Surface-catalyzed Degradation in Storage

A first-class operation all the way through synthesis and purification can still lose out by carelessness in product handling. I have seen cases where wall-catalyzed side reactions in a storage tank, a shipping vessel, or even a long line to marine shipment caused loss of quality for discriminating customers. Common offending reactions are isomerizations and self-condensation to heavy ends. A more subtle version of this problem occurs when a trace impurity, rather than the product, converts in storage to a compound entailing color, odor, or other quality violations (see Chap. 12). Particular metals and basicity or acidity associated with their oxide forms are typical catalytic agents. Modern sub-ppm analytical techniques, using metal coupons to simulate the surface/volume ratio of the real process, make such problems amenable to laboratory solution.

One solution, once laboratory simulations have confirmed the catalytic aspects, is some sort of lining. However, this can sometimes fail, retaining mechanical integrity but allowing permeation of product to the metal wall and of impurities formed back into the product stream. Occasionally the approach of "fighting fire with fire" is cost-effective, adding acceptable impurities (such as trace H_2O) to attenuate the wall chemistry producing the unacceptable impurity.

Reactor Runaway Situations

Catalyst coking or reactor effluent fouling can be self-aggravating in encouraging bypassing effects or pressure-related heightening of fouling side reactions. The following, however, are more striking examples of being felled by a lurking instability activated by the intersection of kinetics with fluid mechanics and/or heat transfer.

Example: Black Snowballs

In ethanol synthesis from gaseous ethylene and water over phosphoric acid catalyst on silica (SPA), conversion is equilibrium limited, so that the exotherm (a large excess of C_2H_4 is used) is modest and large adiabatic downflow reactors can be used. In this case, most of the design ingenuity was expended on corrosion issues related to the phosphoric acid catalyst or mechanical integrity of the catalyst support under the pressure drop involved in tall beds.

From time to time, an exceptionally short run was followed by great difficulty in unloading spent catalyst. It was found that large (black!) "snowballs" of catalyst had formed by consolidation of a large number of individual pellets. The explanation arrived at is as follows: Process humidity equilibrates the aqueous H_3PO_4 catalyst at modest (about 80%w) strength. As long as local temperatures are well regulated, the desired hydration is dominant; its reaction heat is picked up by the gas flow around each catalyst pellet. However, if flow maldistribution and/or agglomeration of fines create a catalyst clump exceeding some critical size, gas flow can no longer adequately cool the center of this clump.

With rising interior temperature, both the thermal effect and the related increase in aqueous H_3PO_4 strength accelerate the minor side reaction of ethylene oligomerization to major proportions (Ono and Sugiura 1972). This reaction is not equilibrium limited. Its added heat effect, combined with the further bonding of pellets by the oligomers, produced the final unobserved snowballing runaway (temperature monitoring was only at the bed inlet and exit). The only outward sign of such an event during the run was decreased hydration of C_2H_4, stemming from flow bypassing, lower Thiele efficiency for hydration of the superpellets, and the adverse effect of temperature on the hydration equilibrium. Curative action taken to counter this type of runaway included a more positive flow distribution for gas flow entering the bed and the use of catalyst pellets of greater strength.

A similar potential exists in trickle-phase hydrotreating, with much more serious safety implications, since the induced hot spots can weaken the wall of a high-pressure vessel. Here a hydrocracking side reaction provides the extra heat effect and it is liquid flow deviations around a pellet clump that initiate the local runaway.

Local Thermal Runaway in Fixed Catalyst Beds

Parametric sensitivity is the technical jargon label for this phenomenon in exothermic packed catalytic tubular reactors. A small increase in feed or coolant temperature or a reduction in feed rate produces a large jump in what was already a "humped" temperature profile (see Fig. 11-3a). It is technically not a runaway in that the temperature jump is bounded, but it is large enough to deactivate catalyst either physically or chemically and to stimulate unwanted side reactions. Chandler Barkelew (1959), a colleague of mine, was one of the first to quantify the parameters (in two dimensionless groups), defining the effective "edge" of such runaways.

Reaction heat simply outpaces the ability to remove heat through the flow reactor walls. High heat of reaction or activation energy promotes the effect. The threat of runaway is a real concern in ox-

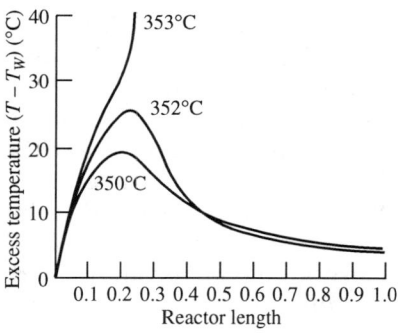
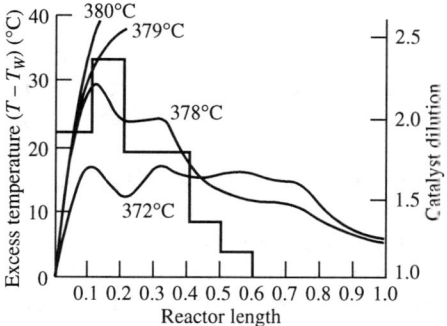

Axial temperature profiles for an undiluted catalyst reactor. O-Xylene oxidation to phthalic anhydride

Axial temperature profiles and catalyst dilution profile for a diluted catalyst reactor. From Calderbank 1969

Fig. 11-3 (a) Axial Temperature Profile: Unit Activity Bed; (b) Axial Temperature Profile: Graded Activity Bed

idation processes for maleic anhydride and phthalic anhydride, in oxychlorination of ethylene, and in other exothermic syntheses. Berty et al. (1989) have described a process combining laboratory tests and theoretical analysis to define the edge of runaway in C_2H_4 oxidation.

Solutions to Incipient Thermal Runaway

Improving the wall heat transfer coefficient would be stabilizing, but there is not much leverage to do so; a more viable approach is shrinking the tube diameter to increase wall surface. However, below a 5-cm diameter, complications from the large number of tubes entailed begin to outweigh the benefits. Surprisingly, catalyst activity per se has little effect in promoting runaway (unless it accompanies raising of the conversion in an integral bed), nor does raising the feed rate.

Reducing per-pass conversion is a defeatist solution to the runaway problem. A high-productivity strategy came from Calderbank (1969a,b) and others: to weaken catalyst activity only in the region where a unit-activity catalyst would produce a peak temperature while actually stepping up activity farther down in the bed where reaction driving force falls off from reactant depletion. These gradations of catalyst activity can be achieved by dilution with inerts or (preferably) by varying impregnation of the support. Calderbank's five-zone dilution pattern over the first 60% of a bed in the oxidation of o-xylene to phthalic anhydride is shown in Fig. 11-3b. Note that the runaway occurring in Fig. 11-3a at a jacket temperature of 353° C now does not occur until the temperature exceeds 380° C. For a jacket at 372° C, a quasi-isothermal path at a gas temperature of 388° C is obtained. A useful graded activity pattern need not comprise more than three or four zones; this approach has become standard technology in a number of exothermic catalytic reactions.

Recourse to a graded bed in troubleshooting is recommended in the proper context, but it may encounter certain counterarguments:

1. Operating people may resist institution of a graded bed in an established reactor, arguing that there is no need to provide the "activity notch"; nature will provide it by burning out activity in the high-sensitivity area by the incidence of a runaway. The rejoinder to this objection (aside from real safety issues) is that fouling and phase-change effects accompanying hot-spot runaway can produce harm not limited to the immediate hot-spot region and lasting throughout the remainder of the catalyst run.
2. Pirkle and Wachs (1987) and others argue that a graded bed is vulnerable to relatively small operating disturbances. For example, a small dip in feed temperature may move the exothermic peak down the reactor tube, where higher-activity catalyst has been loaded. This issue can be adequately addressed via the control system.

Other Catalyst Configurational Issues

Catalyst Packing Issues in Small-diameter Tubes

Suppose that we have used 3.8 or 5.1-cm-diameter tubes to improve heat removal in an exothermic catalytic reactor. Conventional wisdom advises a pellet diameter <1/10th of the tube's internal diameter to avoid channeling effects;[99] here, an opposing constraint appears in the reduction of radial conductivity by small pellets.[100] Picture radial heat transfer as a random walk involving flow division

---

[99] With larger pellets, the voids near the wall restrict radial heat transfer, as well as encouraging flow channeling.

[100] There may thus be an optimal pellet size with respect to heat removal. I am indebted to Bob Hawthorn for raising this issue.

around pellets rather than conduction through the pellets and fluid. The suggested compromise is the use of 0.3-cm pellets. Only with catalysts like 0.5-mm IER beads may radial conductivity decrease become limiting.

Catalyst Loading Precautions

To achieve uniform flow division among a large number of parallel packed 5-cm-diameter tubes, it is worthwhile to take special precautions in catalyst loading to secure uniform packing density. I have seen situations where each of several thousand tubes was tamped, topped off with extra catalyst, and individually flow-tested when a long catalyst run was at stake. Even in large-diameter reactors, loading catalyst by free settling in an inert liquid will secure denser, more uniform loading than free fall from a "sock" feeder.

Stacked Bed Inlets

The insight on the packing size effect on radial heat transfer extends to evening out axial flow profiles as well. Recognizing that early flow maldistribution can be self-perpetuating, the stacking technology of catalyst beds puts coarser pellets on the top to gain a quick initial distribution, switching to finer pellets (for pore efficiency) farther down. Multipoint source feed distribution at the entry is, of course, also essential; there are some innovative ideas (beyond simple notched weirs) for achieving this. It should be appreciated that diffusion of liquid in a bed is not symmetric; the wall gap acts as a sink, so that outward spreading is not reflected inward. Therefore, in tall beds, periodic catch points and redistribution onto the next section are a necessity.

Bed Supports

This seemingly minor factor can be a source of real trouble. For example, a gridwork system like Johnson screens can be an excellent solution for supporting a fine catalyst like IER beads. However, the apparatus is vulnerable to minor commissioning errors[101] that can allow whole beds to percolate away in a matter of weeks.

The traditional stacking of support balls of increasing size downward below the bed (the inverse of the inlet stacking described above)

---

[101] This refers to gasketing and other problems in coping with common defects like out-of-round in a large-diameter column.

is often adequate but is awkward in beds subject to frequent changeout. Another requirement of such support balls is chemical inertness. In one instance, aluminum balls combined with leachings from the process catalyst to cause fouling problems and catalyze exit side reactions. Here as with fillers used as diluents in (e.g., graded-activity catalyst) beds, inertness should be double-checked in a laboratory rather than rest on vendor assurances.

Traditional reactor engineering focuses on major components and the thermal and flow issues that emerge from mainstream chemistry and fluid mechanics. In the next chapter, we turn to the chemistry and physics of trace reactions and the resulting impurities that govern an elusive area of quality and operability whose rules are much less well known.

References (References already cited in Chap. 10 are not repeated)

Alstrup, I. *J. Catal. 109* 241–51 (1988).
Andrew, S.P.S. *CHEMTECH 9*(3) 180–4 (1979).
Astarita, G. *Mass Transfer with Chemical Reaction* Elsevier (1967).
Baker, R.T.K., and Chludzinski, J.J. Paper 18b, AIChE Chicago meeting (Nov. 16–20, 1980).
Baker, R.T.K., et al. *J. Catal. 26* 51 (1972); ibid *77* 74–84 (1982).
Balandin, A.A. *Adv. Catal. 10* 120 (1958); *Dokl. Akad. Nauk USSR 132* 577–80 *133* 1073–6 (1960a); *Kinetika i Kataliz 1*(1), 5–14 (1960b).
Barkelew, C.H. *Chem. Eng. Prog. Symp. Ser. 55* No. 25 37–46 (1959).
Bartholomew, C.H. *Stud. Surf. Sci. Catal. 34* 81–104 Elsevier (1987).
Berty, J.M. *Laboratory Reactors for Catalytic Studies, Applied Industrial Catalysis*, Vol. 1 pp 41–66 Academic Press (1983).
Berty, J.M., et al. *IEC Res. 28* 1589–96 (1989).
Birke, P., and Engels, S. *Z. Chem. 17*(2) 77–8 (1977).
Bjornbom, P.H. *IEC Funds. 14*(2) 102–6 (1975).
Bond, G.C., and Sheridan, J. *Trans. Far. Soc. 48* 651–68 (1952).
Boudart, M. *Chem. Eng. Prog. 57*(8) 33–41 (1961).
Brian, P.L.T. *AIChE J. 10* 5–10 (1964).
Brian, P.L.T., et al. *AIChE J. 8* 205–9 (1962).
Brian, P.L.T., et al. *Chem. Eng. Sci. 21* 551–8 (1966).
Brown, L.F., et al. *J. Catal. 109* 329–46 (1988).
Calderbank, P.H. *Brit. Chem. Eng. 14* 1199 (1969a); *Chimie. Indust. 2* 101 (1969b).
Calderbank, P.H., et al. *Trans. Instn. Chem. Engrs. 37* 173–85 (1959); *39* 363–8 (1961a); *Chem. Eng. Sci. 16* 39–54 (1961b).

Chang, C.-S., and Rochelle, G.T. *IEC Funds.* 21 379–85 (1982).
Chermin, H.A.G., and van Krevelen, D.W. *Chem. Eng. Sci.* 14 58 (1961).
Dadyburjor, D.B. *Stud. Surf. Sci. Catal.* 6(Catalyst Deactivation) 341–51 (1980).
D'Aniello, M.J., Jr., et al. *J. Catal.* 109 407–22 (1988).
Danowski, F., et al. *CFI Ceram. Forum Int.* 65(11–12) 486, 489–91 (1988).
de Bruijn, A., et al. *IEC Proc. Des. Dev.* 20 40–5 (1981).
de Jongste, H.C. *Spec. Chem.* 4(2) 42–45 (1984).
Duda, J.L., and Vrentas, J.S. *AIChE J.* 25 1 (1979); *J. Appl. Polym Sci.* 25 1297 (1980); *ACS Div. Polym. Chem. Polymer Preprts* 22(1) 111–13 (1981).
Evans, K.J., and Palmer, H.J. Paper 63b *AIChE Chicago meeting* (Nov. 1980).
Fiederow, R.M.J., and Wanke, S.E. *J. Catal.* 43 34–42 (1976).
Frusteri, F., et al. *Stud. Surf. Sci. Catal.* 34 197–208 Elsevier (1987).
Gaidal, N.A., and Kiperman, S.L. *Kinet. i Katal.* 15 954–62 (1974).
Geus, J.W., et al. *J. Catal.* 96 454–90 (1985).
Gioia, F., *Chim. Ind. (Milan)* 49 921–7, 1287–93 (1967).
Happel, J., and Walter, E. Paper 84g AIChE Miami Beach meeting (Nov. 1986).
Hayashi, K. *Kagaku Kogyo* 8(9) 722–29 (1957).
Hikita, H., and Asai, S. *Int. Chem. Eng.* 4 332–40 (1964).
Hikita, H., et al. *Chem. Eng. Sci.* 30 607–16 (1975).
Hofmann, H. NATO ASI Ser., E 73 Vol. 2 171–97 (1983a); 73–97 (1983b).
Hulburt, H.M. *AIChE J.* 22 604–7 (1976).
Huq, A., and Wood, T. *Chem. Eng. Sci.* 29 31–6 (1974).
Ihm, S.-K., and Oh, I.-H. *J. Chem. Eng. Japan* 17 58–64 (1984).
Kainer, H., and Reh, H. *Keram. Z.* 41(8) 576, 579–81 (1989).
Karanth, N.G., and Luss, D. *Chem. Eng. Sci.* 30 695–8 (1975).
Katzer, J.R., et al. *J. Catal.* 76 369–84 (1982).
Kolboe, S. *IEC Funds.* 6 169–74 (1967); *J. Catalysis* 13 193–214 (1969); ibid. 27 379–88 (1972).
Krylova, A.V., et al. *Appl. Catal.* 39 325–31 (1988).
Kulkarni, B.D., and Doraiswamy, L.K. *AIChE J.* 21 501–6 (1975).
Léger, G. et al. *Hydrocarbon Proc.* 64(3) 51–9 (1985).
Leitão, A., and Rodrigues, A. *Chem. Eng. Sci.* 45 679–85 (1990).
Levinter, M.E., et al. *Neftepererab. Neftekhim. (Moscow)* 1974(4) 4–6.
Lokras, S.S. et al. *IEC Proc. Des. Dev.* 9 293–7 (1970).
Malek, J., et al. *Appl. Catal.* 7 159–68 (1983).
Malinowski, S., et al. *Can. J. Chem. Eng.* 61 93–7 (1983).
Mathieu, M.M., and Primet, M. *Appl. Catal.* 9 361–70 (1984).
McCarthy, J.G., and Wise, H. *J. Chem. Phys.* 72 6332 (1980).
McGill, W.A., and Albright, L.F. *Oil & Gas J.* 85 (34) 46–50 (1987).

McGown, W., et al. *J. Catal.* 51 173 (1978).

Morrison, S.R. *The Chemical Physics of Surfaces* Plenum Press (1977).

Mross, W.-D. *Catal. Rev. -Sci. Eng.* 25 591–637 (1983).

Ng, K.M., and Chu, C.F. *Chem. Eng. Prog.* 83(11) 55–63 (1987).

Onda, K., et al. *Chem. Eng. Sci.* 25 753–68, 1023–31 (1970).

Ono, R., and Sugiura, T. *Intl. Chem. Eng.* 12 520–5 (1972).

Ono, Y., et al. *J. Mol. Catal.* 43 41–50 (1987).

Peralta, L., et al. *Surf. Sci.* 104 435–47 (1981).

Pereira, C.J., et al. *Appl. Catal.* 42 47–60 (1988).

Phillipp, B., and Dautzenberg, E.H. *Z. physik. Chem.* 229 210–24 (1965).

Pirkle, J.C., and Wachs, I.E. *Chem. Eng. Prog.* 83(8) 29–34 (1987).

Roberts, G.W. *Chem. Eng. Sci.* 27 1409–20 (1972).

Roberts, G.W., and Satterfield, C.N. *IEC Funds.* 5 317 (1966).

Rostrup-Nielsen, J.R. *J. Catal.* 11 220–7 (1968).

Sample, G.E., and Miller, W.B. U.S. Patent 2,434,868 (Jan. 20, 1948) (to Shell Oil Co.).

Sartori, G., and Savage, D.W. *IEC Funds.* 22 239–49 (1983).

Satterfield, C.N. *Mass Transfer in Heterogeneous Catalysis*, MIT Press (1970); reissued by R.E. Krieger (1981).

Satterfield, C.N. *Heterogeneous Catalysis in Practice* McGraw-Hill (1990).

Schuit, G.C.A., and Gates, B.C. *CHEMTECH* 13 556 693 (1983).

Schulman, B.L. *Ind. Eng. Chem.* 55(12) 44 (1963).

Sharma, M.M. *Trans. Far. Soc.* 61 681–7 (1965).

Sheely, C.Q., Jr. "Kinetics of Catalytic Dehydrogenation of Isopropyl Alcohol" *Univ. Microfilms* Ann Arbor, Mich (1964).

Smith, L.A., and Huddleston, M.N. *Hydrocarbon Proc. Intl. Ed.* 61(3) 121–3 (1982).

Smith, L.A., Jr., et al. *Proc. Intersoc. Energy Convers. Eng. Conf.* 19(2), 998–1002 (1984).

Smith, T.G., and Carberry, J.J. *Can. J. Chem. Eng.* 53 347–9 (1975).

Soled, S. *J. Catalysis* 81 252–7 (1983).

Somorjai, G.A., et al. *J. Catal.* 77 439–59 (1982).

Starks, C. *CHEMTECH* 10 110–17 (1980).

Straguzzi, G.I., et al. *J. Catal.* 66(1) 171–83 (1980).

Struth, B.W., et al. U.S. Patent 3,641,190 (to ESSO) (Feb. 8, 1972).

Takezawa, N., and Toyoshima, I. *J. Phys. Chem.* 70 594–5 (1966).

Tanabe, K. *Tetrahed. Lett.* 43 3861 (1976); *J. Org. Chem.* 43 1660 (1978).

Tanabe, K., and Nishizaki, T. *Proc. 6th Intl. Cong. Catal.* 2 863 (1977).

Ternan, M. *Can. J. Chem. Eng.* 65 244–9 (1987).

Thaller, L.H., and Thodos, G. *AIChE J.* 6 369–73 (1960).

Thiele, E.W. *Ind. Eng. Chem.* 31 916 (1939).

Trimm, D.L., et al. *Carbon 13* 189 (1975); *J. Catal. 29* 15–19 (1973).

Trimm, D.L., et al. *J. Chem. Tech. Biotechnol. 31* 311–16 (1981).

van de Vusse, J.G. *Chem. Eng. Sci. 21* 631–43, 645–54, 1239 (1966).

van Krevelen, D.W. *Chemie-Ing. Technik. 30* 553–632 (1958).

Villadsen, J. In *Front. Chem. Reac. Eng. 1* 262–83 (1984), ed. L. Doraiswamy, John Wiley, New York.

Vivian, J.E., and Whitney, R.P. *Chem. Eng. Prog. 43*(12) 691–702 (1947).

Weekman, V.W. *AIChE J. 20* 833–40 (1974).

Weekman, V.W., Jr. *Chem. React. Eng. Proc. 4th Int. Symp.* 615–46 (1976).

Wefers, K., and Bell, G.M. "Oxides and Hydroxides of Aluminum" (Tech. Paper No. 10) Alcoa Research Labs, East St. Louis, Ill. (1972).

Weiss, A.H., et al. *IEC Proc. Des. Dev. 16* 352 (1977); *J. Catal. 72* 185–98 (1981); ibid *86* 417–26 (1984a); *J. Mol. Catal. 25* 131–40 (1984b).

Weisz, P.B. *Z. phys. Chem. 11* 1 (1957).

Westerterp, K.R., et al. *Chem. Eng. Technol. 11* 367–75 (1988).

Wheeler, A. *Adv. Catal. 3* 313–22 (1951).

Whitaker, S., et al. *Chem. Eng. Sci. 35* 10–16 (1980).

Yabroff, D.L. *Ind. Eng. Chem. 32* 257–62, 950–3 (1940).

Yang, R.T., and Chen, J.P. *J. Catal. 115* 52–64 (1989).

Yang, R.T., et al. *IEC Funds. 16* 486–9 (1977).

Yates, D.J.C., and Sinfelt, J.H. *J. Catal. 8* 348–58 (1967).

Yates, J.T., Jr., et al. *J. Catal. 96* 1–11 (1985).

12

Classic Problem Elements: Trace Chemistry

An idealist is one who, on noticing that a rose smells better than a cabbage, concludes that it will also make better soup.

H.L. Mencken

Importance of Trace Chemistry

Trace chemistry, an obscure, even undefined, field 20 years ago, has become a major factor in modern process technology, due to environmental sensitivity, the thrust for product quality, and demands for higher standards of process reliability. More time is spent, or should be spent, in policing components that may amount to 1 ppm to 0.5% of the yield structure than is spent upon the major components that define yields. What is the reason for this disproportionate effort?

Trace Impurities and Product Performance

There used to be a strong distinction in chemical manufacturing between *specification* and *performance* products. The suitability of the former for sale was vouchsafed by their chemical analysis alone; the latter could only be judged by end-use performance. However, product returns and process reruns due to customer dissatisfaction were untenable for both producer and customer over an extended period. Increasing sophistication allowed end users to identify their performance problems with specific compounds, often at trace levels (*trace* here connotes any level between 1 part-per-trillion (ppt) and 0.1%w). Special requirements were set by the manufacturer for shipments to particular end users.

In this way, performance requirements were gradually converted back into analytical specifications, but at the price of sustaining a variety of grades and corresponding product inventories. The logic of the situation has led back to single grades, leveled up to the needs of the most demanding users; the need to maintain a multitiered quality standard proved too disruptive of the manufacturing process. Thus, all ethylene consumers may get a product with <1 ppm of O or CO, even though not all need such purity.

Trace Impurities and Process Performance

Environmental/toxicological issues also brought sensitivity to trace impurities. New attention was given not only to emissions of major components, but also (not always rationally) to known classes of toxicants present as trace impurities in commodity chemicals. Thus, residual styrene monomer in polystyrene and ppm levels of benzene in acetone came under pressure.

The effect of these changes has been to increase attention by technologists to the origins and pathways within their process of a host of ppm impurities. With this has come a recognition of self-interest beyond incremental sales in reducing or excising key trace impurities. Trace chemistry was found to figure in a host of operability issues within the process envelope.

Trace components, via surfactant effects, can markedly affect distillation stability and throughput, as well as reaction pathways. They determine catalyst performance through fouling and functionality changes of active sites. Self-sustaining cycles of trace chemistry can cause corrosion in unexpected places. Coping with odor or color problems from ≤1 ppm impurities in a finished product can force a reduction in production rate. Components thought to have been rejected via fractionation reappear through aliasing chemical transformations. Stabilizers added in one section of the process become poisons to a catalyst in a subsequent section.

Only recently has this awareness spread to management and process development R&D personnel. Therefore plant support people have had to learn or guess the chemical genesis and pathways of the trace impurities in their process, without benefit of the fundamental investigations routinely accomplished for the major components in the original development of an established process. Very little of such knowledge is surfacing in academic work or in the patent literature. Many trace chemistry phenomena are so central to the commercial practicality of a process that companies elect to keep them as trade secrets rather than seek patent protection.

How Trace Impurities Originate

The examples that follow illustrate genesis and transformations of objectionable trace impurities and methods for eliminating them from a process. Occasional vagueness about compound identity has been used to protect the proprietary context. I emphasize in-process origin or transformation of trace impurities, since there is already general awareness of impurities entering with feeds.

In-process Genesis of Trace Impurities

Frequently impurities arise as isomers of desired reaction products. Insight into the mechanism often comes only after we discard organic chemistry rules that are not without exception. A ready example is the formation of small amounts of the normal alcohol in the hydration of olefins, along with the sec-alcohol that the Markovnikov rule leads us to expect will be the sole product. This is a more serious problem in the direct heterogeneous catalytic hydration processes for synthesizing isopropanol than in the earlier indirect hydration process. The reason for this is that the former operate at higher temperatures, and whatever energy barrier resists protonating the primary carbon of C_3H_6 is lessened by rising temperature (in the usual Arrhenius formulation).

Once we see that a rule can have exceptions, we are prepared for the formation of some n-propylbenzene (NPB) isomer in the acid-catalyzed alkylation of benzene by C_3H_6 to make cumene for phenol manufacture. Here the NPB engenders other impurities that end up in coproduct acetone. One reason zeolite processes have had difficulty gaining a foothold in cumene manufacture is that their higher synthesis temperature increases selectivity to NPB.

Another example of bending traditional mechanism rules is the Cannizarro reaction, which deals with aldehydes self-condensing in the presence of base to form esters. (This leads to organic acids as objectionable impurities in several processes.) According to this rule, only aldehydes with no hydrogens on the α carbon (e.g., formaldehyde, benzaldehyde) undergo this reaction. However, in my experience, it can occur at trace levels with other aldehydes, an acidity headache "betrayal" by nature.

A seemingly simple choice among acid catalysts can make a striking difference in by-products stemming from a particular acid. Aqueous H_3PO_4 is used at elevated temperatures to catalyze the hydration of ethylene to ethanol and the alkylation of benzene by C_3H_6, when a lower strength (and/or temperature) of H_2SO_4 would suffice to give the desired reaction rates. This choice avoids the greater oxidizing power of H_2SO_4, which would cause quality and operating problems from oxygenated and sulfurous by-products.

Example: Trace Impurities Formed on Bare Support of a Catalyst

Heterogeneous catalysis also can create problem impurities out of major components. In a process involving vapor-phase hydrogenation of a ketone, a vendor urged the substitution of his "more advanced" Ni catalyst for the "primitive" one in established use. He cited the higher Ni content and greater surface area of his catalyst as making more Ni available for catalysis, with a greater tolerance for surface poisoning by S impurities in the feed. What he did not note was that his catalyst also had a much higher surface of bare γ alumina support. Fortunately, we tested the new catalyst in the laboratory before committing to a plant trial. It was indeed more active initially, but it proceeded to self-poison rapidly by the support-catalyzed condensation of the ketone feed to heavy ends.

Self-engendered Catalyst Poisons: Red Oil

Fouling is not the only situation in which a catalyst can destroy itself by side reactions. Strong base catalysts can engender acidity formation (see the above discussion of the atypical Cannizarro reaction). Strong acid catalysts can create their own basic poisons out of hydrocarbons alone. This process I term the *red oil* problem, from the characteristic color and ionic properties of the hydrocarbons that arise to poison strong acid catalysts.

Conjunct Polymer in Aliphatic Alkylation

Sulfuric acid, used as a catalyst for creating high-octane gasoline components by alkylating isobutane with lower olefins, is discarded as its acid strength drops below the critical minimum (92–94%w) for acceptable rates. While some of this decline is indeed caused by accumulation of water and absorption of feed S compounds, a major part is caused by a class of hydrocarbons loosely termed *conjunct polymer (CP)*, formed by side reactions (Albright 1972). CP appears to have a cyclopentene nucleus, with olefinic side chains; it probably arises by a cationic analog of the free-radical cyclization of dienes and olefins described under postpyrolysis fouling reactions in Chap. 10.

This constellation of purely hydrocarbon groups apparently creates the minimum basicity to interact ionically with a strong acid. A less basic hydrocarbon "polymer" would stay in the alkylate layer and leave the reaction system. A fair amount of CP is present in spent acid. However, I include it under trace chemistry both because CP is formed incrementally from minor hydrocarbon impurities (such as dienes) and because of its mechanistic novelty, little known to practitioners of the technology. Better understanding of CP chemistry should lead to exclusion of key hydrocarbon impurities in feedstocks and perhaps to novel ways to purify and recycle spent alkylation acid.

Trace Impurities in Feeds

The naive technologist looks upstream for the source of impurities. I have deliberately begun this chapter with self-engendered impurities to counter this tendency. However, a critical impurity often does enter with the feed. Elemental O and S or Cl compounds are common examples. Analytical recognition of the problem compound is the first step toward a solution. However, it is sometimes technically difficult to get adequate interception at the outset.

I recall one startup, bedeviled by fouling induced by trace O_2 intrusions, which was not completed until virtually every flange had been encased in a plastic N_2-purged bubble!

In some cases, it is not cost-effective to remove all the impurity in the feed, and interception farther down the process train is sought. The difficulties with this strategy are as follows: The impurity undergoes parallel and sequential chemical transformation within the process, leading to a variety of compounds, of which only a few affect performance; this spreading out renders an already low concentration even lower and further impedes analytical tracking.

S Feed Impurities

For example, product odors are often associated, for good reason, with S feed impurities. In one case, where front-end treatment had removed 95% of 10 ppmw S equivalents in the feed, we were seeking an in-process treatment to ensure good product odor. The 500 ppb of total S entering the process had fanned out, after the primary synthesis step, into dozens of individual compounds. With attenuation in processing, the ultimate malodorous impurity could have been at the sub-ppb level in the finished product! The analytical burden was severe; the nose is much more sensitive than the best GC detector.[102] Only when we concentrated the product impurities in laborious distillations to bring them up to the ppm level, where they were identifiable by gas chromatography/mass spectrometry (GC/MS), were we able to devise an effective tail-end treating system (but see GC/MS limitations in Chap. 13).

Trace Impurities with Leverage on the Main Reaction

Free-radical (FR) Reaction-inhibiting Effects by Trace Feed Impurities

The scope of FR chain reactions has been dealt with in the last chapter. Trace components can have strong leverage in chain-stopping transfer reactions, such as the effect of trace C_3H_6 in converting active propagation radicals into the allyl radical, which is in a resonance energy well and cannot continue the chain reaction.

Trouble from an Accelerating Impurity

But a trace impurity need not figure simply as an inhibitor.

---

[102] Building upon this, clever investigators install diversion valves so that GC peaks can be passed to an "olfactory" detector, their *noses*, to verify which are odor contributors.

Example: Finding an Unseen Villain

One of the more thrilling troubleshooting ventures of my career came in a several-year pursuit of a trace impurity that was causing damage by accelerating an initiation reaction substantially, even at levels well under 100 ppm. This compound, chlorine monoxide (Cl_2O), was surmised to be an impurity in Cl_2 feeds to reactions, not on the basis of analyses (we never obtained a direct analytical method at the low levels involved), but largely by inferences from thermodynamics. Cl_2O splits much more readily than Cl_2 into a $Cl\cdot$ and a $ClO\cdot$ radical, which initiate chains as does the fission of Cl_2 into atoms, but at much lower temperatures.

This trace activator caused harm by initiating a FR reaction at an unfavorably low temperature, where selectivity was impaired. The insight afforded by quantitative estimates of dissociation energy and kinetics was enough to overcome our analytical blindness and sketchy understanding of how it originated as a trace impurity in the chlorine feed. We found a workable method inside the Cl_2 finishing train to remove from crude Cl_2 a compound that we had never directly detected, except through its final effect on synthesis selectivity, and one unknown even to the industry grouping: the Chlorine Institute (Saletan and Chun 1981)!

Transmutability of Impurities in the Course of Doing Damage

S Compounds

I was nonplussed, early in my career, to find caustic soda and sulfuric acid both in use to remove S compound. We must make distinctions among S compounds on the basis of acid/base behavior and state of oxidation before proceeding to technical judgments. Consider as classes H_2S, mercaptans (RSH), thioethers (RSR), disulfides (RSSR), CS_2, COS, and thiophene. The first two have definite acidity (hence capture by caustic) and move up to elemental S and RSSR, respectively, when oxidizing conditions are present. Yet they are nucleophilic enough to add to double bonds in basic media to create new and different odor bodies.

The thio ethers, RSR and RSSR, are basic enough (see the discussion of the H_2SO_4/diethyl ether system in Chap. 9) to cleave to and interact with other moieties in the presence of a strong acid like H_2SO_4. RSSR can also cleave under thermal/free radical conditions.

Thus, a mercaptan converted to a more innocuous disulfide by a front-end treatment like the MEROX process (see Chap. 10) may come back to bite us if exposed to cleavage chemistry later in the process. In typical refinery-generated feedstocks, at least three different classes of S impurities may be present. This is one reason why removals under 99% are obtained in most front-end treatments.

Iron as a Trigger for Undesirable Trace Chemistry

Fe is a ubiquitous problem source. Its role in tube wall coking has been treated in Chap. 11. Chap. 2 described the roles of Fe_2O_3 and Fe_3O_4 in coke formation on catalysts. In Fe oxide catalysts for ethylbenzene dehydrogenation to styrene, components such as K are added to counter coking by promoting the water-gas reaction.

The ability of trace Fe to reach the overhead in distillations has puzzled me for some time. Entrainment explains some instances. Volatile compounds like triallyl iron are a more remote possibility. In most cases, however, it is likely that HCl, generated locally, is responsible by attacking ferrous metallurgy in the overhead.

The transmutations of Fe from the simple oxides through mixed oxychlorides to $FeCl_3$, the catalytic form involved in much trace chemistry, is subject to classic thermodynamic equilibria in terms of the partial pressures of HCl and H_2O involved (Schäfer 1949, 1959, 1952). $FeCl_3$ is often aggregated as the dimer Fe_2Cl_6, which is relatively nonpolar, is freely miscible with hydrocarbons, and has just enough vapor pressure to allow trace Fe to reach unexpected places.

In chlorinated hydrocarbon systems, $FeCl_3$ can be a nuisance, catalyzing unwelcome dehydrochlorination/rehydrochlorination rearrangements. It can complex with double bonds (particularly as $H^+FeCl_4^-$, with HCl formed in reboilers), causing unwanted polymerizations and color bodies. Chelating agents[103] can often tie up Fe if suppression at the source is not feasible (Alternative metallurgy can be an expensive solution.) A supplementary tactic is to use HCl-scavenging compounds; epoxybutane, volatile enough to act in vapor spaces, can be particularly useful.

---

[103] A table of chelation stability constants should be at hand to plan sequestering of trace metals. Note, however, that these agents have a reputation for picking up new metal from wall deposits.

Aliasing

Aliasing is a special vulnerability of distillation from reversible conversions of impurities on the trays, which thwarts their expected rejection. For example, in stripping light carbonyl impurities overhead from a crude reaction product, reversible formation of an acetal or hemiacetal or hydration to a diol seems to be involved (Buschmann et al. 1980). Such forms will have a different (generally heavier) volatility, which permits the carbonyl compound to continue to travel with the product, later reverting to its original form as an impurity in the finished product. The phenomenon is a broad one; halogenated carbonyls in particular furnish striking examples. Chloral/chloral hydrate equilibrium is labile in the 60°–100°C range (depending upon water activity), permitting aliasing under normal distillation conditions. Water or alcohol is usually involved to shift the chemistry toward the alias species. Sørenson and Jencks (1987) have reviewed the acid- and base-catalyzed reversion of the hydrate and hemiacetals of acetaldehyde.

Disproportionation Reactions Affect Distillation Volatilities

Chemistry-altered volatility can stem from disproportionation reactions with a reagent present in a distillation column. Consider the case of trace mercaptans RSH reacting with NaOH to form the mercaptide salt (NaSR); both mole weight and the distillation medium affect the outcome. In an aqueous environment, volatility of higher RSHs is enhanced by a lowered solubility that outweighs their lower vapor pressure. Their acidity is too low to reduce volatility by the proportion of the RSH tied up as NaSR. But in a hydrocarbon environment, the higher RSHs will be retarded both by lower vapor pressure and by greater compatibility of the NaSR with a hydrocarbon medium. The effective stripping factor is KV/L for the free RSH times the proportion of RSH *not* tied up by the NaOH. If this factor >1, the RSH will still move up the column, even though 90% of it is tied up as NaSR at a given point.

Metathesis Reactions

Metathesis reactions can also create a volatile unwanted version of a trace impurity we expected to be rejected as a heavy end in a distillation column. Examples include weak acid phenolics attacking the Na salt of a strong light organic acid (e.g. formic acid) to regenerate the volatile free acid (RCOOH) and metal chloride corrosion residues reacting with lower alcohols to form metal alcoholates and send volatile organic chlorides to locations where they can cause further complications.

Transmutation of Silicones

In Chap. 3 we described how a startup team, stymied by an unexpected foaming problem in a primary synthesis step, introduced a silicone antifoam agent with little prior thought and almost immediately killed catalyst activity in a downstream conversion step to the final product. In the major crisis this caused, I was called in as part of an R & D investigatory team and found myself hurriedly learning silicone chemistry.

Example: Documenting Unexpected Transmutation of Silicone Additive

The silicones were fed at ppm level to counter the native surfactants present at somewhat higher levels. It was expected that the polymeric silicones would be rejected in stripping the crude primary product overhead, although no real attention had been given to the ultimate fate of silicone in the system. High Si content in the scale fouling a sample of dead catalyst convinced us quickly that Si was volatilizing with the primary product in some fashion. At the same time, consulting organic chemistry texts convinced us that we had the conditions for substantial depolymerization at the bottom of the primary product stripper to a cyclic dimethylsiloxane tetramer that would distill overhead.

Laboratory catalytic conversion tests with pure primary product, with and without tetramer spiking, and current plant primary product confirmed the tetramer as a plausible vehicle for damaging the catalyst. In the laboratory, we went on to devise a leaching procedure for regenerating the Si-encased catalyst and to find an alternative defoamer that could be used on an interim basis. The leaching procedure developed was soon found to be a useful alternative way of regenerating the catalyst even under normal operating conditions.

Learning from a Trace Chemistry Episode

From such an episode, several global conclusions can be drawn about trace chemistry skills. The first is that it took interlocking weaknesses in three stages of the process to let a simple error like silicone usage bring the operation to its knees. In each of these areas, a trace chemistry fault was the kernel of weakness. Second, in a mature process like this one, trace chemistry should have been at the top of the support agenda as the "new frontier" of the process. Technical tools to explain how native surfactants were formed (and destroyed) were not hard to find; the hurdle was in believing that components inconsequential from a yield standpoint could be so important to operability.

Finally, the importance of stepping outside our particular process rut for our information sources becomes apparent. It turned out that an unanticipated volatilization of a silicone antifoam to kill an unrelated catalyst had already occurred in the company in another process. There was no need to repeat the mistake if we had had not just communication, but a means to disseminate the technical information so as to catch the imaginative attention of support people in all process areas via a shared disciplinary "home."

A crisis such as this can send threads of curiosity about the trace chemistry into the various phases of the process affected. If followed up with insight, it can serve to shore up and strengthen the various shaky substructures of the process. For this particular process such curiosity follow-ups included (1) better management of the elements in primary product synthesis that had helped form the native surfactants; (2) seeking a purge path for removing key impurities from the primary product before it reached the catalytic conversion to the final product; and (3) examining the vulnerable surface chemistry of this converter catalyst to learn more about what poisons it and how to regenerate it with more subtle methods.

Countering the Deleterious Effects of Trace Chemistry

Surface Reactions in Product Storage

Metal oxides in the surface of storage tank linings at ambient conditions can be sufficiently basic or acidic to catalyze isomerization and condensation problems which constitute serious quality issues for supposedly finished products. Zn, Sn, and Fe are common culprits. Some countermeasures to such problem have been presented in Chap. 11.

Attempts to Fight Trace Chemistry with Counteradditives

Sometimes we fight trace chemistry indigenous to the process with an additive that proves to cause significant new problems. An example is the use of heavy amines to counteract HCl, cited in Chap. 2 as potentially causing fouling. Those who believe that "less is more" prefer deletions as less likely to cause unwelcome repercussions than additions.

Methods for Removing Trace Impurities

The above account establishes a preference for deletion of trace impurities or their formation chemistry in PTS work. However, an operation like distillation is less effective in rejecting objectionable trace impurities than it is in major component separations. Since the trace component cannot be boosted sufficiently in concentration for convenient discard, its removal in a purge stream will require further processing (see below). The problem of aliasing has been discussed above. Another difficulty with fractionation rejection is that the trace components to be rejected may be a class of compounds with a range of volatilities. Conversion to a more innocuous set of impurities (reactive purification, or RP) or adsorption will often make more sense.

Reacting Away Trace Impurities (RP)

Conversion of a troublesome impurity can be done either in tandem with fractionation or separately. In the first case, it would for example, be reacted with a major component to a readily rejected heavy end. The second case covers situations where removal of the converted impurity is not needed. For example, in an odor problem involving a ppm-level acrylate contamination of an alcohol product, it proved more effective to catalytically add alcohol across the double bond, destroying the odor-contributing functionality, than to try to remove the impurity.

Another example of the second case concerns an unstable impurity, such as an unsaturated (allylic or vinylic) alcohol impurity in a finished alcohol vulnerable to autoxidation, dehydration, or other sequential quality-impairing reaction. Such an impurity can originate from an enolizing catalyst site in hydrogenation/dehydrogenation reactions associated with the main synthesis train. A finishing hydrogenation of such a liquid alcohol product can prove valuable for its stability in further storage and shipping by converting such unstable intermediates either to the product itself or to an impurity that engenders no quality problems.

Missing: A General-purpose Reducing Reagent

However, a full-scale hydrogenation is a bit extreme for reducing perhaps 50 ppm of a stream. Yet, in my experience, sodium borohydride, hydrazine, and other potent reducing agents leave much to be desired as robust reducing agents for troublesome trace precursors. Many incipient quality problems could be averted by a selective reducing agent that does not introduce new complications into the process stream. Rohm & Haas made an abortive attempt some years ago to introduce a regenerable redox exchange resin. It is not clear whether the problem was performance or the lack of a sophisticated market for such a treating agent.

Purge Criteria for Impurity Buildups

An impurity buildup requiring a purge may occur in several distinct ways. The CP buildup in a strong acid system discussed earlier represents cumulative retention because of chemical interaction with the catalyst. More innocently, nonreactive impurities may concentrate as diluents in a reactant recycle loop. Finally, bulge buildups of impurities in distillation columns may reach magnitudes that compromise operability. In each case, a purge is the prime means of limiting such buildups, but what sets a reasonable rate for such a purge?

In the first and last cases, the onset of trouble is likely to be abrupt at a certain level of buildup; hence purge becomes essentially a control issue. The middle case, typified by benzene impurities building in an alkylbenzene synthesis recycle loop, represents a more difficult choice. The flow rate of "drag" (i.e., purge) benzene is usually set by a balance between the cost of processing the drag benzene and the energy and reactor occupancy disadvantages of allowing the diluent impurities to ride around the reactor/distillation circuit. If some of the impurities can be construed as mild catalyst poisons (see Chap. 10), it may significantly alter the judgment of an optimum purge rate.

Removing Impurities by Sorption/Reaction

Invoking a creative heterogeneous reaction can frequently improvise a way out of a difficult problem in removing a trace impurity from a sensitive major component. Removing the last of a S- or phenolic-based inhibitor used in storage or shipping of dienic monomers is often better done in this way than by redistillation. A recent example from my experience was a proposal to use N-alkylation (via a bed of t-amine IER) for heterogeneous removal of organic chloride impurities.

Color and Odor Problems Involving Trace Impurities

Chromophore Arithmetic and Chemistry

My interest in chromophore definition began when I first encountered visible brown, yellow, green, or red color in process streams that clearly came from organic, not inorganic, contaminants. In one case, relatively simple C_3–C_5 molecules were involved. A little reading revealed that the presence of multiple conjugated double bonds was a necessary condition; carbonyl group participation strengthened the effect. Later, color problems in phenolic systems gave me an appreciation of quinoidal systems with much stronger effects. Clearly, resonance capabilities strongly affect the visible light absorption required to create a color problem.

How do we plausibly implicate a particular trace impurity at a particular concentration level as causing an observed color effect? We use the kind of three-coordinate triangulation a small boat skipper uses to establish position from known landmarks sighted. A visible absorption spectrum will yield the Beer's law extinction coefficient ϵ at various wave-lengths λ. For a component to give an absorption value $A = 0.05$ (a level that will register on the unaided eye) at a concentration of 1 ppmw ($\approx 10^{-5}$ molar) in a cell with a 10-cm path length requires an ϵ of about 500, realizable by a variety of compounds. A second "coordinate" for screening candidates is that the absorption peak be at the wavelength λ appropriate to the color. The final coordinate for color body identification is, of course, a chemical structure derivable from plausible precursors present in the process stream.

My most exciting experience with chromophores involved potent biquinoidal color bodies. These can have extinction coefficients ⩾5000. The identification difficulty with such a potent chromophore is that far less than 1 ppm can be significant. With the low volatility also involved, we were up against the vagaries of liquid chromatography linked to mass spectrometry (LC/MS) in attempting analytical definition of the molecular structure of possible color body components at very low concentrations. The outcome was a half-blindfolded (but ultimately very successful) attack on guessed classes of precursors, with the exact chromophore chemistry remaining obscure.

Malodorous Trace Impurities

The nose is more nonlinear than the eye. Butyl mercaptan, the critical ingredient of a skunk's protective technique, is vile at the ppm level but is reputedly an ingredient of perfumes at the ppt level. Hence, there are even more difficulties in quantitative identification than indicated above for color impurities. We have already cited the linkage of S compounds to foul odor. Meilgaard (1975) gives odor threshold values (TV) for various S compound classes. This, with GC/MS, can narrow the list of odor compounds we should consider. For example, if we are trying to implicate disulfides (RSSR) in an odor problem, we must show a fairly generous (⩾1 ppm) concentration; a make/break analytical confirmation should be readily devised. For a t-mercaptan, on the other hand, with a ppt TV,[104] our strategy will be quite different and will involve a lot of nose power and reaction mechanism thought.

The most important activities in odor problems are microanalytical work[105] and a discerning literature search. The ubiquitous "cat's urine" compound discussed in Chap. 9 was identified in one of our products via such a search. Another reputedly raunchier S species is

---

[104] The TVs are for a vaporized sample. How we sniff a liquid sample, the volatility of the odor body relative to the major component, and the masking effect of the latter greatly affect ultimate sensitivity.

[105] The virtues of devices like flame photoionization detectors for extending capillary GC sensitivity cannot be overemphasized. Distillative concentration to build up to GC detectability has been cited above. GC/MS is useful only for identifying compounds at the level of several hundred ppm.

thioketones, particularly the lower aliphatic unconjugated ones (characterized by a red color as well) (Mayer 1964). For a non-S example of a strong odor body, consider acrylates. Their conjugated double-bond system appears to explain why their odor is intense compared to those of other esters. Use of a reaction that destroyed the conjugation and odor without their physical removal has been described above.

Beneficial Uses of Trace Impurities as VPMs (see also Chap. 11)

C_2 chlorides are commonly used as vapor-phase modifiers to improve selectivity of AgO catalysts in ethylene oxide synthesis. Berty (1989) claims that dichloroethane, vinyl chloride, and ethyl chloride are equivalent vehicles for Cl (the fate of the organic moiety was not established). Since Cl is scavenged (as, e.g., C_2H_5Cl) by ethylene or ethane, continuous addition of the Cl VPM is required.

Another system with a similar issue of interchangeability in Cl VPMs is regeneration of Pt/Al_2O_3 catalysts. CCl_4 or $CHCl_3$ would be more convenient than HCl, and the literature suggests that they are valid alternates. Yet, one of my great surprises was finding strong differences between H_2S and CS_2 as VPMs in hydrogenation/dehydrogenation systems. Perhaps the distinction in the case of Cl-containing VPMs is that they are applied in catalyst systems with strong oxidation/reduction cycles that level the modifiers.

Structuring Trace Chemistry Kinetics

Modeling Autoxidation of a Stored Product as a Chain Reaction

The following illustrates structuring an investigation of a quality impairment related to FR reactions in storage. It does not report an actual episode.

Suppose that we are confronted with oxidative degradation of a common solvent like methyl ethyl ketone (MEK) in storage under atmospheric O_2 and are trying to decide if blanketing with N_2 or addition of a liquid-phase inhibitor is warranted. We seek rough kinetics to guide laboratory interpretation of long-term storage trace chemistry analyses. A quick search of the literature, building upon the notion that this is a FR reaction, uncovers a useful article (Zaikov et al. 1961), which identifies attack on the 3-carbon, followed by chain branching via fission of the intermediates 2,3-keto-hydroperoxide (KHP) and 2,3-butanedione (diacetyl, or DA) into two radicals, as important features, and gives a pseudo-first-order rate expression (for a complex chain) that extrapolates to 1300 ppm/day at 40°C and 1 atm of air pressure.

This is large enough to create concern, but more probing on the initiation reaction creating KHP is needed before taking any action. The experiments of Zaikov et al. were at elevated temperatures and large MEK conversions, with KHP at its quasi-steady-state concentration (see the discussion of FR and QSSA in Chap. 10), whereas in the early stage of oxidation KHP will just be starting to form via the following reaction:

$$O_2 + C_4H_8O \rightarrow \cdot OOH + C_4H_7O\cdot$$
$$C_4H_7O\cdot + O_2 \rightarrow C_4H_7O_3\cdot$$
$$C_4H_7O_3\cdot + C_4H_8O \rightarrow C_4H_8O_3 \,(KHP) + C_4H_7O\cdot$$

In effect, early autoxidation rates would be well below Zaikov's QSSA rate, and a separate laboratory study of this phase would be needed.

This use of reaction mechanisms gleaned from the literature is but one way of probing into a system for which we lack clear evidence. In the next chapter we canvass other tools for inferring what goes on inside process piping not amenable to direct observation.

References

Albright, L.F. *IEC Proc. Des. Dev.* 11 447–50 (1972).

Berty, J.M. *Chem. Eng. Comm.* 82 229–32 (1989).

Buschmann, H.-J., et al. *Ber. Bunsenges. Phys. Chem.* 84 41–4 (1980).

Mayer, R. *Angew. Chem. Intl. Ed.* 3 277–86 (1964).

Meilgaard, M.C. *Tech. Quarterly, Master Brewers' Assn. Amer.* 12(3) 151–68 (1975).

Saletan, D.I., and Chun, H.W. U.S. Patent 4,247,532 (to Shell Oil Co.) (Jan. 27, 1981).

Schäfer, H. *Z. anorg. Chem.* 259 53, 75, 265 and 260 127, 279 (1949); 261 142 (1950); 264 249 (1951); 270 304 (1952).

Sørenson, P.E., and Jencks, W.P. *JACS* 109 4675–90 (1987).

Zaikov, G.I., et al. *Dokl. Akad. Nauk SSSR* 140 701–3 (1961).

III

Finishing the Job

13

Diagnostic Aids: Seeing the Unseen

Aristotle would have avoided the mistake of thinking that women have fewer teeth than men if he had simply asked Mrs. Aristotle to open her mouth.
                                                                Bertrand Russell

About Inference Gaps

We all feel uncomfortable about the opaqueness of the processes we deal with. On an introductory trip around the chemical plant on my first job, the old hand escorting me spat on a bare hot pipe to illustrate how an operator could gauge temperature without an instrument. Even with the sophisticated sensors available to us today, there is still a veil between us and the process, which we pierce only by inferences that bridge our connection to the process via the sensor. The series arrangement is really

Process ↔ Inference ↔ Sensor ↔ Inference ↔ Technologist

This means we must consciously supply *two* judgments before we can speak of what the "process" is doing. We tend to focus on the second inference gap only: how we manipulate the instrumental outputs to derive an understanding of the process. But how the instrument is coupled to the process can also be critical to success.

Process Coupling Criticality

For example, suppose that we are doing counts of gamma ray transmission (GRT) or neutron backscatter (NBS) to assess flooding at particular tray locations in a distillation column. The statistics of the counting process matters, but more important are elements like accuracy and reproducibility of the vertical alignment of the source and receptor (for GRT) or the presence of moisture in the external insulation (for NBS). If we are trying to verify flooding with NBS, we must be aware of its limitations: that it registers primarily H-rich fluids and only mass within a few centimeters of the vessel wall. For example, with a center downcomer, careful peripheral positioning is needed to get downcomer backup via NBS.

Doppler devices coupled to a pipe wall, to gain information on internal axial flow, exemplify the relatively direct high-tech exploratory instrument most of us embrace readily. Here the inference gap we should be considering is, for example, whether the two-phase system inside the pipe is really amenable to the technique. We must be certain that the process we couple to is the process we care about. This seems a trivial issue, but before me as I write is an article (*The Economist* 1990) describing the very different perspective given by an infrared (IR) scanner beam traversing the exhausts of cars moving down the road past it from that of a motor laboratory "rack" determination of which auto engines cause the most air pollution.

High-tech Sensors

Schneider's review (1983) of newer sensors for mass flow measurement contrasts instruments where emphasis is on correlative and other mathematical digestion techniques with those, such as magnetic flow-meters, vortex-shedding elements, or vibrating elements, where a somewhat unusual physical property is harnessed. In the latter class of instruments our major attention should be on the inference gap between process and instrument.

We should aim high. If radio transmitters can traverse the human circulatory system, why not go for the idea of embedding a transmitter in a pellet in a fluidized bed to gain understanding of its circulation pattern? But while keeping an eye on high tech, we must remember the older simple devices that infer composition from viscosity, refractive index, or other simple measurements.

Inferential Raw Data Processing

The other extreme (where the inference gap is between the sensor and ourselves) is represented by the use of online autocorrelation and cross-correlation techniques to estimate the mean and variance of residence time distribution. This might be done in a trickle phase reactor, utilizing the noise of flow or temperature. Here the emphasis is on the power of the mathematical/electronic manipulation once location, sensitivity, and stability standards for the individual sensors have been met. There is considerable information to be extracted via these dynamics-derived techniques (see the discussion in Chap. 8), but care is required in filtering and in choosing among working in real time or the frequency domain or in summing everything up as moments or Fourier transforms.

For example, acoustic emissions are used online to test overall structural integrity in lined pipe and vessels (Fowler 1987). While skill is, of course, needed in positioning the sonic emitter, the emphasis is on subtleties of filtering data to avoid losing the acoustic "signatures" of particular kinds of material flaws. In another dynamic assessment method, characterizing the onset of flooding in a pulsed extraction column, spectral analysis was more informative than correlation functions (Strand and Garlid 1975).

By contrast, radiography and ultrasound work over small areas, as in defining localized thinning of a metal wall from internal corrosion, with little mathematical data manipulation. Here effective preventive maintenance requires intelligent selection of areas to concentrate on for regular inspection by careful review of process chemistry and hydraulics. We are back to the criticality of the process-sensor inference gap. Mindless attempts to cover everything will lead to careless, low-quality work (see Chap. 16).

Use of Tracers

The use of tracers (radioactive, dye, salt) to elucidate flow patterns in closed systems is of broad interest and falls somewhere between the above extremes. That is, equal emphasis must be placed upon the spectral analysis of the raw data and the proper choice of tracer and sensor. Tracers can be used to advantage not only for matters like residence time distribution (RTD) or for evaluating the quality of interphase contacting (Mills and Dudukovic 1981), but also for purposes as basic as leak detection (White 1987).

Here are a few examples of cases in which RTD information deduced from tracer experiments provided key insights in my own troubleshooting:

1. Dynamic response of amperage efficiency of a Cl electrolysis cell to pH lowering, achieved by a pure "pulse" input arranged conveniently by adding a quart of aqueous HCl in one dollop;
2. Preferential phase holdup of the heavier (aqueous) phase in an open tubular upflow reactor, done with two different markers, LiCl and radioactive tracers, with similar results;
3. Use of a nonvolatile dye to measure heavy-end entrainment overhead as a sensitive function of boil-up in a column;
4. Use of radiotracers (via a commercial service) to establish the extent of short-circuiting of reactor effluent by slippage under a baffle separating two reaction stages;
5. Use of a UV-absorbing dye tracer to establish the extent of axial centerline plunging in a tubular flow reactor for polymerization, caused by wall drag and strong radial viscosity gradients.

Bridging Gaps Between the Process and Sensor

Remote Sensing

In most cases, we are separated from the process primarily by a pipe or vessel wall needed simply for fluid containment. However, in some cases, the process environment is so aggressive that we start one degree more removed in accessing information. An example involves sensing tube wall temperature in process furnaces. Grantom (1985) describes an evolutionary program that, with no change in the quality of the IR pyrometers used, improved the overall accuracy of sensing by using secondary standards, background refractory temperatures for correction, and a preferred wavelength window in the CO_2/H_2O spectra. Such patient yet imaginative method development must be followed even when we install some touted new sensor, rather than viewing the instrument substitution as a painless shortcut to success.

A clever calculational/inferential measurement for flow and liquid-level detection in aggressive environment is based upon heat loss from precisely heated resistance temperature detectors.

Potential Errors in Gamma Ray Traverses (GRT)

GRT scanning of distillation columns can be very valuable for troubleshooting (Fulham and Hulbert 1975; Jones and Jones 1975) but has several pitfalls besides the vertical positioning issue already mentioned (Severance 1985). These include random fluctuations of source intensity; overlong path or loss of source intensity; and background calibration with the wrong phase. When it is done well GRT can go beyond identifying isolated pinch points or flooded trays to provide a true hydraulic profile of a column, including spray height and jet flooding phenomena that depend on surface tension and other effects not encompassed in the original design calculations (Harrison 1990). Special preparatory procedures like a "dry" scan with the column empty (to calibrate out all metal structural peculiarities) can (as in the pyrometry example above) add to the interpretive value of a GRT scan (Bouck et al. 1986).

Nondestructive Metallurgical Testing

We have mentioned metal walls as screening us from process knowledge. Sometimes the story is *in* the wall, even though derived from process events (e.g., carburization of alloy from pyrolysis coke). Dressel et al. (1991) reviewed nondestructive testing methods (ultrasonic, radiographic, eddy current, liquid penetrant, hardness) to infer a variety of materials flaws in NH_3 plants.

The Power of On-line Analytical Information

The above discussion has dwelled on understanding the physics of system internals. It is timely here to emphasize the other dimension of process knowledge we can gain by paying the entry cost for reliable process analyzers, operating either continuously or at short enough intervals to give a running picture of drifts and other shifts in process performance. These can be either continuous analyzers like refractive index, one-wavelength spectrophotometry, or electrical conductivity or discrete, short-time devices like flow injection analysis. The key consideration is to follow analytical trends with good time base and amplitude resolution. Accuracy is secondary. In this way, we can infer the catalyst's response to shifts in feed quality or flow rates and other effects that take us well beyond textbook predictions.

Chap. 8 pointed out the fallacy of a large computer control system regulating with only flow and temperature information. It was an eye-opener for me to see, in an early steam cracking computer control installation, what a bank of fallible process GCs on the effluent from the pyrolysis furnaces could do for process visualization, trends, and time-base resolution of events.

Pitfalls of Laboratory GC Identifications

Since emergence time is an ambiguous identifier, it is common to couple GC to MS or some other device to complete species identification. Even such coupled systems can fail, however. One example from my experience involves confounding a higher alcohol with an olefin oligomer of slightly higher carbon number, caused by the dehydration tendency of conventional mass spectrometry. Newer techniques in GC/MS may soon eliminate such confusion.

Other Screens Between the Process and Us

Shutdown Procedures

There are other ways to overcome screening of the observer from the process. If, for example, we need to infer the process by which a catalyst fails, we may be well advised to spend effort on decontamination techniques for process shutdown, in order to get an uncompromised "failure specimen" for convenient bench-scale examination, rather than on elaborate online, through-the-wall sensing approaches.

Process Noise and Drift

However, there are also time sequences and drifts that can separate us from the critical characteristic or trend. There are two basic responses to such time screening: (1) filtering to display a true overriding trend, as in the CUSUM plots of statistical process control or with leading economic indicators to infer the start of a business recession; (2) deliberate perturbation of the process, overcoming process noise either with a single large pulse or with low-level, repetitive perturbation (see Chap. 8).

Elusive Chemistry

The cloaking of a compound by chemical aliasing has been discussed in Chap. 12. Online analyzers may be inappropriate for following trace chemistry, very rapid reactions, or systems where the act of sampling itself disturbs the system. In some of these situations, a laboratory environment may be able to re-create the effect. In others, a theoretical "spine" to our observation (as in the optimal quench path of Chap. 10) can pull a coherent story together.

This chapter has dealt with rational methods of inference in problem diagnosis. But metaphoric imagery can also enrich diagnostics and the pool of prospective solutions. We return to this topic in Chap. 14.

References

Bouck, D.S., et al. Paper 86e *AIChE Miami Beach Meeting* (Nov. 1986).
Dressel, M., et al. *Chem. Eng. Prog.* 87(4) 30–5 (1991).
Fowler, T.J. *Chem. Eng. Prog.* 83(5) 25–31 (1987).
Fulham, M.J., and Hulbert, V.G. *Chem. Eng. Prog.* 71(6) 73–8 (1975).
Grantom, R.L. *Chem. Eng. Prog.* 81(7) 41–4 (1985).
Harrison, M.E. *Chem. Eng. Prog.* 86(3) 37–44 (1990).
Jones, D.W., and Jones, J.B. *Chem. Eng. Prog.* 71(6) 65–72 (1975).
Mills, P.L., and Dudukovic, M.P. *AIChE J.* 27 893–904 (1981).
Schneider, H.J. *Chem. Ing. Tech.* 55 767–74 (1983).
Severance, W.A.N. *Chem. Eng. Prog.* 81(9) 38–41 (1985).
Strand, M.J., and Garlid, K.L. *Can. J. Chem. Eng.* 53 677–85 (1975).
The Economist p. 64 (July 7, 1990).
White, R.L. *Chem. Eng. Prog.* 83(5) 33–8 (1987).

14

Thinking Metaphorically

Nonsense is good only because common sense is so limited.
                                                    George Santayana

Making the Familiar Strange

In Chap. 5 we linked metaphor to the imaginative constructs used in lateral thinking (LT) to generate unusual and creative approaches to defining and solving problems. Here we pick up this thread and expand the notion of metaphor as making the strange familiar and the familiar strange (MSF + MFS). There is a practicality to this mode that goes beyond generating unusual ideas; it can strengthen motivation and communication as well.

Like many professionals in process improvement, I had come to bemoan the rather lethargic attitudes of operators toward the (to me, thoroughly exciting) processes they control. In roaming control rooms in process support work after the day shift had gone home, the real animation I encountered in these men, whose roots were in the woods of East Texas, came in conversation about the last or the next hunting or fishing trip. It came to me that the way to enlist the enthusiasm and ability of such operators was to install the magic and excitement of hunting *in the control room*. While changing work practices (i.e., job enrichment) is a task for management, I have used this kind of imagery in communicating on process matters with operators. An unmastered reactor became a wild beast; a control chart oscillation became a bird call to be recognized.

Poetry frees our imagination and frees the right side of our brains—holistic, intuitive, nonverbal, diffuse—to help us in our work. Technical people tend to be left-brained, using a verbal, analytical, rational, sequential, bits-and-pieces approach to problems; they have been conditioned not to acknowledge an emotional component in their everyday technical work. Of course, we have all found ourselves making jokes, puns, and Freudian slips. But, by and large, we tend to look askance at anyone who shows emotional enthusiasm. I recall a heated planning session on vinyl chloride process improvement where there seemed to be as many different technical idea threads being advanced as there were participants. I asked the group to precede their advocacy remarks with a description of the emotions and visions of accomplishment from which their proposal flowed. The silence was deafening.

The participants were driven by strong feelings; but expecting them to start injecting poetry into the workplace in a large meeting, when they were constrained by "the banality of the rational" even in their musings alone in their offices, was rather naive. And there is, admittedly, a certain disrespect in harnessing poetry for prosaic, utilitarian ends. In embracing metaphor for technical work, we are not aspiring to the artistic completeness of a well-turned poem, just turning loose the child-like creative side of our psyche in a flight of fantasy and an inversion of our normal values.

The Values of Ambiguity

One of the characteristics of the high achievers whom A.H. Maslow termed *self-actualized* is that, despite efficient perception of reality and authenticity, they do not separate reality and unreality into mutually exclusive compartments. The physicist Nils Bohr supposedly brought forth the complementarity version of quantum mechanics after reading Kierkegaard, who believed that two opposed ideas can exist simultaneously and both can be true! (Prausnitz 1988)

The dream or fantasy world and the poetic process are means by which medium-achieving, relatively staid individuals can learn to use mind-stretching techniques in problem solving. To loosen up, I suggest learning to coin and enjoy puns and spending some time reading the Alice works of Lewis Carroll.

Puns

The pun epitomizes the kind of double-edged openness we seek in problem definition and in enlarging our choice of solutions. It is language turned on itself and experienced as foreign.[106] It springs from the same combination of wit and imagination that feeds the poetic process. It makes the strange familiar. The pun uses assonance (sound-alikes) as a sort of solvent or glue to convey an initial air of affinity, somewhat like the use of rhyme in poetry.

A pun thrives on ambiguity. We tend to think of the grosser kind of pun as groan-provoking, the kind of wit used in humorous birthday-card greetings. The problem with such efforts is that there is nothing except assonance to unite punning words; they are not even decent opposites. But the creative pun as used by Shakespeare (over 3000 new-minted ones in his plays) and many other writers is much more powerful and serves us in two ways:

1. It deflects, bifurcates, or diffuses our attention; this heightens perception and enriches the field of possibilities.
2. A pun provides an instantaneous perception of analogies between apparently incongruous things. These enter the mind before they can be absorbed into consciousness or processed in any other (e.g., filtering or censoring) way. A good pun seems both accidental and inevitable.

The defocusing of a pun is like Zen techniques that get our consciousness out of the way of an action like shooting an arrow or the tennis player's trick of bouncing the ball before serving it. The double meaning in a pun gives depth to our perception, much as the two slightly different photo views in an old stereopticon device conveyed a three-dimensional image. The duality/linking of even near-opposites reaches its highest form in the death/life reconciliation that we experience in humanity's great myths.

---

[106] I am indebted to Walter Redfern's fine book for some of the material in this section (Redfern 1984).

Alice in Wonderland

My attention was drawn to this resource in technical work when I found Tom Baron, one of the great personalities at Shell Development, using quotes from it to preface sessions at which he was selling a prototype process computer control concept to a less than enthusiastic operating management. I urge you to get a copy and refer to it often, not for bedazzling your colleagues but as a launching pad for your own metaphorical experiences.

Much of Lewis Carroll's humor rests on the incongruity between the literal meaning of words and the figurative meanings we often give them in common usage. Alice is anxiously conservative in sticking to the literal meaning, as children are just before they take off into love of riddles and appreciation of ambiguity. This take-off is important for developing creativity. Humpty-Dumpty is a case at the other extreme that enables us to strike a balance. He assigns quite arbitrary and unusual meanings to words; language is turned loose from its moorings to "spin upon the waters."

Provocations, jumps, and reversals, introduced in Chap. 5 in our discussion of LT, are part of the value in Humpty-Dumpty's way with words (Fig. 14-1). So is *densification* (assigning additional meaning to existing words), which can enrich a language more than adding new words to describe new things or concepts. Best of all are fusion words juxtaposing two ideas, like *slithy* in the *Jabberwocky* poem of Alice. The value here is both in the hybrid vigor of a new coinage and in its quality of holding off premature facile differentiations and rejection of words/ideas.

Dreams

The world of Lewis Carroll is not nonsense; rather, it is the kind of "countersense" we experience in dreams, where we hold everyday realities and macabre imaginings together in our minds without rejecting either. There is a double consciousness in a dream reminiscent of the stereopticon alluded to above. The dream world echoes Alice in a further sense: In the everyday world, we assume words have fixed meanings that control how we must use them. But in dreaming, we are not in control; we are like Alice confronting

"When I use a word," Humpty Dumpty said, in rather a scornful tone, "it means just what I choose it to mean - neither more nor less."

Humpty-Dumpty, for whom "a word means just what I choose it to mean, neither more nor less." In a sense, the dream is dreaming us. In dreams, metaphor becomes reality and loss of control becomes, paradoxically, the key to a better shaping of reality. When a person faced with a difficult decision decides to "sleep on it," there may be no expectation that a clearer head will come tomorrow or that a revelation will unfold in sleep. Rather, it may be that a transformed version of the problem can be worked on in a dream and a more imaginative answer reached than in the cold world of "reality." This is the hoped-for outcome of the problem-solving procedure called *metaphorical conversion* by McKinney (1987).

Why Metaphor?

A pun is a first step away from the "transparent" word without depth. It is not a metaphor per se, but it carries us toward a metaphor. A pun becomes a metaphor when it conveys that two different concepts/expressions also share deep affinities. By holding ideas that are at once like and unlike before us, metaphor makes us think, see, and hear on more than one level concurrently. The simultaneous distancing and closeness of MSF + MFS is what gives metaphor its power in many instances to open up a problem.

The following are some examples of taking a metaphorical trip that enables us to examine our problem with fresh insight and return to our prosaic state to begin solving our problem:

1. McKinney's example concerns an obscure leak of process gas, with economic/environmental consequences, which defied solution until the technical people brainstormed it as a jail breakout problem for prisoners. In identifying with the would-be escapees, they generated a large number of options. When these options were translated back into the context of the real problem, they opened up new lines of investigation that found the leak in a defective "blind" supposedly sealing off a dead leg in the piping.

2. Suppose that we encounter a problem, like that developed in Chap. 2, of unequal flows through seemingly identical manifold branches. We visualize a sow nursing a litter of seemingly equivalent offspring and see how a trivially small differential among the piglets jostling for position at the sow's teats leads finally to the "runt of the litter" phenomenon. The insight of how a small difference is magnified in a self-sharpening feedback cause/effect loop is then transferred back to our real work.

3. A deeper metaphorical level of the same problem was presented late in Chap. 2. The image of a "seed of its own undoing" to characterize how coke caused more coke, finally to plug reactor tubes, helped us transcend our initial image of Fe seeds creating coke. It allowed us to see that it was the very procedure for removing coke (air/steam oxidation) that was creating the Fe seeds.

There is no sharp distinction between metaphorical conversion and LT; both enlist the right side of the brain and the dream world. Such dream devices as working backward from the wished-for result, using ephemeral bridges ("intermediate impossibles"[107] of LT) to link needs to solutions, are valid in both methods.

---

[107] A famous example (deBono 1970) is imagining cars with square wheels to elicit radically new concepts for attacking traffic congestion problems.

Example: Building a Theory to Fit Observation

Consider corrosion caused by water penetrating acid-containing process streams through minor defects in heat exchanger tube walls when a pressure differential is supposedly maintained to prevent this intrusion. A prosaic explanation of such corrosion is operator neglect so that the pressure differential specified was in fact not maintained. The metaphorical MFS approach sees the water penetrating against the pressure gradient. This gets us to the correct inferential starting point: that a partial pressure driving force is involved.

The remaining act of imagination brings to bear the second vehicle of metaphor: MSF consists of visualizing a micro-crevice in the tube wall, one fine enough that hindered diffusion,[108] rather than bulk fluid flow, is involved. In this context, it becomes credible that water molecules, driven by partial pressure, can work their way against a slow outward flow of major components from the process side. The tube wall is, in a manner of speaking, a membrane. The corrosion from nucleating an aqueous acid phase on the process side (see Fig. 9-6) creates new crevices; a self-aggravating sequence results.

In this chapter we have shown how better to enlist our imagination in solving problems. Now we turn to a more positive use of the insights gained: turning the "alligators" from the problem "swamp" into a team to help us achieve our goals.

References

de Bono, E. *Lateral Thinking: Creativity Step by Step* Harper and Row (1970).

McKinney, P.T. *Chem. Eng.* 94(21) 131–3 (1987).

Prausnitz, J.M. *CHEMTECH* 18(8) 466–8 (1987).

Redfern, W. *Puns* Basil Blackwell (1984).

Ternan, M. *Can. J. Chem. Eng.* 65 244–8 (1987).

---

[108] The strong diffusivity difference between molecules like H_2O and larger molecules near crevice diameter in size (Ternan 1987) assists counterdiffusion in these liquid-filled "pores."

15

Turning the Tables: Making Phenomenology Work for You

The art of our necessities is strange, that can make vile things precious.
Shakespeare's King Lear

When we have been sensitized by hard knocks to all the nonlinear boobytraps embedded in a mature process, defeatism can begin to infect all of our activities, as with a swimmer who always enters the surf worrying about sharks. This chapter provides examples of two reversal strategies to use in overcoming this attitude. The first strategy uses the grasp of phenomenology we have acquired in troubleshooting to try to make nature work for us. It invokes reversals whereby a terrain tilted against us is turned into an upbeat enhancement (see Fig. 15-1). It is summed up by the adage "If you can't lick 'em, join 'em." The second strategy, discussed later in the chapter, consists of grafting new technology onto the old to create vigorous hybrids.

Phenomenology Reversal

Example of Undoing a Callus

Consider a case where a homogeneous catalyst in a reactor train is deactivating much faster than premised in the design due to a rise

Fig. 15-1 Turning Slippery Downslopes into Up-tilt Opportunities

in reactor temperature settings over the years that has made possible more throughput without capital expenditure. If we do not allow this capacity/catalyst use trade-off to be forgotten as a callus (see Chap. 3), we will take advantage of a downturn in the business cycle that reduces throughput to lower the temperature again, offsetting lost sales with catalyst savings.

Foaming Away Troublesome Surfactants

The presence of trace surfactant impurities is a common problem. In one situation involving tight emulsions and phase separation problems in a multistage L/L reaction system, we attempted to use these same surface-active properties to selectively purge such trace impurities. The hope was that the surfactants would concentrate in the interfacial film created by sparging fine gas bubbles into the liquid and become enriched in the "foamate" overhead. Our laboratory experiments, using a "foam-over" column, copied from *foam fractionation*[109] already in commercial uses like effluent detoxification, were unsuccessful.

Later, with a stronger grasp of surface chemistry from other troubleshooting activities, we realized that the quasisolid "carbon" particles that made up our surfactants had polyfunctional wetting properties. These required an extended interface with a nonpolar *liquid* phase to give interfacial enrichment (see Levine et al. 1989 in Chap.

---

[109] The generic concept has been termed *adsorptive bubble separation* (Grieves 1982; Lemlich 1972). Other subcategories are flotation, distinctive in that rather large particles of a solid phase are carried at a froth (G/L) interface, and solvent sublation, in which a supernatant liquid phase acts as a collector for the surfactants from the rising bubbles. There are also analogous liquid injection techniques termed *adsorptive droplet separation*.

6). This understanding did not take form as a totally new emulsion fractionation scheme. But it led us to increase the agitation and interfacial area in the final reaction stage to facilitate rejection there to the residual raffinate layer effluent of the surfactants from the reaction recycle loop.

Other Types of Reversal

Following are more aggressive examples of reversal, the LT device cited earlier for enlarging the field of possibilities.

Reversal of Catalyst Structure

We have cited several side reactions on the bare support surface of metal hydrogenation catalysts. One of the bolder development ideas inverts the structure so that the metal is a film under a metal oxide exterior. The concept is for H_2 to diffuse in ahead of the reactant, dissociate to H at the metal, then diffuse back, counter to the reactant, for a more selective hydrogenation.

Deliberate Poisoning of a Catalyst

Example: Reversible Use of a CO Moderator

We are using a tubular vapor-phase hydrogenation reactor in dual service on unsaturated compounds A and B. A has a reaction rate 300 times that of B on the catalyst used. If catalyst activity is adequate to achieve 98% conversion in the flow reaction of B, we will have thermal runaway problems (see Chap. 11) when A is fed. Much higher space velocity in A service is not practical. Instead we introduce CO with the H_2 as a temporary poison during the conversion run on A. After a run of a week or so on A, we suspend CO addition and switch to feedstock B. This simple change in operating mode avoids the need for separate reactor systems.

"Just-Right" Acidity

An even more striking use of reversal for innovative proposes introduces catalyst poisons to forestall deactivation. This can occur in the following context: Use of extremely strong acid catalysts can breed (see Chap. 12) out of feedstock impurities a class of hydrocarbons (conjunct polymer in the case of aliphatic alkylation) basic enough to interact ionically with the acid catalyst. In this way, a catalyst initially stronger than needed for its task is weakened to the point where it is deficient in activity. There are a number of ways to do this. One of the more innovative embodiments is by feeding a mild hydrocarbon poison (preferably via a recycle loop) that is not basic enough to bond ionically. In this way the acid catalyst can be brought to "just-right" strength where the strongly basic hydrocarbons are scarcely formed.

S to Restrain Activity

A similar line of thinking has led to presulfiding of re-forming catalysts to obviate rapid deactivation. Other uses of ostensible poisons (VPMs) such as S, Cl, and CO at ppm levels to improve selectivity have already been cited in Chaps. 11 and 12.

Reversal in Heat Transfer Innovation

Reversal was surely involved in duPont's development of Teflon as a heat transfer material. One would think that the relatively low thermal conductivity of such a plastic could only be a detriment in such applications. But as soon as we vault over the hurdle of our disbelief, the positives come flooding in: less fouling, flow slip at the wall because of lack of wetting by the process fluid, and dropwise condensation of vapor for a better U value.

Cost Reversal

Least-cost thinking can be a block to creative ideas. Often reaching for the more expensive ingredient exposes advantages that more than compensate for the extra cost. Examples from my experience include using K rather than Na compounds for solubility advantage and using LiCl as a substitute for HCl in reaction systems. The pharmaceutical industry often uses the more costly tert butyl alcohol in preference to other C_4 alcohols in formulations for solubility and other advantages.

Another version of cost reversal is the use of expensive electrical energy to fill separation "niches" where the effect obtained is superior to that achievable with cheaper mechanical or thermal forms of energy and more massive equipment. Examples involving the use of microwave power include removing minor water inclusions in hydrocarbons, batch distillation of fine essences, and regeneration of powdered activated carbon. Electric pulses are useful for (electroacoustic) dewatering of sludges and L/L contacting.

Retrofitting High Tech onto Mature Processes

Our second strategy for climbing out of the swamp of adversity involves a hybridization whereby relatively new technical developments are grafted onto a mature process to remedy key embedded defects or enhance its capabilities at disproportionately low incremental cost. The stereotype of processes being born and dying over a 10- to 15-year life span has persisted long after any brief validity it may have had in the heyday of the petrochemical era. As profitability in the chemical industry has shrunk, the period required to recoup the original investment has risen far above the 6-year horizon once used to plan expansion and innovation. Even after 20 years of life, an older plant may be protected, by the high cost of new grassroots investment, against displacement by newer, more efficient alternative processes; the qualifier is that the older process must be kept alive by timely pruning and by splicing in new elements at minor capital cost.

Technically oriented vice presidents have expressed surprise that there could be anything new to learn or change in a mature process that has been operating for 30 or more years. Yet it is in just such systems, which took shape before the modern era in chemistry and chemical engineering, that rich opportunities lie for grafting new technology onto the rootstock of the old. This usually leaves the chemistry more or less intact[110] while adding modifications involving separations, fluid mechanics, or possibly a new catalyst. Such grafting mimics most invention, which tends to combine a novel technique with a well-established operation rather than springing whole from the brain of the inventor.

[110] Entrenched quality specifications of a commodity chemical from an established process are an added hurdle for any new substitute process. The additional fractionation needed to remove carbonyl impurities in a new process for methyl ethyl ketone synthesis by direct oxidation of butylene is a recent example (Newman et al. 1989).

Reconfiguration of Reactors

Consider a catalytic reaction constrained by pressure drop: dienes synthesis by olefin dehydrogenation. One idea for grafting in this area was the use of radial-outflow reactor layouts, which increase flow cross section as the number of moles increases with reaction. This exemplifies the shallowest level of innovation, proceeding almost directly from implications in the differential equations themselves. Such a graft improvement can be done inexpensively because the cost of adiabatic reactors, even complex shapes like this one, is minor compared with the capital investment in the plant as a whole.

A deeper level of insight in pressure drop–constrained situations came from discriminating necessary wall drag from wasteful form drag in momentum transfer,[111] seeking to minimize the latter as not central to the needed contact of gas with catalyst surface.

This line of thought led to the monolith or honeycomb catalyst that permitted insertion of an afterburner converter in the automobile exhaust system without adding significant pressure drop.

A physical chemistry–oriented graft improvement promises new levels of efficiency in the already well-worked MeOH synthesis process (Westerterp and Kuczynski 1986). A solid adsorbent, flowing down over fixed catalyst, countercurrent to a rising synthesis gas mixture, removes MeOH and shifts chemical equilibrium to permit 100% conversion.

Many investigators are pursuing reactor configurations in which membranes are added, somewhat as in the above system, to diffuse away a component that blocks reaction completion. This approach tends to dictate the use of inorganic membranes (to cope better with temperature) and the choice of light components for removal. Moser (1991) reports some success along this line in securing high conversion in ethylbenzene dehydrogenation to styrene (surprisingly, with enhanced selectivity) by abstracting H_2 as it is formed through a membrane wall.

Catalytic Distillation (CD)

CD, introduced in Chaps. 7 and 11, a broader version of the fusion of separation with reaction embodied in the above adaptations of

---
[111] Momentum loss does not parallel the well-known identity of the j factors for mass transfer and heat transfer because of form drag.

MeOH and styrene synthesis. In addition to displacement of the chemical equilibrium by selective removal of product from the reaction zone, CD offers related options to return excess reactant to the catalytic zone to complete its conversion, either by stripping up from the bottoms or by reflux from the overhead. Prompt removal of product from the catalyst zone can also minimize an unwanted sequential side reaction.

CD requires a mix of reactive (catalyst plus packing) and nonreactive (tray or packed) zones. A transform system developed by Doherty may be useful in configuring a CD system (Barbosa and Doherty 1988). Naturally, the volatilities of reactants and products need to be in a certain range and order to make best use of CD. But changing the nature of the third phase is also an option, as in the flowing adsorbent cited above that removes MeOH downward more feasibly than liquid reflux could at the 200°–250°C temperatures involved.

A conscious acceptance of the graft nature of technical improvements can guide the technologist and chemist to a more effective groundwork for a CD process improvement (see the discussion of planning below). For example, the limiting factor in promulgating wider use of CD is finding catalysts that provide a reaction rate equal to or slightly greater than the G/L mass-transfer rate achievable in a distillation column. The latter rate is roughly 0.1/sec with respect to liquid-phase residence time (consistent with 1 to 2 transfer units in a 15-sec residence on a distillation tray).

At normal liquid-phase molar concentrations of reactant and roughly equal masses of catalyst and liquid per column volume, this means that we must seek reaction rates of about 0.1 gmol/L sec, two to three orders of magnitude faster than common liquid-phase catalytic reactions operating at near-ambient temperature. MTBE synthesis by CD over ordinary CERs is an exception made possible by the ease of protonating isobutylene.

Reversal of CD

The above process makes distillation the handmaiden of reaction. A simple reversal uses a (reversible) reaction to help the separation of isomers and other close-boiling binaries (Terrill et al. 1985). A strong base or acid, utilizing a differential in the acidity or basicity of hydrocarbons, cresols, and so on, is a common agent. The preference is to operate like extractive distillation, with the reagent going down the column, reacting preferentially with one isomer. Of course, if the reagent catalyzed isomerization, the reaction could also be used to drive to 100% yield of one isomer.

Reconfiguration of Contactors

Membrane Contactors

As in monolith catalyst development, one line of high-tech thinking in mass-transfer operations consists of working from the recognition that form drag on bubbles and droplets contributes little to the mass transfer mission relative to pressure drop expenditure. It seeks to replace chaotic tumbling of one fluid against another by a one-way pattern of tubes whose walls are *porous membranes*, one phase flowing inside, the other countercurrently outside. Such systems are a growing factor in gas separations and in G/L and L/L contacting applications. In another area of innovation, we now see cartridges that work on wetting or surface tension principles in common use in oil/water separations to hold back a minor phase.

Unsteady-state (USS) Separation

In Chap. 6 we noted the utility of the unsteady state (USS) for special performance needs. Adsorptive fixed-bed separation is inherently USS. However, the advent of pressure-swing adsorption (PSA) gave an extra transient twist to the regeneration step. The productivity of this graft has led to an almost autocatalytic series of improvements: ingenious layouts of multiple beds in various cycles, breakthroughs in automatic switching valves, and so on. It has become a dominant technology in O_2 and N_2 concentration from air, CO production, and many other small- to medium-scale fractionations.

Planning for Breakaway Technology

A narrow viewpoint might argue that such departures should not even be considered by technologists in an operating plant until the materials innovation side of, say, producing porous tubes or membranes is well advanced. However, innovations such as this realistically require both a push and a pull aspect. The technologist who, sensitive to process needs, helps to provide pull can significantly advance the commercial availability of, say, a porous-tube contactor as a graft to improve the performance of an established process.

The marriage of new technology and need comes most easily when both reside within one organization. Union Carbide was in this position with their Linde Division, seeking process applications for O_2 or O_2-enriched air where this more costly form of air had advantages. Through Union Carbide's contact with biotreating, as a chemical process operator with effluent treating needs, they recognized the potential of O_2 to augment the capacity or effluent quality of existing effluent biotreaters; this took shape as the UNOX system.

Powdered activated carbon is a useful adsorbent for removing heavy toxicants from aqueous effluents in terms of the favorable equilibria and variety of compounds covered. However, handling complexities due to its shape and the difficulty of regeneration are a common obstacle to its use. It is not surprising, then, to see research in the advanced materials sector on the synthesis of (weavable) C fibers, despite significantly higher cost.

Retrofitting a Novel Separation System

Supercritical fluid extraction (SFE) is a relatively new, much-touted unit operation, the sort of technology we need to consider for grafting to an older process. Although SFE provides unusual solvency for heavy molecules,[112] it is likely to be of interest only for solutes that justify treatment costs on the order of $1/lb. This criterion is likely to be met only in special situations like removing low levels of toxicants from a low-volume effluent. How can we find an opening for this technology in regular process operations? The example below shows such an opening, arrived at simply by keeping the potential of SFE in mind and waiting for the right partner.

Example: SFE for Regeneration of an Expensive Catalyst

SFE was investigated on a laboratory scale in a process development seeking to extend the life of a noble metal catalyst. This catalyst was on a support such that thermal regeneration was not feasible; hence each time it fouled (from polymer by-products of the

---

[112] Introduction of cosolvents is providing another dimension of flexibility in tailoring SFE to applications (Walsh et al. 1989).

reactant), it would, in a commercial realization of the process, have to be sent to an off-site chemical reclaimer, at a cost of 10% of the metal value, plus processing charges. Laboratory results indicated that batch SFE of fouled catalyst in an on-site treating vessel, geared to timely dissolution of the tarry polymers, restored sufficient catalyst activity to extend the overall catalyst usage time between reclaiming processings by a factor of 5. This was, unfortunately, not enough in itself to allow the development to go forward to commercialization.

New Horizons for Processing

Suppose that an interesting new concept surfaces in the literature.[113] How can a technologist structure its possible intersection with her process needs, in a way that also sharpens and focuses the concept itself, without going overboard in what might be wasted effort? For example, suppose that she suspects that unsteady-state operation (USSRO; see Chap. 8) can confer significant selectivity or productivity advantages on her reaction system. She also recognizes that much of the infrastructure of USSRO needs development before it can be a commercial option. She decides to go just far enough to delineate the transient catalyst surface chemistry in a cyclic operation and to rough out the time base of an advantageous USSRO cycle.[114] Then she puts its aside, awaiting concrete developments by others in the broader technical community.

This may seem disconnected and drawn-out to those steeped in project mentality. But it must be recalled that we are talking of process support over a 5- to 30-year period. There is plenty of time to nurse ideas along until the moment is ripe.

This chapter has emphasized using our acquired knowledge of shoals and currents to sail our process ship purposefully and prof-

---

[113] Uncritical enthusiasm is not suggested. For example, ultrasound is perennially touted for process chemistry, but I would assign a higher likelihood to e.g., laser-initiated reactions (see Clark et al. 1984) for commercial utilization in our time.

[114] I became aware of the surprisingly long time constants for gas/surface dynamics in considering the decay and renewal of VPM action on a heterocatalyst (see Chap. 11). For example, in CO or S moderation, where as many as half of the sites on a metal catalyst are to be covered by a trace component that is present only at the ppm level in the vapor, the arithmetic of the traffic gives a time constant of many hours for renewal or removal. Typically, adsorption/desorption of such modifiers will involve hysteresis as well.

itably. In the next chapter, we address two strong tides of events that command our attention, whether we will it or not.

References

Barbosa, D., and Doherty, M.F. *Chem. Eng. Sci. 43* 1523–37 (1988).

Clark, J.B., et al. *SPIE 458 Appls. Lasers to Ind. Chem.* 82–8 (1984).

Grieves, R.B., *Treatise on Analytical Chemistry* Part I, Vol. 5, ed. P.J. Elving, Chap. 9, Wiley (1982).

Lemlich, R., ed. *Adsorptive Bubble Separations* Academic Press (1972).

Moser, W.R. Paper 79f AIChE annual meeting (Nov. 18, 1991).

Newman, S.F., et al. U.S. Patent 4,885,399 (to CATALYTICA) (Dec. 5, 1989).

Terrill, D.L., et al. *IEC Proc. Des. Dev. 24* 1062–71 (1985).

Walsh, J.M., et al. *Chem. Eng. Comm. 86* 125–44 (1989).

Westerterp, K.R., and Kuczynski, M. *Chem.-Ing.-Tech. 58*(12), 929–33 (1986).

16

Safety and Health: A "Must" with a Price

We are absolutely certain only about things we do not understand.
                                                            Eric Hoffer

Troubleshooting in our time is involved with the obligations of safety and health (S&H) in several important ways:

1. As considerations forcing plant changes (like closing off vents) that entail new problems of process performance (yield, reliability), which troubleshooting must counter.
2. As considerations that may inhibit plant tests, sampling, and other activities needed to define a process problem.
3. As opportunities for troubleshooters to use their skills to provide least-cost operating solutions rather than a capital-intensive total rebuilding of a unit.

Below are vignettes from my experience with the qualitative and semiquantitative judgments that we can bring to hear on safety and health problems. I stress developing problem-opening insights and avoiding major blunders rather than complete assessments. "Loose talk" (i.e., order-of-magnitude thinking) will often be *useful and appropriate* in sorting out the significant factors in S&H, just as it is in the areas of troubleshooting already discussed.[115] We all wish that safety were an exact science; we all know that it is not.

The reader is challenged here to bring global value issues into his or her technical decision making. This is not just a matter of personal ethics. It is a survival issue for technologists because of the waste and distortion that cripple their everyday work if they avoid such issues. Because of the broad nature of S&H claims, it is particularly important here *not to expend a great deal of technical energy on the wrong problem.* Trevor Kletz, a shrewd writer on safety, suggests that time is better spent looking for all the sources of hazard in a situation rather than quantifying with greater accuracy the subset already identified.

Cow 1 is not Cow 2 (B)

This covers situations where we are working on the wrong problem through misdefined categories.

One example is the so-called ozone (smog) problem in the airsheds of many major cities. Smog is now understood to result from the chemical interaction of reactive hydrocarbon emissions with natural and man-made oxidants, principally the NOx from automotive and other engines. Bureaucratic conformism has applied hydrocarbon emission control programs designed for the Los Angeles basin to many other urban areas. In Los Angeles, NOx is abundant and hydrocarbons limit smog formation. However, in other cities, such as Atlanta and Houston, hydrocarbons are abundant and it would probably be more effective to attack NOx formation as the limiting factor.

As another example of the wastefulness of misplaced emphasis, we seems to be spending most of our regulatory energy on the 5–10% of toxicant deaths that are reasonably attributable to synthetic chemicals. Ames' (1989) story of the many *natural* toxins that exist and the value-of-a-human-life argument given below will, I hope, persuade the reader of the need for a more balanced approach.

---

[115] I make no apology to those for whom safety is an absolute, in either a quantitative or a moral sense or because they feel compelled to follow orders. But in order not to mislead, I have labeled the sections that follow in terms of whether we are dealing with (A) general social philosophy; (B) a loose sorting of technical issues; (C) examples of order-of-magnitude quantitation; or (D) a situation where exact quantitation is achievable and should not be shirked.

A Balanced Agenda for S&H Activities (A)

Much of the technical work at operating plants over the past several decades has been in response to S&H issues. This (often mandated) activity has displaced profit-seeking process improvement projects, in some cases completely off the agenda. But in the years ahead, S&H projects are likely to rejoin the mainstream in the sense that they are subject to economic criteria (albeit with the once-neglected social costs included).

Economic Realism and Limited Resources (A)

The reason we need to attach (large but) finite values to life and health is that this is the only way we can put any kind of bound on extensive measures to minimize risks to life, limb, and property. Without such a bound, we misdirect our limited energy and money resources even within the domain of S&H. As we are slowly learning in the medical care system, overvaluing one need leads inexorably to slighting a competing claim that may be equally or more worthy.[116] It is one thing for emotional laypersons to assert that life is beyond price; but when responsible technical people, in the glare of media publicity, acquiesce, they do a disservice not only to their organization's balance sheet, but to the cause of S&H itself.

Both sides of the balance pose challenges: the realistic valuing of injury to life or health and the probability that countermeasures at a particular cost will be effective. In the earlier days of the National Aeronautics and Space Administration, we in the petrochemical industry thought that we could not aspire to the 99.9999% assurance that we believed characterized the manned space program. After the self-destruction of the Challenger spacecraft on its launching pad, it became clear that finite budget and calendar constraints were limiting the safety achievable in the manned space program as well.

Benchmark Values for Deaths and Risk of Death (B)

General societal acceptance of benchmark monetary criteria for loss of life or health is the first step in making judicious technical work

---

[116] As I write, the state of Oregon is proposing to make Medicaid available to *all* the indigent, at the price of making sharp judgments among medical procedures on cost effectiveness.

in the S&H area once again possible. This has been done to some extent in the past but usually quietly and discreetly, without explicit adoption of standard methods. The resumption of a 65 mph rural speed limit, with a well-established consequence of thousands of added traffic fatalities, is sure evidence that American society does in fact put a finite value on human life. A recent newspaper article cites hesitancy by the civil aviation authorities to make airlines install the SAVE system of pressured water tanks in jetliners, shown to give passengers critical extra minutes to evacuate before flames engulf after a crash landing. The reason for the hesitancy is that the fuel and passenger-capacity wastage of carrying the extra weight on every flight works out to $18.6 million per prospective life saved.

Risk Criteria (B)

An alternative approach is to work only against an accepted risk of a fatality, for example, per 10^8 hr of exposure (FAR) (Gibson 1980). There would be a near-uniform standard for the exposed public, based upon common risks that are accepted with equanimity. For example, the annual risk of injury from automobile collision or home accidents is on the order of 10^{-4}, accepted willingly by most of us. There might be a different, looser, standard for risk to an employee, set to an industry norm, on the basis that there is a voluntary acceptance of risk in taking employment.

This second approach fails to provide an allocation criterion for technical resources in a situation where such resources are torn between safety and economics-driven activities. Even with the FAR approach, however, it is possible to do some societal allocation based upon simple ranking. For example, given the readily accepted automobile collision risk cited above, how much effort should society spend on further cleanup of drinking water containing trace chloroform at the Environmental Protection Agency limit, where the risk factor is more than two orders of magnitude less?

Matching the Tools to Responsible Concern (B)

Despite what may be perceived above as a less than compassionate attitude, I was an early community activist trying to force considerations of fault trees[117] and other risk assessment techniques onto the permitting process for new plants. In earlier days, securing pro forma balance of risk against investment would have made us very happy. Even without quantification, the imaginative construction of several dire-event scenarios promotes a priori safety thinking by designers and reduces the exaggerated anxiety of the lay public by giving formless fears a name.

Quantitative Assessment of Risk to Surrounding Population from Cl_2 Emissions

The early efforts of ICI (Sellers 1976) in quantifying hazards to the surrounding population of potential Cl_2 emissions from a chloralkali plant encouraged me in efforts to secure an appropriate safety review mechanism in the permitting process. Theirs was a landmark study, both in setting an acceptable (small but) finite fatality risk, over the lifetime of the plant, from an accidental Cl_2 release and in relating this criterion to plant layout and operational safety strategy.

Excessive Sophistication Too Early (B)

As noted earlier, when events bring a new tide of needs for technical conceptualization, the first response often comes from mathematical specialists using elaborations that would be more appropriate to a mature art than one just taking form. Instead of computerized fault trees and discrete-event simulations, what we need at this early stage is research on factors like failure modes, life expectancies of devices, the "ergonomics" of presenting information to an operator at a control panel or video display, or an understanding of detoxification mechanisms in the human body. The ubiquitous tendency in our society to address new problems with what we know how to do, rather than what the problem requires, is a misallocation of intellectual resources as profound as working on the wrong problem.

---

[117] A fault tree systematically sketches cause-effect chains that connect primal failure events to the ultimate accident, with its safety, performance, or environmental damage implications. There is a temptation to computerize such fault trees (Powers and Tompkins 1974), moving from probability of each of the initiating failure events to a composite probability of occurrence of one or more ultimate accidents. In most situations, it is more appropriate simply to sketch a fault tree to ensure that we have included all the events and pathways that may be operative rather than to proceed to a full calculation (see Kletz's remark above).

Paralleling Part II of this book, we now shift from a general orientation to a discussion of some classic S&H risks and the tools available to deal with them.

Atmospheric Dispersion of Toxic Spills and Emissions (C)

Consider dispersion of a toxic vapor released near ground level. Even with a steady point source emission, there are numerous vagaries affecting three-dimensional spreading to the downwind boundary of concern: the logarithmic air velocity gradient above the ground, the scale of atmospheric turbulence, and density and temperature factors. It obviously makes considerable difference when an inversion layer gives effectively two-dimensional spreading. But when these uncertainties are superimposed on a once-in-5-years undefined event of accidental release, we must be cautious in what we claim for a computer model's outputs.

> I have seen able but overworked technical people pursuing a fill-in-the-blanks approach to risk assessment using a computer model designed for relatively light, completely vaporized sources, when the risk they were addressing involved a liquid mist whose properties were not at all embodied in the mathematics used to generate the computer program. Much of the formalized attention to environmental risks, whether elicited by anxious managers or staid bureaucrats, has this characteristic: seeming to address a problem, with usage of appreciable technical time, when in fact nothing useful is being accomplished. A more down-to-earth approach described by Blewitt et al. (1988) to a particularly nasty mist emission—HF from an alkylation unit, pipeline, or tank car—included not only mathematical modeling but field tests and the use of positive-response devices as well (e.g., water spray curtains).

Responses to Accidental Releases (B)

There are only three elements in addressing accidental releases with environmental implications: prevention, containment, and alerting those likely to be affected once the event takes place. *Prevention* includes activities like corrosion surveillance. Here it is worth noting that the use of lined pipe in lieu of solid metal, while it may well be appropriate to the situation, involves demanding design considerations. Solid metal tends to tolerate unforeseen changes in service conditions; composites like steel/glass or steel/plastic can provide high performance but require careful specifications and a relatively narrow arena of use. Lack of early attention to thermal expansion, possible periodic vacuum conditions, and other factors may lead to a failure much more unexpected than with a metal of known slow-corrosion behavior.

In addition to spray curtains, *containment* can comprise ditches, ponds, and chemical scrubbers that are activated only at need (e.g., for emergency shutdown of a Cl_2 line). As for event *alerts*, dispersion modeling (e.g., for a variety of wind conditions) should be viewed as an illustrative exercise, not a recipe for planning. The release may occur under quite unexpected circumstances; the emphasis should be on flexible and rapid dissemination of warnings and information to those with a right to know, including the surrounding community.

Control of Vents and Fugitive Emissions (C)

In this area, the search for quality devices has had rewarding results. Double-sealed pumps, new diaphragm materials, and catalytic combustion or adsorptive cartridges on unavoidable vents are examples of rapid progress under regulatory duress. Of course, flares (with associated scrubbers) are still a mainstay. Here we emphasize sensitivity to seemingly minor elements in the scheme whose reliability may be critical. For example, pulsing flow from a compressor may cause erratic combustion, loss of ignition and other problems.

Testing and Redundancy (B)

One common method for increasing reliability is redundancy. We may use twin instruments to reduce greatly the likelihood that instrumental failure will provoke an incident. For example, if one instrument has a failure rate of once in 5 years, use of two instruments promises 25 years of coverage (i.e., both instruments would have to fail simultaneously to create a problem). This reasoning has two potential flaws. The first is that we may not be able to tell which instrument's reading is wrong. The second problem is that if the instruments are truly "identical twins," they may well both fail in a single episode. Even when we spare an electric drive pump with a steam-driven one, we must address the likelihood that both steam and electricity utilities may fail together due to a single event.

Shutdown and startup are considered dangerous periods because the terrain is unfamiliar compared to steady-state operation. There is a tendency to provide some form of automated shutdown on large units (e.g., with battery-operated instrumentation) and to back this up with testing exercises. In one case, a clever control engineer demonstrated that testing such a safety device too often could increase the risk! This was because human error was a major element in executing the test and could shut down the plant when it occurred.

Layout Factors in Effects of Emissions

Series Interactions of Plumes (B)

Occasionally, a relatively inoffensive plant emission interacts chemically with a downwind emission from another unit to produce a much more offensive air contaminant. An unusual eye irritant formed by interaction of styrene and halogen plumes, with assistance from sunlight UV activation, was vividly described by Adams and Schneider (1952). Not only was the irritation threshold down in the ppb range, but the interaction was rediscovered in two other locations over an 8-year period! Similar plume interactions have been noted in other areas where the terrain channels wind movements, as in the Kanawha Valley of West Virginia.

Turbulent Wakes and Other Seeming Oddities (B)

When air flows over a nonstreamlined intervening object, the stream breaks into a turbulent back eddy. One example, for those of us pumping self-service gasoline into our autos, is the difficulty of avoiding inhalation of fumes, try as we will to face away from the pump. I first came across this phenomenon in the modeling of hydrocarbon emissions reaching a nearby road in the event of loss of ignition at a plant flare. The presence of a large intervening structure, far from helping, gave higher ground-level concentrations at the section of the road in the wake of the structure.

Emissions from Ponds and Other Effluent Surfaces (C)

Organic chemicals can be lost to the atmosphere undetected from treating ponds, water-cooling towers, and other situations marked by extended surface and convective conditions. This may come to our attention via poor mass balance closure on trace components in liquid effluent. There is nothing mysterious here; the technologist is encouraged to use standard boundary-layer mass-transfer correlations, with driving force corrected for what may be high activity coefficients in the liquid phase (Mackay and Yeun 1983). For aqueous streams, a clever expedient for predicting transport effectiveness is to link the estimation to its converse: reaeration pickup of O_2, known in a wide variety of contexts (Smith and Bomberger 1978).

Upper Atmosphere Environmental Issues (C)

The significant but indirect threat to the earth's protective ozone layer posed by chlorofluorocarbons (CFCs) and other halohydrocarbon emissions has recently come into technical definition (Chemical and Engineering News 1987; the estimation of Rowland 1989). In many respects, risk here is cleaner and more quantitative than that in estimating the health impact on downwind individuals of low-level toxicant plumes. The issue of ozone shield attrition involves a number of the methodologies we have addressed in earlier chapters:

1. The use of skin cancer incidence in the general population as an index of damage has gained broad acceptance.
2. The technologist is given a social cost adjunct that provides a more informed basis for complying with or contesting a regulatory requirement for tightening up on the plant's vents of relevant compounds.
3. The FR kinetic estimation methods and literature discussed in Chap. 10 are very useful in the search for more innocuous compounds to replace the Freon family.

Chemistry of Ozone Depletion

For example, it has been known for some time that Cl· atoms initiate a chain reaction that converts ozone (O_3) to ordinary oxygen (O_2):

$$Cl\cdot + O_3 \rightarrow ClO\cdot + O_2$$
$$ClO\cdot + O\cdot \rightarrow Cl\cdot + O_2$$

The ozone and singlet O are formed in the stratosphere by solar UV bombardment of molecular O_2. However, the ozone accumulates in a layer somewhat lower in the stratosphere, where it constitutes a shield that protects us at the surface of the earth from high levels of UV radiation. This shield also protects the CFCs from UV-stimulated dissociation (e.g., $CCl_2F_2 \rightarrow CClF_2\cdot + Cl\cdot$) on their turbulent diffusional path into the stratosphere. However, when the CFCs reach a level about 30 km above the ozone layer, they do become dissociated. The mystery has been how the $Cl\cdot$ released at such an elevated altitude descended to launch the ozone depletion cycle.

Research on the "ozone hole" over Antarctica reveals that descent occurs when $ClO\cdot$ combines with NO_2 to form $ClONO_2$, which is inert but heavy. Heterogeneous clouds form in winter over Antarctica, where molecular reactions with HCl and water release Cl in active form from the reservoir $ClONO_2$:

$$HCl + ClONO_2 \rightarrow Cl_2 + HNO_3$$
$$H_2O + ClONO_2 \rightarrow HOCl + HNO_3$$

The first sunlight of spring releases $Cl\cdot$ from these species, with massive ozone depletion. FR chemistry also suggests a practical remedy for ozone depletion: substituting hydrogen-containing analogs of CFCs as refrigerants and propellants; they react with $\cdot OH$ radicals, abundant in the troposphere, to intercept the Cl content before it can ascend to the stratosphere:

$$CHCl_3 + \cdot OH \rightarrow H_2O + CCl_3\cdot$$

Mistaken Judgments When One Model Element Is Wrong or Missing (B)

My own environmental thinking in this area went wrong (I was promoting a process application of CCl_4) in reasoning that vented heavy molecules like CCl_4 do not reach the stratosphere. It turns out that large-scale atmospheric turbulence, not molecular diffusion, controls this upward transport to altitudes of 100 km and that vented CCl_4 would indeed reach the stratosphere.

Fires, Explosions, and Detonations (B, C)

Factor Elimination in Minimizing Explosion Hazards (B)

We begin by recognizing the three antecedents for an explosive energy release: the coincident presence of a fuel, oxygen,[118] and a source

---

[118] Cl_2 can replace O_2 as an oxidant. Rozlovskii (1982) explores the similarities and differences between the two. Dokter (1985) reviews typical fire/explosion hazards in Cl_2-containing systems. Degnan (1983) describes a curious hazard from wet Cl_2 attacking steel, via H_2 generation and H_2/Cl_2 reaction, that raises the metal temperature until Fe combustion by Cl_2 vapor results.

of ignition.[119] This triad leads in several directions for further safety thinking. The first might be to consider whether we need to eliminate two or only one of the three factors in order to be safe in a given situation. For example, inerting the atmosphere with N_2 seems robust enough[120] in itself to block combustion, but we need to recall the long series of fatalities occasioned by lone operators entering, say, a N_2-blanketed pellet storage tank without an air-support system.

There are explosive regions involving little more than rapid pressure rises, where we are tempted to settle for containment by high pressure rating, as in the "cool flame" region of a H_2 flammability envelope. There will always be an explicit or implicit economic pressure to settle for the one-factor-elimination solution; a qualitative tension between such pressure and a professionally satisfying assurance of safety must be maintained.

Causes of Accidental Release of Combustibles (B)

A common source of fires is a release of hydrocarbon or other flammable agent due to sudden failure of an elbow or section of pipe. Once this release ignites, other complications may rapidly ensue. But let us pause to consider the initial pipe rupture. We cannot expect materials selection to be omniscient. How can we improve preventive maintenance to catch such failures before they happen?

Unexpected localized internal corrosion is one frequent antecedent of pipe failures. The thinning of the pipe wall is in principle detectable by ultrasonic or other external inspection devices. However, it is not practical to scan every square inch of piping surface. I suggest inspection programs[121] that give preferential scanning to fluid mechanics situations that are classic sources of localized cor-

---

[119] There was previously an implicit assumption of confinement in a vessel. The Flixborough explosion, cited below, demonstrated that confinement is not necessary, given a sufficient mass of released fuel, mixed with air by the mode of its emission. This massive explosion drew worldwide attention (see Warner 1975) to unconfined vapor explosions but was by no means the only industrial event of its kind (Davenport 1977). Military use of such "deflagrations" by and to Iraqis has become known since the Gulf War.

[120] The atmosphere need not to be devoid of O_2; bringing the O_2 content down below 8–10%v usually eliminates an explosive event. Additional discussion of fuel/O_2 proportions defining the flammability envelope is presented below under static electricity hazards.

[121] Online inspection techniques are necessary for the many processes that go several years between shutdowns for maintenance.

rosion, for example, boundary layer thinning at the entry region to a pipe, or at elbows and the piping just beyond, where new flow profiles are established with considerable high-frequency "flutter". Rapid alternation of regimes may be as significant for enhanced corrosion as boundary layer thinning per se. A dramatic example of catastrophic corrosive failure from slug flow effects in a CO_2-containing pipeline on the Alaska North Slope was recently described by Jepson et al. (1991).

On the chemistry side, we should look for unusual corrosional combinations of two or three ingredients that would not have had the attention of the original design team. Before me as I write is an article suggesting that (contrary to some earlier judgments) the critical event in the 1974 Flixborough cyclohexane vapor cloud explosion may have been rapid embrittlement of a stainless steel pipe section via *external* impact of molten Zn droplets released from nearby galvanized equipment by an earlier fire.[122] While this might seem an improbable composite event, a full preventive fault tree for a variety of internal and external corrosion scenarios might well have indicated how a trivial fire could launch a sequence of events with a disastrous culmination.

In another episode, a small, light hydrocarbon leak from an aluminum pipeline conveying powdered polymer product ignited and burned up the Al from the outside. The expectation that the critical event will come from inside the process is the kind of "domination" that LT helps us to overcome.

Hazards without Oxidants (B)

Another important situation is where only two factors can constitute a hazard, as in an exothermic reaction of a compound that can provide destructive force without the need of an oxidant. Fragmentation to produce more molecules, as with peroxides or nitro compounds, adds more explosive punch, but runaway polymerizations are hazard enough, as in tank-car exotherms during shipping that lead to rupture and a fire or toxic gas release. There is much activity on (mathematically and experimentally) screening quasiadiabatic runaway hazards of reactive commodity chemicals in storage vessels (Kohlbrand 1985). But it seems to me that other, seemingly minor,

---

[122] Interestingly, the reactor failure that led to the use of this temporary bridging piping and the ultimate explosion was itself a result of stress corrosion of a mild steel jacket from external spraying with nitrate-treated cooling water.

factors merit increased attention. These include inhibitor packages (and means for removing them) and minimizing exterior heat pickup (from fire and sunlight). It is certainly true that societal pressure, justified or not, is forcing many companies to adopt just-in-time manufacturing sequences, where a dangerous intermediate is made and consumed on the same site, thereby minimizing both inventories and handling.

Compounds such as ethylene oxide can deflagrate without O_2. Even without polymerization, opening the cyclic ether structure, (e.g., to acetaldehyde) is an energy-releasing transition. Less well known is the exothermic "analytical decomposition" of compounds such as ethylene via pipeline walls that have been inadvertently heated to a danger point, with CH_4 and C as the chief products. (Normally C—C bond breakage is endothermic.) I once pursued the possible parallels to this phenomenon with 3- and 4-carbon olefins. The conclusion was that the phenomenon is much less significant, presumably because of the higher proportion of endothermic cracking reactions and the higher molar heat capacity of the higher-carbon-number olefins.

Explosions from a Metastable Phase Transition (D)

An interesting pseudoexplosion is a so-called superheat-limit vaporization (see the earlier presentation in Chap. 7). No chemical energy is involved. It occurs in situations like liquefied natural gas (LNG) leaking into water or extremely fine water droplets suspended in a heated oil stream. Surface tension resists vaporization of a fine droplet until it is well above its boiling point. Then, at a temperature typically 0.88 of the critical temperature of the liquid in question, there is a microsecond transition from liquid to vapor, releasing all the stored sensible heat of the droplet (Katz 1972; Reid 1976). Although the publicized cases have involved LNG tanker leaks, I believe this phenomenon underlies many mysterious pressure surges inside plant distillation columns that end with all of the trays being found in a heap at the bottom upon inspection after weeks of poor fractionation.

Detonations (C)

The energy release per unit mass of an exothermic decomposition is of high importance, both in terms of speed of propagation (e.g., detonation-type velocities in pipes) and ultimate damage. A *1 kJ/g energy release* appears to be a useful rule of thumb for initial screening of the detonation risk, particularly for liquids. I have seen this used on ternary diagrams to delineate roughly a detonation region inside a broader explosive envelope. Distance, however, is also a factor; it takes some pipe length before a marginal case can build to detonation velocities. Deflagrations may propagate at a rate of 0.1 m/s or less, while detonation velocities are typically 2000 m/sec.

Organic peroxides can provide energy for a nasty explosion, particularly in laboratory work. Since peroxides formed from trace autoxidation chemistry, as in storage of an ether, are likely still to be a minor component, safety concerns should focus on the concentration mechanisms needed to reach energy release potentials approaching 1000 J/g. This can, of course, happen in distillation, with the peroxide as a heavy bottoms residue. However, what is involved in many laboratory incidents of uncapping an old reagent bottle is a discrete peroxide phase that has formed under the cap, either by organic vapors meeting intruding air there or by precipitation of a polymeric polyperoxide.[123]

Minimizing the Impact of Flammables Release, Fires, or Explosions (B)

Finally, we consider the measures we can take to mitigate a fire or explosion once it occurs. Examples are fire walls to prevent involvement of other units and relief panels for timely venting of dust explosions in a harmless direction (Bartleknecht 1977). Some of these ameliorative approaches verge on the preventive, as in spraying of quenching halogen compounds into vessels within microseconds of a sharp pressure rise,[124] pioneered by Fenwal, Inc., and others (Rae and Thompson 1979). Another example is the use of dilution, small-diameter pipes and sections packed with (preferably wetted) metal or ceramic packing to mitigate acetylene decomposition, developed by the Germans during World War II (Ivanov and Kogarko 1964). Here the principle seems to be attenuation of the chain reaction by thermal sinks and perhaps also surface trapping of FRs.

---

[123] Alternating polyperoxides from dienes, for example.
[124] As in a dust handling explosion, where the rate of pressure rise is slow enough to permit sensors to react.

Risk with No Obvious Ignition Source

Although ignition sources of carefully controlled energy (e.g., sparks) are used to determine experimentally flammability or explosive envelopes, the absence of a putative ignition source should not be relied upon for safety in a problem situation. For example, a jet of hydrocarbon from a pipe or flange leak may self-ignite *from static electricity effects alone*. Or fluids may *autoignite*, either by unexpected release of a very hot liquid into the atmosphere or via the impact of a liquid routinely exposed to air upon a hot surface which is above its characteristic autoignition temperature (T_{ai}) (Hilado and Clark 1972). This temperature is presumably set by an autoxidation that becomes sufficiently rapid to overcome heat dissipation available at the exposed surface of the liquid.

Example: Eliminating an Autoignition Risk

I was surprised when we found that toluene had a major advantage over *n*-octane as a candidate heat-recovery fluid due to its high T_{ai} (482°C vs. 206°C). I have not tried to explain this difference in terms of chemistry fundamentals.

Post Facto Investigations (C)

Although our safety thinking must emphasize prevention, the after-the-fact reconstruction of the source and energy of a blast calls for many troubleshooting-type skills. Patient collecting of fragments and logging their mass and distance from the blast center can reveal the nature of an explosive event. For example, in the 1969 Texas City vinylacetylene (VA) distillation explosion (Buehler et al. 1970; Freeman and McCready 1971; Keister et al. 1971), this type of investigation estimated the mass of VA concentrated in an accidental total reflux operation. Blast energy can make or break a detonation scenario, for example. Such findings can be central to revising startup or shutdown procedures to avoid future incidents.

Explosions of Acetylene and Acetylene-derived Compounds (C)

A long series of explosions and/or detonations is associated with the decomposition of acetylenic compounds. Rather than recount incidents, let us try to pull a thread of quantitation and general understanding through these phenomena. This will allow us to use troubleshooting-type thinking in addressing proposed preventive design and operating measures in handling these compounds.

The energetics of acetylene C_2H_2 decomposition parallels the analytic decomposition described above for ethylene. Mechanistically, a C_2H_2 explosion probably is a chain reaction carried by $C_2H\cdot$ radicals. However, the heat evolution from C_2H_2 (to $3C + CH_4$) is about 10 kJ/g vs. 4.5 kJ/g from C_2H_4, 0.8 kJ/g for propylene, and 1.0 kJ/g as the rule of thumb cited above for detonability. If we are dealing with substantially diluted acetylenic streams, the hazard is much less. A compound like vinyl acetylene is harmless as a minor component, but its inadvertent concentration in a column led to the explosion cited above.

For C_2H_2 gas, defined[125] boundaries exist for deflagration and detonation as functions of pressure and tube diameter (Sargent 1957). For example, acetylene can be handled in a 2.5-cm-diameter tube without deflagration at up to 33 psia and up to 47 psia without danger of detonation; the respective limits are 24 and 31 psia in a 5-cm-diameter tube. If we must use larger-diameter pipe, we can invoke the dilution or packing cited earlier.

Copper Acetylides (C)

The danger in the formation of cuprous acetylides can be viewed as an alternative way to create a discrete phase[126] that meets the energy/g criterion for detonation out of a dilute initial acetylene concentration, analogous to the solid peroxides around a flask stopper described above. Presumably chemical equilibrium lends itself to this kind of concentration in the case of Cu.[127] However, the applicability of thermodynamics to these phenomena points the way as well to forestalling them without necessarily abandoning Cu metallurgy. Dilution effects, either in metallurgy (brasses containing less than 50–65% Cu are reportedly relatively immune) or in competitive occupancy of the Cu surface have been shown to prevent explosivity.

---

[125] Boundaries differ slightly for vertical vs. horizontal pipe.
[126] The acidity of the α-H of acetylenes, as well as the insolubility of the Cu acetylide, probably also plays a role.
[127] Cu acetylide is used as a catalyst in Reppe chemistry, the combination must be reasonably labile. The catalyst form is presumably made safer by the heat capacity of the support.

An example of the latter is surface blockage by sulfide ion (Nosetani and Sato 1971). Here the free energies of formation of the species confirm that the equilibrium $Cu_2C_2 + S^= + 2H^+ \rightleftarrows Cu_2S + C_2H_2$ is very far to the right.

Static Electricity (C)

Early in my career, I became involved in systematizing the investigation and prevention of fires and explosions originating from electrostatic discharges (Saletan 1959). This field has strong similarities to troubleshooting generally: multiple highly interactive causative factors, which need be quantified only to order-of-magnitude precision to permit judgments to be drawn. For example, the streaming current in a flowing liquid represents charge detached from the pipe wall boundary layer and will range from 0.01 to 1 μcoulomb/gal of flow. The lower end of the range will apply with low velocities and low-conductivity fluids, the upper end with high velocities and two-phase mixtures. In many flow situations, the electrostatic accumulation can be adequately estimated by balancing incoming coulombs of charge (μcoul/gal × gpm) against prospective conductive leakoff of charge.

There are many explosions in which static electricity's role as the ignition source is in dispute. Consider a flange leak spraying out a liquid/vapor dispersion of a light hydrocarbon into the atmosphere. If the emission were simply a vapor cloud, it might drift along the ground until it found a flame to set it off. But with the two-phase situation, I believe self-ignition occurs from static electricity almost coincident with flow out the leakage point. A comparable dispute on grain elevator explosions[128] is unresolved; the insurance industry seems to prefer attribution to thermal or flame ignition, perhaps because the liability situation is more favorable; I believe they are caused mostly by static electricity.

Hazard Factor Elimination in Electrostatics (B)

The three factors in an electrostatic problem are an explodable medium; a discharge path through that medium; and a charge accu-

---

[128] Reeves (1978) reviews the Goodpasture grain elevator explosion on the Houston Ship Channel of February 1976. This episode was followed by a series of four similar grain-handling explosions in 1 month (in December 1977) in other cities.

mulation mechanism, usually involving an interrupted flow path. How do we modify one or more factors to provide adequate safety?

Discharge Path (B)

It is generally unwise to base security solely on the absence of a path for an electrical discharge. For example, in the case of a hydrocarbon stream free-falling into a storage tank, a strong coulombic charge accumulates near the middle of the free liquid surface. Discharge is to a conducting object projecting from the tank ceiling (e.g., a sampling chain). In my experience, the absence of an obvious "lightning rod" does not provide security.

Flammability Envelopes (C, D)

The use of experimental explosion envelopes for vapors, as in the excellent work of the U.S. Bureau of Mines (Coward and Jones 1952; Zabetakis 1965), gives the envelope as a yes/no boundary. In the context of electrostatic discharges, however, much greater spark energy will be required near the edge than along the stoichiometric spine of the envelope.[129] If we are inherently at the margin (e.g., with stored toluene, whose vapor pressure is barely into the explosive region) or in a dust explosion situation, where the ignition energy required may be 100 mJ (vs. 1 mJ for a near-stoichiometric vapor mixture with air) counting on the absence of a discharge "target" may sometimes be an acceptable approach.

Spark Energy in Solids-handling Operations (C)

There are several order-of-magnitude discriminations to be made in an operation that is questioned: bulk conductivity of the contents and surface conductivity of a sack for draining charge or capacitance effects in inducing a charge on a nearby conductive collector body.

---

[129] Lewis and von Elbe (1961) give the effect of spark energy and geometry on explosive boundary. In gases, there is always some offset between the envelope for a deflagration, with a significant but slow pressure rise, and one for an explosion, but the offset is significant only for "cool flame" regions, as with H_2.
With fine particulates (dusts or mists), the lower explosive limit (we rarely deal with an upper limit here) is about half of the stoichiometric fuel/air ratio, paralleling vapor mixtures. However, as particle size exceeds 40 μm, other effects enter: For mists, sedimentation lowers the lower limit. With dusts, larger particles increase the lower limit (only the outer surface can participate in combustion on the time scale of an explosion).

The energy of a discharge is given by $0.5\, CV^2$. It is difficult to muster a capacitance term C much greater than 200 pf[130] unless we are dealing with large dust clouds or liquids in tanks with a thin insulating lining as dielectric. Therefore the voltage V will have to be about 7000 volts to have any prospect of giving a spark discharge with an energy of about 10 mJ, needed for most dust explosions. Much higher voltages than this may be observed on frictionally handled resin pellets but, in the absence of a conductive collector, the only effect may be to ionize the surrounding air, giving a corona-type electrical discharge. The energy in such a discharge is too small for an explosion unless sorbed combustible vapors are involved.

Minimizing Charge Buildup (B, C)

Conducting footwear is recommended when a spark energy ≥ 10 mJ appears possible in an operator/collector solids-handling situation. A resistance to ground R of <10 megohm is sufficient and is readily checked in the field. A final consideration is charge drainage dynamics, given by the $\tau = R \cdot C$ time constant. If the operator is inductively acquiring charge in handling a sack, with $R = 10^7$ ohms and $C = 2 \cdot 10^{-10}$ farad to ground, the time constant of 0.002 sec means that charge will drain as fast as she gains it.

In seeking to minimize charge buildup, we can try to lessen charge generation; for example, in refueling an airplane, we reduce pipe velocity or employ additives that reduce the zeta potential at the fluid/wall boundary. Or we can eliminate flow path interruption, as in avoiding free-fall entry of streams into a tank. Or we can try to enhance dissipation of charge back to ground, as in humidifying a plastics conveying belt[131] or inserting a grounded "Christmas-tree" grid below the liquid surface in a tank.

Grounding (B)

Grounding is good practice but grounding everything in sight does not always contribute to safety. Suppose that we have free fall of a

---

[130] The capacitance to ground of a human operator, whose body serves as charge collector in many electrostatics situations involving batch solids handling, is ca 150 pf [Haase 1977].

[131] I think it significant that General Electric sustained two silo explosions in the transport of powdered Bisphenol-A during a March in Indiana, when motive air would be expected to be dry (Yowell 1968). By contrast, I know of no such incident in the humid Gulf Coast environment.

relatively polar, conductive liquid into a lined tank, where the lining is an insulator. If we ground both the outside wall of the tank and the feedline to the tank, we have guaranteed the maximum voltage, and hence stored charge, across the electrical condenser made up by the wall and liquid in the tank, sandwiched around the lining as dielectric.

Specific Compound Toxicity and Safety Problems (B, C, D)

Pyrophoricity (B)

Many metal catalysts, particularly Ni, are pyrophoric in their active state. This is less of a problem with shipping (where CO_2 passivation is relatively standard) than with in-plant handling during reactor unloading and recharging.

Metal Carbonyls (C, D)

A catalyst toxicity issue is formation of the extremely toxic metal carbonyls, which are particularly threatening because of their volatility, when CO is present in gas feeds to a metal catalyst system. Formation of, say, $Ni(CO)_4$ is described by a conventional chemical equilibrium constant above 70°C; formation decreases strongly with increasing temperature (Brief et al. 1965, 1971). Below 70°C, kinetics of formation governs (Goldberger and Othmer 1963). Shutdown of a reactor with cooling probably represents the worst occupational exposure risk. The metal carbonyl may form during shutdown and then decay in a matter of hours after the reactor is opened. The accepted threshold level for affecting health with sustained exposure is only 1 ppb.

Precisely because the thermodynamics and kinetics of metal carbonyl formation and reversion are reasonably robust, we should not allow ourselves to be driven by their undoubted toxicity into adopting extreme preventive measures that exclude areas of profitable and defensible process activity. Quartic dependence on the partial pressure of CO, for example, means that there will be an enormous difference between hydrogen containing, say, 1 ppmv CO vs. a 50 ppmv concentration, in terms of the risk of a metal carbonyl health issue. Likewise, oxidation of a significant portion of a Ni metal surface with S or O substantially reduces the formation of $Ni(CO)_4$, as might be expected from the reduced activity of Ni.

Interactions and Nonlinearity in Toxicity (A, B)

I was working in a related area when the *vinyl chloride (VC)* occupational health issue came to national attention. Unfortunately, it sometimes takes exotic tumors, in a small work group exposed to a toxic compound, to elicit action when rat tests might have left us arguing for decades. Exotic manifestations overcome the statistical difficulty of inference from small populations. I wish to address here, not VC per se, but two toxicity issues that this episode highlighted for me.

The first was the strong interactive effect of smoking or drinking upon which individuals were most affected in a shared group exposure to VC. The second was that the tumors were found in operators who received high, short-term exposures in cleaning out PVC reactors. To what extent is a long-term, low-level exposure equatable to this situation? And even if it is equatable (i.e., there is no threshold below which the body detoxifies and excretes the offensive compound as quickly as it is ingested), are we talking about linear or exponential kinds of impact from varying concentration levels?

The alcohol effect illustrates the need to understand better detoxification in the body, particularly the liver. To a nonlinearity buff like myself, a *finite threshold* exposure level is appealing, but only in-depth biochemical detoxification research can settle the issue. Lacking this evidence in most cases, multiple bureaucratic agencies tend to impose successively tighter constraints. If chronic exposure to 100 ppm is proven dangerous, a prudently chosen 1 or 2 ppm workplace exposure limit can become a legislated 10 *ppb*. This is unfortunate. Since linearity is almost surely conservative as we work downward in concentration, I would argue for linearity plus placing a finite value on a death as benefit inputs (against regulation costs) for choosing workplace limits. This pattern (albeit without the dollar value) is beginning to take hold in governmental cost-benefit analyses.

Detoxification (B)

Neural toxicity of n-hexane provides some insights into the biochemical mechanisms of detoxification in the body. I was among those who advocated voluntary early displacement of n-hexane as a process chemical by substitutes. Yet why the similar compound *n-heptane* was free of a neurotoxicological onus remained unclear until health concerns were expressed about methyl n-butyl ketone. Only then did engineers like myself get a glimmer of 6-carbon centrality in toxicity and the likelihood that an intermediate oxidation product of body processing of n-hexane intake was the toxic species.

Along this line, I continue to encounter technical as well as laypeople who analogize from benzene to toluene on toxicity issues. Here, as with n-heptane, *one carbon can make all the difference*. Here it is easier to intuit, from FR chemistry, the facility in body detoxification of toluene by abstracting H from the terminal methyl group of toluene, vis-à-vis the benzene ring hydrogens.

Biotreater Operation and Layout (B, C)

The first (thing to understand) about even well-run biotreatment is that we are not getting rid of waste but rather converting it from liquid to solid form (Moores 1972). The solid is the sludge or incinerated form thereof, embracing spent biota, hopefully freed of the C they assimilated from organic compounds in the effluent. This *exchange of problems* may be desirable, but heaps of incinerated sludge around plants attests that biosolids disposal, as landfill or road base, is not always simple or cheap.

The second point is that the leverage on system performance of operating skill is at least equal to that of installed capital equipment, rather like process problems as a whole. Know-how in sludge handling and recycle is particularly important. Surveys have shown that inadequately trained supervision hurts performance as much as poor layout/design. On the capital side, poor cold-weather performance can be improved by an original design using deep ponds of limited surface to approach adiabatic operation (at a cost in extra internal circulation). Graft-on retrofits like UNOX O_2 supplementation may be less expensive than a major expansion to cope with more client units or an enlarged plant. Poduska (1979) describes some operating chemistry tools to enhance performance.

Biotreater Tests and Modeling (B)

There are two of pitfalls in laboratory modeling of biotreater kinetics: (1) the laboratory must run at modest conversions to fit kinetic rate constants efficiently (by the Monod equation or another roughly suitable expression), while the plant must run at very high conversions, a severe test of any model's fit; (2) the organisms used in the laboratory will never be quite the same as the "acclimated" multispecies biota of the plant. If laboratory studies are carried out, they can most appropriately concentrate on qualitative features.

A difficult substrate like phenol, which may be poorly biodegraded in the laboratory, does reasonably well in the plant, with organisms that have adapted to cope. Such substrates are a serious problem for a biotreater handling multiple process units. When one unit is down for a turnaround and comes up again, the biota population may have protracted difficulty in handling effluent, until "reassimilation" takes place. In studies of the dynamics of acculturation for successful biotreating of chlorinated oxyorganic compounds, lags on the order of 10 days were found unless inoculation of acclimated cultures was practiced (Shamat and Maier 1980).

Anaerobic Treating (B)

Aerobic biotreating is, of course, not the only option. Anaerobic digestion merits consideration in many cases as an alternative or auxiliary treatment, early in project concept development.

Other Clean-Water Issues (B)

Biotreating is mostly a matter of removing the "oxygen demand" from a stream; that is, the components treated are toxic only in a narrow sense. Other issues that arise in reuse of effluent water are pH, heavy metals content, bacterial counts, and the presence of "named" toxicants such as $CHCl_3$ (some of which arise in the process of water chlorination itself). New, innovative approaches to removal of named toxicants and heavy metals are constantly appearing.

Toxic Waste Disposal (B)

Let us examine briefly four schemes for disposal of toxic wastes:

1. *Incineration*: Both land and sea incineration are effective and acceptable. Proper atomization of viscous feeds and overall

maintenance are critical elements. Heavily chlorinated hydrocarbons require supplemental fuel to sustain complete combustion. Most protests come from people who are technically uninformed and driven by "not in my backyard" alarms. More legitimate concerns attend the transportation of wastes to the incinerator.

For concentrated low-volume streams, exotic variants like Na metal dehalogenation of polychlorinated biphenyls (PCBs) may merit consideration.

2. *Deep-well disposal*: At this writing, some respected chemical companies are still defending the technical viability of this method, with all the implied responsibilities of perpetual care in injecting wastes into selected strata. When a cheapest-outlay solution such as this is advanced by industry, the burden of proof must lie with its advocates. On the other hand, when the mandated treating cost mounts to a disproportionate level compared with normal processing costs, as in so-called remediation of gasoline service-station subsoil from minor fuel leaks over the years, the social agencies should bear the burden of proof of substantive harm from the prior practice.

3. *"Burning" with supercritical water*: This is a space-age solution from the same people who are planning the "hydrogen energy economy". The costs are ludicrously high except for very special cases.

4. *Reprocessing for lower-level use*: This is the method recommended by Habermehl (1988) for spent catalysts containing metals like Ni, Cu, Cr, or Zn. Landfill is the worst alternative:

Quality of People (A)

We have almost, but not quite, left the era when safety and waste disposal problems were called important but not staffed with high-caliber people. They are indeed important activities. However, honesty compels us to admit that the combination of tedium (safety is the long interval between accidents) and personal visibility only when things go wrong still deflects many able people. If we could cut out some of the waste motion imposed by ill-considered regulations and inject just a bit more technically exciting material, the battle would be won.

The above has dealt with honest and efficient allocation of human resources in a societal context. We now return to the flow of communication, with management and others, which is essential if our hard-won findings in troubleshooting are to be implemented.

References

Adams, E.M., and Schneider, E.J. *Proc. Air Poll. & Smoke Prev. Assoc. of Amer.* 45 61–4 (1952).

Ames, B.N. *CHEMTECH* 69(10) 590–7 (1989).

Bartleknecht, W. *Chem. Eng. Prog.* 73(9) 45–53 (1977).

Blewitt, D.N., et al. As reported in *Oil & Gas J.* 86(41) 58–62 (1988).

Brief, R.S., et al. *J. Indl. Hygiene* 47(1) 72–6 (1965); *Arch. Envir. Health* 23(11) 373–84 (1971).

Buehler, J.H., et al. *Chem. Eng.* 77(9) 77–86 (1970).

Chemical and Engineering News 65(33) 10–13 (1987).

Coward, H.F., and Jones, G.W. *U.S. Bureau of Mines Bull.* 503 (1952).

Davenport, J.A. *Chem. Eng. Prog.* 73(9) 54–63 (1977).

Degnan, T.F. *Matls. Perf.* 22(11) 45–7 (1983).

Dokter, T. *J. Haz. Matls.* 10 73–87 (1985).

Freeman, R.H., and McCready, M.P. *Loss Prev.* 5(5) 61–6 (1971).

Gibson, S.B. *Chem. Eng. Prog.* 76(11) 46–50 (1980).

Goldberger, W.M., and Othmer, D.F. *IEC Proc. Des. Dev.* 2 202–9 (1963).

Haase, H. *Electrostatic Hazards* Verlag Chemie (1977).

Habermehl, R. *Chem. Eng. Prog.* 84(2) 16–19 (1988).

Hilado, C.J., and Clark, S.W. *Chem. Eng.* 79(9) 75–80 (1972).

Ivanov, B.A., and Kogarko, S.M. *Intl. Chem. Eng.* 4 670–9 (1964).

Jepson, W.P., et al. Paper 20ld AIChE annual meeting (Nov. 21, 1991).

Katz, D.L. *Chem. Eng. Prog.* 68(5) 68–9 (1972).

Keister, R.G., et al. *Loss Prev.* 5(5) 67–75 (1971).

Kohlbrand, H.T. *Chem. Eng. Prog.* 81(4) 52–6 (1985).

Lewis, B., and von Elbe, V. *Combustion, Flames and Explosions of Gases* Academic Press (1961).

Mackay, D., and Yeun, A.T.K. *Envir. Sci./Technol.* 17 211–17 (1983).

Moores, C.W. *Chem. Eng.* 79(26) 63–6 (1972).

Nosetani, T., and Sato, S. *Sumitomo Keikinzoku Giho* (Engl.) 12(3) 119–23 (1971).

Poduska, R.A. *Proc. 34th Indl. Waste Conf.* (May 1979), pp. 167–83, 215–28.

Powers, G.J., and Tompkins, F.C., Jr. *AIChE J.* 20 376–87 (1974).

Rae, D., and Thompson, W. *Comb. & Flame* 35 131–8 (1979).

Reeves, M. *The Houston Post*, Sept. 25, 1978, p. 8A.

Reid, R.C. *Amer. Sci.* 64 146–56 (1976).

Rowland, F.S. *Amer. Sci.* 77(1) 36–45 (1989).
Rozlovskii, A.I. *Fiz. Gor. i Vzryva* 18(1) 8–16 (1982).
Saletan, D.I. *Chem. Eng.* 66(11) 99–102 and 66(13) 101–106 (1959).
Sargent, H.B. *Chem. Eng.* 64(2) 250–4 (1957).
Sellers, J.G. *Inst. Chem. Engrs. Symp. Series* 47 127–34 (1976).
Shamat, N.A., and Maier, W.J. *J. WPCF* 52 2158–66 (1980).
Smith, J.H., and Bomberger, D.C. Paper 23a *AIChE Philadelphia meeting* (1978).
Warner, F. *Chem. Eng. Prog.* 71(9) 77–84 (1975).
Yowell, R.L. *Chem. Eng. Prog.* 64(6) 58–62 (1968).
Zabetakis, M.G. *U.S. Bureau of Mines Bull.* 627 (1965).

17

The Friendly Arts of Persuasion

Tell them what they need to hear when it doesn't count. Then tell them what they need to hear when it does count.

Anonymous

As one who has been often too immersed in the technical detail of defining and opening up a plant problem, failing to keep open communication with management and other people critical to ultimate implementation of a solution, I speak here as a student rather than as a teacher. It is absolutely essential to spend the time to keep such people current with progress and the approach being taken. Failure to do so will be self-defeating, even though it may stem from past wounds due to bureaucratic obstructionism or a fear that exposing our uncertainties in the awkward problem definition phase may detract from our credibility. Aside from the tactical issues, there is actually a payoff to our own understanding from being obliged to communicate the essentials of a subject to nonexperts, as all novice teachers can attest.

The less decision makers can follow what the troubleshooting team is doing in the first phases, the more likely they are to develop an image of end running and instability of the whole team effort. I have seen good support even in an authoritarian organization by showing a key person the succession of trial scenarios for problem definition that we had to run through before we could submit a definitive proposal for solution; that the payoff from ultimate success made the intervening game worthwhile.

A matrix organization, where concurrence is also needed from a number of skill/specialists, can be more difficult in terms of communication/persuasion. This seems to be due more to territorial self-protectiveness than to the multiplicity of communication required. On the positive side, I have participated in an all-day "bargaining session" with a distillation specialist, attempting to persuade him to agree to a long-shot but inexpensive retrofit of column internals. Once his design-oriented preference for a surer but expensive solution was set aside, he became a powerful contributor to a successful conclusion.

For those who need encouragement in dealing with a bureaucracy, I recommend Drucker's book on management (Drucker 1974), which makes some shrewd distinctions between communication and the mere passing of information. Another useful experience is a course in team problem solving that provides familiarity in working with others who operate from different perspectives.

The rational elements in bureaucratic procedure include exposition, familiarization, and logical sequencing. An initial slowdown to conform to these precepts may be rewarded later by a "flywheel" of continuity that gets you past difficult junctures because your work has achieved project status. I saw this in the corrosion problem discussed in Chap. 4, where the team that was deployed seemed much slower and more ponderous in their efforts than I would have liked, yet displayed bulldog tenacity in seeing the investigation through a number of plant tests and implementation over a 2-year period.

There are, of course, nonrational elements as well in the bureaucratic response that must be dealt with: Fear of the unknown needs to be soothed away by slow familiarization. (And what a sweet moment it is when one hears played back words from the organization demonstrating that unorthodox concepts have become conventional wisdom!) The seemingly tedious tooting of the same horn referred to in the epigraph of this chapter is therefore needed and appropriate. This should be sharply differentiated from the ceaseless tooting of his own horn used by many a consultant to secure new business or as a "cover" while he gets familiar with a current problem to which he came without preparation or expertise.

In most organizations there is a formal chart structure, but also a shadow organization that actually does the day-to-day business. In troubleshooting there is a temptation to work only through the shadow organization. In terms of gathering information on the problem this may be appropriate but, sooner or later, the formal organization must be enlisted. The message of this chapter is: Make it sooner.

Reference

Drucker, P.F. *Management: Tasks, Responsibilities, Practices*, Chap. 38 Harper and Row (1974).

18

Summing Up

Science is the meeting place for two kinds of poetry: the poetry of thought and the poetry of action.

G. Agosthino da Silva

    This book is inherently diffuse. Indeed, a considerable part of the methodology taught encourages its practitioners to be diffuse in their approach to problem solving. Nevertheless, I would like to leave the reader who has come with me this far with a creed for a troubleshooting craft:

1. Problem solving in industrial-scale chemical processes is a profession that can be rich and rewarding.
2. A straight line is normally *not* the surest connection between perceived problem and desired resolution. The holding of mutually incompatible possibilities in our minds *at the same time* is a cornerstone of creativity in problem solving. Nonlinearity is not merely an unfortunate possibility; it is the normal (and exciting) state of the world.
3. A mature operating process deserves the same respect in terms of data gathering and remediation that we have come to give ecological and other living systems.

4. The role of trace components in terms of reaction mechanisms and the behavior of fractionating systems merits a major and systematic research effort in the universities and research laboratories of the world.

5. Most progress comes not from the brand-new but rather from the hybridization of the old and stable with new elements, with major attention to the detail work of their marriage. Many brilliant inventions have lain unused for long periods because their creators lacked the temperament to seek such fusions.

I hope and believe that this book will not merely draw a nodding appreciation from the fishermen out there in the chemical process unit ponds, but will also enlist them as apostles of a new craftsmanship embracing the lowly and the sublime, the data gathering and the inspired surmising of the Problem Solver.

Index

Accidents
  fault trees of, 332, 339n
  primal fault of, 55
  fires and explosions
    auto-ignition, 342, 347; see Static electricity detonation, energy release criterion for, 341
      discrete-phase aspect of, 341, 343
      of acetylenes, 341–44
      of ethylene, 343
      fuels, flammability envelopes of, 341, 345n
        as unconfined vapor clouds, 338n, 339
    oxidant, chlorine as, 337n
      explosions in absence of, 339–40
      hazard of nitrogen inerting, 338
    post-facto investigations of, 342
    safety, through factor elimination, 338, 344
    static electricity as ignition source, 342, 344–47
      charge accumulation in, 344–47
      discharge path for, 345–46
      spark energy criterion for ignition by, 345n, 346
    superheat explosions of liquids, 109, 340
  handling of hazardous intermediates, 340
  layout factors in plume dispersion, 335
  releases, containment of, 333–34
    mitigation of, 333–34, 341
  reliability, by equipment improvement, 334
    by redundancy, 334–35
  risk, quantification of, 9, 330–32, 336
  social costs, 330, 336
  value of a life in, 330–31, 348

Calculation
  of vapor pressure, extrapolation errors in, 40, 193–94
  of material balance, 5, 109–10, 191, 335
  of mathematical transforms, 150n, 155, 166, 305, 323

Catalysts:
  acid, heterogeneous, 265–67
    homogeneous, 229
  acidic sites on, 29
  anion exchange resin as basic, 266
  bifunctional, 254, 267
  bimetallic, 252
  capillary condensation in, 264
  correlations of kinetics on, 247
  deactivation of, 14–15, 27, 57–60, 267–68, 293
    by metal sintering, 268, 272–73
    by poisons, 268, 319
    by pore mouth blockage, 59
    forestalling of, by hydrogen, 257
    of acidic, by ionic hydrocarbons, red oil, 288
    conjugate polymer, 288
  dilution patterns of, in fixed beds, 277–78
  hydrogen spillover effects on, 255
  life tests of, 29
  liquid-phase supported (SPLC), 224–25, 234–35

359

360 Index

metal, supported, 246–47, 251
  coverage of, 251, 253
  crystallite size effect on, 250
  novel shapes for, 259, 322
  pore efficiency of, 258–59
    improvement of, 259
    in macroreticular ion-exchange resins, 264–65
  preparation of, 252–53
  regeneration of, 14, 15, 21–22, 28, 47, 138–39, 166, 293
    by in situ methods, 270–73
    by metal sintering reversal, 253, 272–73
    cyclic, in ion-exchange resins, 271–72
  retrofitting variants of, 28–29, 246–47
  selectivity enhancement of, 252, 254–57, 299
    by retention of heavy ends, 265–66
    by vapor-phase moderators, 319–20, 326n
  silica-alumina, 254
  supports for 252–53
  topology of, 59, 250–51, 253
  vendor support of, 29, 253, 267
Catalytic reactions
  of chlorine/ethylene addition by Fe, 243n
  hydrocracking, 167
    for tar removal, 271
  hydrogenation, 24, 247, 251, 262, 319
    of acetylenes, 227, 251–52, 255–56
    of dienes, 251
  in distillation columns, 322–24
  models of, gas-solid, 248–49
Chemical behavior, scales of
  acidity, of hydrocarbons, 231
    Hammett index of, 229
Chemical compounds and elements
  CO, role of, in hydrogenation, 254–56, 319
    in formation of metal carbonyls, 347–48
  COS, 34, 290
  Fe, as ubiquitous problem source, 15, 291
  olefin(ic) 10, 47
  oligomers, 43
  precursors, 19, 52, 269, 298

S, as irreversible catalyst poison, 268, 272
  role of, in minimizing coking, 257–58
  in odors, 289, 298–99
SO_2, 42
sulfuric acid, 10, 42
Chemical reactions:
  base-catalyzed, 229–30
  chain, via free radicals, 205–06, 215–20
    autoxidation, model of, 299–300
    branching of, 218
    chlorination, model of, 210
    cyclization, in tar formation, 219, 230
    initiation of, 215, 217–18
    inhibition, by radical traps, 218
    liquid-phase, prediction of, 221
    quasi-steady-state assumption in, 214–15
    termination of, viscous slowing of, 218
    vinyl chloride synthesis by, 205–06, 215
  ionic, in the liquid phase, 228–29
  chlorination, selectivity of, 217, 221–22, 262
  coke formation, 14–15, 18–19, 24, 51, 221, 275–76
    plugging by, 14, 16, 18–19, 24, 40
    reversal of, 256–58, 270–71
    role of wall metals in, 256–57
  cyclization to graphitic coke, 269
  dehydration, 123–24, 247
  dehydrogenation, 14, 47, 247, 257
  depolymerization, 27
  hydration, 10, 124
  interaction with diffusion, 260–61
    membrane separation in, 322
    selectivity, effects on, 262–63
      enhancement strategies for, 222–27
  ionic, effect of medium on, 231–32
  mechanisms of, 10, 37, 213–14
    acetylenes hydrogenation, 255–56
    acid-catalyzed hydration, 213
    activated intermediates in, 214, 222
    aliphatic alkylation, 228–29
    aromatic alkylation, 222–24
    COS hydrolysis by bases, 34, 245

Index

steam effect in pyrolysis, 221
transalkylation of aromatics, 224–27
side-reaction of, 19, 28, 38, 107, 122–24, 205 207–08, 255, , 266, 276, 280
degradative, 22, 84, 123–24, 204
cracking, 19
dehydrochlorination, 22, 27, 109, 123, 291
fouling, 16, 27, 55, 57, 82, 84, 101–02, 122–23
in reactor effluents, 269, 280
troubleshooting of, 268
in product storage, 204, 274–75, 294n, 299–300
tar formation, 205, 220, 256
removal, by hydrocracking, 271
by solvent washing, 272
Control, types of
cascade, 150–51
feedback, 4, 5, 10, 38, 145–49, 160
instability of, 52, 68, 82–84, 109, 260, 340
phase lags in, 155
dead time in, 71–72, 146–50, 164–66
dynamic lags in, 146–147, 156–57, 159
maximum loop gain in, 146–48
feedforward, 4, 38, 146, 149, 161
fuzzy, 151–52
level, 112, 152–54, 161–62
of centrifugal compressors, 174
of heat exchangers, 174–75
of pH, 164–65, 171–73
of reactors, 165–72
at metastable equilibrium, by cold-shot injectiojn, 169
by reflux cooling, 169
feedback of heat and mass in, 167
in catalytic cracking, 169–70
instrument recfonfiguration in, 170–73
limit cycles in, 170
multiple steady-states in, 168, 190–91, 260
phase plane plots in, 170, 222
stability analysis of CSTRs, 167–69
wave phenomena in fixed beds, 85–86, 166, 270
optimal, theory of, 154
shutdown/start-up, 174, 335

via computer, 143–44, 308
via sampled data, 150
Corrosion, 27, 47, 197, 210–11, 265, 285, 315–16, 333, 339n
early detection of, 305, 338–39

Distillation, control of, 110–14, 159–63
material balance schemes for, 161–62
pressure schemes for, 162–63, 193–94
quality control schemes for, 161–62
temperature profile in, 111
fluid mechanics of,
douching for ice removal, 24, 187
entrainment, carry-over in, 42–43, 47–48, 50, 58–60, 97–99, 221–22
flow regimes in, 96–97, 102
loading, flooding of, 48, 52, 97–100, 112
plunging of second liquid phase in, 55, 60, 103, 107–09
rivulet flow in, 90, 103–04, 127, 131
generation of acidity by side-reactions, 20, 57
internals of,
blanking strips, 101–02
bubble-cap trays, 96, 100
downcomer(s), 53, 99, 102, 125
dual-flow trays, 52, 101
knockback section, 51, 116–17, 129–30
layout errors in, 121–22, 131
packing, 102–03, 106–07
sieve trays, 102
valve trays, 46, 101
reaction during, 124–25, 234, 241; see catalytic distillation

Equipment, configuration of,
fallacies of symmetry, 16, 121
fixed beds, layout of, 18–19, 70–71, 266, 277–80
linings, fallibility of, 275
retrofits, 27, 29, 46, 50, 325
grafting technology in, 321–26, 358

Failure specimen, 17
Fluid flow,
division of, 69, 71, 102, 212, 279
by gas spargers, 89
by sonic flow orifices, 16, 69–70

inequality of, by channeling, 67–68, 70, 103
residence time variation from 66, 305
segregation effects in, 53, 68, 84–85
voids at wall, effect in fixed bed, 18, 70–71
manifolds for, 16
interaction with reaction, turbulent, 227–28
in downer reactors, 68
in jet-stirred reactors, 208
in riser reactors, 68
in wall boundary layers, 210
preferential phase drop-out in, 233–35
patterns, by tracer injection, 305–06
of axial flow, 66–68
of idealized stirred tank, 67, 71
derivations from, 71–73, 206–07
effect of power input on, 71
entrance effects, 210–11, 339
gas/liquid mixtures in pipe, 84–90
in jet-stirred vessels, 72–73, 208–09, 212
in recirculated tanks, 73
non-axial flow, 211, 322
plug flow, deviations from; see axial flow patterns
mixing, by momentum exchange, 212
pressure drop in, 84n
downstream/upstream communication in, 37–38, 47
sonic orifices re, 69–70
slip, between phases, 78–79
at wall of pipe, 67, 320
wall effects, in reactors, 210
pellet size effect, in fixed beds, 70–71
Fractionation, in distillation columns:
azeotropic distillation, 56, 113–14, 149, 158, 190
bottoms product from, 29, 43, 57
concentration profiles in, 11
"bulges," 24, 46, 55, 109, 116, 124, 184–88
entrapment; see "bulges"
extractive distillation, 47, 51, 115–19, 129–30, 188–89
volatility enhancement in, 188
light ends from, 21, 123–24
overhead from, 21
pasteurization sections in, 119–21
side-draws from, 110, 119–20, 185
stripping in, 21, 39, 42–43, 57, 106, 116, 125, 146, 148
temperature/pressure effects on, 192–94
trace impurities, limited rejection in, 120–21, 295
thwarting, by aliasing reactions, 124, 188, 292
by metathesis reactions, 293
under vacuum, 98, 102–03
Fractionation, in extractors, reaction during, 234
Fundamental relationships:
Avogadro number, 59, 130
Brownian motion, 7
Einstein viscosity equation, 80
SI system of units, xi

Hazards, cumulative:
detoxification mechanisms in the human body, 332, 348–49
exposure of surrounding population, 332
general population, societal constraints in, 11
atmospheric chemistry in, 329, 336–37
risk yardsticks for, 331
in-plant toxic exposure in, 347–48
Heat effects in energy integration, 28, 166
in endothermic reactors, 14, 19, 167, 249
in exothermic reactor dynamics, 167, 276–77
Heat transfer
by boiling and evaporation
film regime in, 83–84, 122
forced-circulation mode of, 82
instability in, 82–83
nucleate regime in, 83
spray evaporation, rates of, 82
thermosiphon circulation mode of, 82–83
by conduction and convection
in direct liquid-vapor contact, 131–32
radial, in fixed beds, 18, 70
temperature differential in, 18, 83–84, 122

Information sources, from the literature, 54–55, 255n, 298
 dossiers of plant technology, 55
 exchange across product lines, 294
Instrumentation:
 diagnostic tools in, 33
 for data acquisition and filtering, 11, 16, 309
 inference chain in, 145, 304
 layout errors in, 154
 radiation testing, of columns, 304
  online, for material flaws, 305, 307
 sampling and sample loop design, 57, 61, 145
 sensor problems in, 154, 157
  composition, 158, 160, 305, 307–09
  level, 158, 307
  mass flow, 304
  temperature, 158–59, 306

Mass transfer
 by axial diffusion, 211
 diffusion, in polymers, 80, 265
  in resin catalyst beads, 265
  in zeolites, 265
  inside porous catalysts, 263–65
 enhancement by reaction, 243–44
 interphase, film vs surface renewal models, 242n
 combining resistances of phases, 105–06, 135, 336
 in membrane contactors, 324
 interface resistance in, 42, 237–38
 resistance within gas phase, 7, 42, 104–07
 transfer units for dynamics of, 6, 42, 50, 105–06
 tray efficiency of, 6, 42, 55–57, 101, 103–08, 114, 126–27, 191

Non-rational thought, 11, 355
 ambiguity in, 313
 holding opposed ideas, as part of, 46, 311–14, 357
 arbitrary word meanings in (Humpty-Dumpty), 313–14
 brainstorming in, 12, 13, 51
 counterintuitive conclusions from, 19, 98, 181, 193, 215, 220
 diffuseness, depolarization in, 12–13, 312

 "intermediate impossibles" re, 61, 315
 reversal, as tool in, 12, 61, 245, 313, 317, 319
  of cost thinking, 320–21, 325
 holistic envelopes in, 10, 13, 15, 20
 imagery in, 11–13, 17, 52–53, 237, 278, 311, 315
  metaphorical, embodiments of, 13, 15, 20, 52, 310–11, 314–15
  conversion, as tool in, 13, 314–15
  MSF + MFS, making the strange familiar and the
   familiar strange, as tool in, 20, 310–12, 314, 316
 lateral thinking in, 4, 12, 51, 61–62, 123, 245, 339
 less is more, as principle in, 21, 98, 131, 295, 299
 patterning, 11, 40, 45, 47
 physical self-identification with the system, 51–52, 84, 204–05, 210, 249, 315

Organizational cultures, 35, 41
 brokers of technical interaction, 41
 commercial secrecy policies, 55
 communication gaps, 35–37, 41, 144, 333, 349–50, 354
 cubbyholes of specialists in, 42
 disciplinary divisions, 3–4, 31–32, 42, 52, 61, 294
 emotional driving force in, 311
 generalists, role in, 42
 immaculate disinvolvement stances, xi, 35
 impedance matching of, 41
 jargon in hindering communication, 39–40, 44, 152, 276
 matrix organizations in, 42, 355
 misplaced sophistication in, xi, 143, 332
 staffing quality for health and safety, 351
 "standard" remedies in troubleshooting in, 36, 43–44, 49
 triangular network of disciplines in troubleshooting, ix, Fig. 4-2
 virtues of overlap and redundancy, 145

Phase contacting, gas-liquid

CO_2 capture by hindered amines, 244
scrubbing, 7, 28, 42–43, 77, 237
assisted by two-step reactions, 244–45
swamping effects in, 244
in venturi devices, 43, 77
liquid-liquid, counter-current extraction, 48–50, 119, 132–37
extract phase in, 50
raffinate phase in, 49–51
counter-current, in porous tubular arrays, 325
cocurrent, in laminar contactors, 137–38
sorption, adsorption on solids, 22, 114, 132, 138, 274
Phase equilibrium
systems,
HCl, adsorption on alumina, 199
binary with water, 196–98
scavengers for, 291
ion exchange resins as sorbents, 138
supersaturation effects, 91–92, 109
thermal waves in sorption in fixed beds, 199–200
non-linearity in liquid mixtures, 178–98
in adsorption on fixed beds, 198–200
predictions and correlations of,
activity coefficients, of trace components, 181
in negative deviation systems, 194–98
for major-component binaries, 179, 181, 182, 191–92
temperature effect on, 183
Henry's Law constants, 180–82
general prediction of, 194
vapor pressure, extrapolation of, 40, 193–94
Phase separation, coalescence in liquid-liquid, 76–77
hydroclones in, 50
Plant operating methods:
innovation in, as new-old hybrids, 321
maintenance, preventive, 305
purges in, 10, 110, 120, 184, 185
optimal rate of, 296–97
testing of, ix, 40
unsteady-state in, 34, 93, 326
cyclic operation as deliberate variant of, 175–76, 324, 326

Process, characterization
dynamic, by filtering or time series analysis, 176
by perturbation, 156–57, 309
with random noise inputs, 157
inverse response in, 155, 160, 166
noise in, 112, 156–57
of disturbances entering a process, 159
oscillation, 52, 69, 78, 86, 89
static, by cross-correlation, 11, 156–57
by regression, 10, 304–05
treatment of extreme values in, 74
drifts, trends of, 11n, 157, 307–09
texture of, xi, 5, 37, 39, 42, 44, 54, 84, 100, 146, 205
layering of phenomena, 4–5, 20, 51
organic nature of, 9, 24, 35
phenomenological "corners" in identification, 4
the quality process in, 141–42
Process, flaws in, 24, 54
band-aids for, 21, 24–25, 34
calluses covering, 24, 46, 318
dislocations revealing, 25–26
localized effects in beds, 53, 167, 276
remission of symptoms of, 25, 54
resonating interactions of, 5, 10–11, 27–28, 206
root cause of, 5, 21, 26, 38, 122
syndromes of, 26–27
troubleshooting of,
as a distinctive discipline, viii–x, 33–44, 53, 357–58
relation to TSO function, 31–32
asking the right questions, 220
design function, relative to, viii–ix, 4–7, 16, 24, 32, 60, 355
discovery phase in, 19
enlarging context of, 38–39, 48–52, 220, 312
Vincent's theorem for, 50
hypothesis generation in, 33–34, 37, 47, 53–54, 59
leap from problem to project, 38, 46
problem definition in, 11, 19–20, 33, 38, 45, 48–50, 52–61
scenarios for, 53
closure of, 60–61
problem-solving techniques, 48

scale-down for experimentation, 67, 241, 249–50
scale-up, 66, 69, 76, 241
technology grafting for, 29
suspending judgment in, 45–46
team structure in, 10, 31–37, 39, 47, 205, 355

Reactors, configurations of,
guard beds, 273–74
sponge concept in, 274
loop reactors, 74, 211
micellar, 235–36
phase transfer-catalyzed liquid systems, 233, 245–46
polymerization reactors, 68, 167
slurry reactors, 240–41
staging of reactants to, 227
trickle-flow reactors, 241, 305
zero-slip requirement in, 209–10, 251–52
heat effects in,
hot-spots, local, in beds, 275–76
quench, for post-reactor cooling, 219–20, 269
thermal runaway in beds, 24, 33, 167, 319
graded-activity to control, 29, 277–78
of reactive chemicals in tankcars, 339–40
Reasoning, discrimination
analogizing, 4, 54
blocks, to information and concepts, 10, 59, 294, 307–08, 320, 339
catch-all designations as, 44, 80, 268–69
conventional wisdom in, 16, 355
dominance in, 12–13, 96, 339
circumstantial evidence in, 40
induction, inference, 3–4, 40, 46, 208, 300, 303–05, 308–09, 316
order-of-magnitude sifting, 4, 55, 57–61, 328–29, 344
post-rationalization of observed effects, 42–43, 214, 223
simulation by computer, 35, 55, 144
models, lacking key element in, 55, 329, 332n, 337
parochial vs. global, 249

vertical (project-type) thinking, 13, 38, 326

Trace impurities, 13
as chromophores in color problems, 291, 297–98
conversion, by reducing agents, 296
reactive purification for, 274, 295–96, 297, 299
to innocuous compounds, 295–96, 299
counteradditives for, 295
effects on process performance, 285–86, 294, 296
acceleration/inhibition of main reactions, 289
beneficial, 299
on product performance, 284–85, 295–96
excision of, 295
in odor problems, 185, 201, 295, 298–99
origin, by rule-breaking chemistry, 286–87
in feedstocks, 288–90
on catalysts, 287–88
pathway through the process
analytical problems in defining, 289n
confusion by mutations, 289–93
Two-phase systems
agitation of, 73, 74–76
droplets, coalescence of, 76–77
"daughter" droplets in, 77
electrical effects in, 92
dispersion of, 74–76
by static mixers, 67, 76
size distribution in, 74–75
mists and fogs, 42, 85, 91, 99, 197
HCl/water nucleation of, 196–97, 337
emulsions, 10, 68, 80–81, 91–92, 321
microemulsions, 81, 92
"rag", nature of, 80
spontaneous, in extraction, 92
encapsulation, of Fe in coke, 15

Waste, disposal of
by biotreater, design and operating issues in, 349–50
by deep well, 251
by incineration, 350–51